T0332845

Sex and Behavior

STATUS AND PROSPECTUS

These essays are presented in honor of
Frank A. Beach

Sex and Behavior

STATUS AND PROSPECTUS

Edited by

Thomas E. McGill
Williams College
Williamstown, Massachusetts

Donald A. Dewsbury
University of Florida
Gainesville, Florida

and

Benjamin D. Sachs
University of Connecticut
Storrs, Connecticut

PLENUM PRESS · NEW YORK AND LONDON

Library of Congress Cataloging in Publication Data

Main entry under title:

Sex and behavior.

"Presented in honor of Frank A. Beach."
Includes bibliographies and indexes.
1. Sexual behavior in animals. 2. Sex. 3. Psychology, Comparative. I. McGill, Thomas E. II. Dewsbury, Donald A., 1939- III. Sachs, Benjamin D. IV. Beach, Frank Ambrose, 1911- QL761.S46 599'.05'6 77-17840
ISBN 0-306-31084-8

A Division of Plenum Publishing Corporation
227 West 17th Street, New York, N.Y. 10011

Printed in the United States of America

Contributors

Norman T. Adler *Department of Psychology, University of Pennsylvania, Philadelphia, Pennsylvania*

Gordon Bermant *The Federal Judicial Center, Dolley Madison House, 1520 H Street, N.W., Washington, D.C.*

Lynwood G. Clemens *Department of Zoology, Michigan State University, East Lansing, Michigan*

Donald A. Dewsbury *Department of Psychology, University of Florida, Gainesville, Florida*

Richard L. Doty *Monell Chemical Senses Center and Department of Otorhinolaryngology and Human Communication, University of Pennsylvania Medical School, Philadelphia, Pennsylvania*

G. Gray Eaton *Unit of Primate Behavior, Oregon Regional Primate Research Center, 505 N.W. 185th Avenue, Beaverton, Oregon*

Lawrence V. Harper *Department of Applied Behavioral Sciences, University of California, Davis, California*

Benjamin L. Hart *School of Veterinary Medicine, University of California, Davis, California*

Jerome Kagan *Department of Psychology and Social Relations, Harvard University, Cambridge, Massachusetts*

Burney J. Le Boeuf *Crown College, University of California, Santa Cruz, California*

Thomas E. McGill *Department of Psychology, Williams College, Williamstown, Massachusetts*

Benjamin D. Sachs *Department of Psychology, The University of Connecticut, Storrs, Connecticut*

Karen M. Sanders *Department of Applied Behavioral Sciences, University of California, Davis, California*

Leonore Tiefer *Department of Psychiatry, Downstate Medical Center, State University of New York, Brooklyn, New York*

Steven G. Vandenberg *Institute for Behavioral Genetics, University of Colorado, Boulder, Colorado*

Richard E. Whalen *Department of Psychobiology, University of California, Irvine, California*

James R. Wilson *Institute for Behavioral Genetics, University of Colorado, Boulder, Colorado*

Foreword

Discussion of the precise nature and position of boundaries between disciplines is nearly always counterproductive; the need is usually to cross them not to emphasize them. And any such discussion of the distinction between ethology and comparative psychology would today seem patently absurd. While there may be differences in outlook, no boundaries exist. But when Frank Beach started in research, that was not the case. Comparative psychology flourished in the United States whereas ethology was unknown.

Beach started as a comparative psychologist and has always called himself either that or a behavioral endocrinologist. Yet, among the comparative psychologists of his generation, he has had closer links with the initially European ethologists than almost any other. He was indeed one of the editors of the first volume of *Behaviour*. That this should have been so is not surprising once one knows that his Ph.D. thesis concerned "The Neural Basis for Innate Behavior," that he used to sleep in the laboratory so that he could watch mother rats giving birth, and that in 1935 he was using model young to analyze maternal behavior. Furthermore, for nine years he worked in the American Museum of Natural History—in a department first named Experimental Biology and later, when Beach had saved it from extinction and become its chairman, the Department of Animal Behavior.

It was in 1938, during Frank's time at the American Museum, that he was first introduced to Niko Tinbergen by Ernst Mayr. Of his impressions at the time, Niko Tinbergen has recently written:

> Although at that time I was not at all attracted to Frank's "lab-orientedness," nor to his inclination to measure simple components of the "causal web," he on his side was open-minded enough (I am sure not merely polite enough) to express interest in what we were then doing. And he did influence us, much to the good, by his insistence on the need for measurement. I think that it was a combination of our respect for his research, for his open-mindedness and for his friendly and truly cooperative attitude that made us (W. H. Thorpe and N.T.) approach him for representing "*Behaviour*" in the U.S.

Beach's editorial position with *Behaviour,* and later his membership in the group which organized the international ethological conferences permitted him to play a large part in the liaison between ethology and comparative psychology. I do not think it would be true to say that his work has been very much influenced by ethology—he was already using some of the experimental techniques, and it could not be said that fieldwork would ever have really become Frank's scene. However, a number of his students wrote theses that were ethological by anyone's criteria, and in setting up the Field Station for Research in Animal Behavior at Berkeley he provided research opportunities for a number of ethologists from outside his own Department, such as Peter Marler and Thelma Rowell.

Beach's success in keeping the Department of Animal Behavior in being after Noble's death was no small achievement and had important consequences for ethology in a curious way. Beach persuaded T. C. Schneirla to join the Department, and Schneirla became one of the growing ethology's harshest critics. Fortunately, perhaps, Schneirla's comments were at first little known, in part because the journals in which he wrote were not widely read by ethologists and in part because Schneirla's style was not always as lucid as it might have been. However, Schneirla's message was taken up, emphasized, and supplemented by his pupil, Danny Lehrman, in his "Critique of Konrad Lorenz's Theory of Instinctive Behaviour." Although Lehrman was persuaded by a number of colleagues, including Frank Beach, to soften the tone of that article somewhat before publication, it was both hard-hitting and clear. Ethologists gradually came to accept some of its points and rejected others, and it was from that time on that the boundaries between ethology and comparative psychology began to dissolve.

Not only was Frank, by saving the Museum's Department of Animal Behavior, ultimately responsible for the Schneirla–Lehrman combination, he was also closely involved in the discussions. On the one hand, his model of a "central excitatory mechanism" provided valuable material for Tinbergen's (1951) "The Study of Instinct." On the other, Frank's "The Descent of Instinct" (or Taking the Stink out of Instinct) (*Psychological Reviews,* 1955, *62,* 401–410) took up some of the same issues as did Lehrman's.

But Frank is an experimentalist rather than a theoretician, and it is his example and his presence that has had the largest effect on ethologists. For a long time it seemed almost impossible that there would be an International Ethological Congress without him. What was especially appealing to the scientifically youthful ethologists was that he came not as an advocate for psychology, but to learn. And apparently he saw in the ethologists one of the same virtues that Tinbergen had seen in him, for after one ethological conference he was heard to comment "I have never attended any scientific meeting at which participants were so willing, indeed keen, to listen to criticism as at the ethological conferences."

If I am right in believing that Frank has made a major and often underestimated contribution to ethology as a liaison man, I hope I may be permitted a personal note. During the fifties, several small meetings were organized with the explicit aim of integrating ethology and experimental psychology. The first of these that I attended was organized by Bill Verplanck at Harvard, and my friendship with Frank started in Scollay Square, where by chance or good judgment we remet in the evening after the first day's session. Frank took several of us to see his laboratory in Yale afterward. This was only a few years after I had written my thesis, but Frank could not have been kinder. A year or two later, he organized what was undoubtedly the most important single academic event in my life—a conference in the Behavioral Sciences Center at Stanford for a few comparative psychologists (I remember especially Harlow, Hebb, Lehrman, Rosenblatt, Washburn, and Beach) and a few ethologists (Tinbergen, Baerends, van Iersel, Vowles, and myself). What was unique was that we had five or six weeks to talk. If you didn't finish what you had to say today, there was always tomorrow, and discussion continued round the dart board or in Dinah's Shack on El Camino Real. I made and cemented good friendships at that meeting, and it certainly affected the direction of my work. I don't think it had the same importance for Frank—he did not even mention it in his chapter in Gardner Lindzey's *History of Psychology in Autobiography*.

Another occasion was the conference in Berkeley which led eventually to the publication of *Sex and Behavior*. Frank organized the meeting to take place in two separate periods a year apart and, characteristically, he arranged for participants to be able to bring a student the second year. One incident deserves to be put on record. I asked if I might bring Patrick Bateson, now Director of the Madingley Sub-Department of Animal Behaviour. Frank, ignorant of the Cambridge collegiate system, sent out invitations to him addressed to my Cambridge college, St. John's. Pat belonged to another college, and the St. John's porters had never heard of him. However, with unusual efficiency they forwarded the letter, within 20 minutes of its arrival, to Gregory Bateson, who had been a Fellow of St. John's 20 years earlier. Gregory is a relation of Pat, but at that time they had never met. Gregory was then living in California, but he and Frank had not been on speaking terms for about ten years. He was therefore a little surprised to get the invitation. But Frank carried off the situation with great aplomb. Both Batesons came, and met for the first time. Frank and Gregory spoke to each other, and Gregory gave a superb lecture with his fly open.

If these reminiscences take up undue space, it is because I am gratefully availing myself of this opportunity to acknowledge my debt to Frank. And I am only one of many ethologists who would acknowledge how their work has been enriched by contact with him.

It is too easy, in a book such as this, to remember a man's achievements and to forget where he has failed. Frank has had one academic goal that he has clearly failed to achieve. About five years ago, he announced his intention of giving up research and trying his hand at undergraduate teaching. The latter he has done—I understand with great success. But giving up research is something that Frank can't quite seem to manage.

Robert A. Hinde

MRC Unit on the Development and Integration of Behaviour
University Sub-Department of Animal Behaviour
Madingley, Cambridge, Great Britain

Foreword

For nearly 40 years Frank Beach's contributions to the field of hormones and behavior—which he has recently called Behavioral Endocrinology—have been of exceptional significance. They include many of our basic concepts for understanding the interrelationships among hormonal, neural, and experiential factors in the organization, development, and evolution of reproductive behavior among mammals.

An early contribution was to introduce methods of studying reproductive behavior from the field of comparative psychology to the field of endocrinology. As recently as 1948, when I was a graduate student studying endocrinology, reproductive *behavior* was considered a rather poorly defined secondary sex character in comparison to the cock's comb and plumage color. W. C. Young described the scientific atmosphere in which Frank Beach's studies were initiated in the 1930s when he wrote in 1961:

> It is of . . . interest that endocrinologists, physiologists, biochemists, and anatomists . . . have given little attention to problems centering around the hormonal regulation of reproductive behavior. Much of the work in this field has been done by psychologists and much of the support that has encouraged effort in this direction can be attributed to their foresight, activity, and persistence. Endocrinologists, physiologists, and anatomists have done little with behavior . . . because of . . . the . . . belief that behavioral endpoints are so vague to be of little value in experimental studies. This circumstance . . . is reflected by the absence of any discussion of hormones and behavior in most textbooks of endocrinology.

As a comparative psychologist, Frank recognized that an accurate description of mating behavior of male and female rats was a prerequisite for quantitative endocrine studies. The prevailing concept of hormonal "erotization," proposed by Steinach (which incidentally strongly influenced Freud's concept of libidinous energy) was based upon behavioral observations that did not distinguish between motivational and performance aspects of mating behavior. With more refined observations, Frank developed methods of distinguishing between the arousal or initiation of mating and its performance or maintenance and termination. This distinction has proved extremely fruitful for further experimental analysis of the influence of hormones and sensory stimuli on mating in both males and females. The current status of these concepts has been recently reviewed by Ben Sachs and Ron Barfield in a chapter in *Advances in the Study of Behavior* (Volume 7, 1976) with respect to the organization of copulatory behavior in the male rat. As they indicate, Frank has frequently revised these concepts to accord with new findings, and he has been willing to reevaluate the assumptions underlying them, as, for example, in his recent contrast between the concepts "receptive" and "proceptive" pertaining to female mating behavior, and his introduction of the concept of "gender role" as an intervening variable to account for many aspects of human sexuality.

The study of sensory factors led Frank to propose multisensory control of mating behavior. Arousal or initiation of mating is dependent upon a combination of sensory stimuli and in the execution of copulation both male and female are dependent upon a variety of stimuli to elicit the sequence of synchronized mating responses. An important discovery by Frank was the greater versatility of experienced compared to inexperienced males with respect to the role of sensory stimuli in mating. Experienced males were less affected in their mating behavior by the surgical elimination of any one or two sensory systems: they were able to substitute one sensory system for another in the control of mating while inexperienced males were unable to do so.

Frank's zeal for studying the sensory basis of sexual behavior has sometimes resulted in unusual methods—the cuffing of males rats about their cages is a mild example compared to his adventure into the world of music. Frank observed one day, serendipitously, that the inert crocodile housed upstairs in the Department of Animal Behavior, which was unresponsive to all of Frank's attempts to arouse it sexually, began to move and seemed to become romantic when the sounds of a French horn reached it from the musicians rehearsing on a lower floor of the museum. Frank brought the French horn player to play before the crocodile to confirm his observation that this was the stimulus to which the crocodile had responded.

The study of hormone–behavior relationships was carried out by Frank using classical endocrinological procedures of castration and hormone replacement, and by taking advantage of developmental endocrine anomalies. He investigated the relationships between the gonadal hormones and the male and female mating patterns and between level of gonadal hormone and completeness of the mating pattern. Frank established the basic relationship between homotypical gonadal hormones and mating patterns: he was equally aware of heterotypical gonadal–behavior relationships, the underlying bisexuality they presupposed, and the somatic determination of hormonal action that was implied. Subsequent work by Sachs and by Barfield and their students had made even clearer the bisexuality of the female where evidence was scanty in earlier studies. Several important principles emerged from these studies on the functional relationship between hormones and mating behavior: the completeness of the mating patterns is based upon the level of circulating hormone but once the hormone level is above threshold and the full pattern is exhibited, the level of sexual behavior is dependent upon somatic factors characteristic of the individual.

Discovery of these principles was the culmination of a series of developmental studies that laid the foundation for later studies on the influence of perinatal hormones on the organization of mating patterns in rats. What had to be clarified before these later studies could be conceived was whether the appearance of components of the mating pattern in prepubertal male rats (e.g., palpation and pelvic thrusting) was evidence of the developing organization of the mating pattern. By administering gonadal hormones to weanling males and females Frank showed that the mating patterns were already fully organized before puberty but they lacked the activational support of gonadal hormones. With prescience he wrote in 1942:

> It appears that the neuromotor mechanisms mediating copulatory behavior are well organized and ready for functioning long before puberty and they do function at a relatively low level prior to the occurrence of postpubertal testicular secretory activity. That the resultant behavior is incomplete and infrequent may be due to the presence of high thresholds in the circuits in question. The function of androgen is thus interpreted as a lowering of these thresholds. The hormone does not organize the behavior, but does facilitate its appearance by reducing thresholds in the essential neuromotor circuits,

and again with reference to the development of the female's mating pattern:

> Apparently this response, like the male's copulatory pattern, is organized well in advance of puberty and needs only the influence of ovarian hormones to make it elicitable.

Subsequent work, of course, has shown that Frank was on the right path and he himself came very close to demonstrating the organizing influence of perinatal androgen on male mating behavior in his study with Marie Holz-Tucker in 1949.

The larger view of mating behavior derived from studies on the neural basis, which guided Frank's thinking at this time, led him to focus on the sensory and performance aspects of the deficiencies in mating resulting from early castration rather than on the hormonal aspects in the above study. Under the influence of Karl Lashley, Frank had undertaken studies on cortical control of mating in males and found that mass action was the predominant mode of cortical participation. This suggested a hierarchical neural organization and emphasized the regulation of lower-level neural functions by inhibitory and facilitory influences of higher ones. Translated into behavioral terms, this means that a sensory stimulus has an effect at all levels of the nervous system and does not simply trigger a mating response. It contributes to both the arousal and the performance of mating behavior. Since some of the sensory stimuli that regulate mating are the consequence of sensory feedback from motor responses, failure of a motor response, as occurred in many of the early castrated males in the above study, had consequences beyond the immediate effect on performance.

Frank believed that hormones influence mating behavior in several ways: through their effects on sensory input, on adequacy of the structures required for performance, and by their action on central neural structures. He was interested in analyzing each of these hormonal effects and many of his discoveries can be fitted into this overall conception. Of particular importance, I believe, is his conception that sensory stimuli many compensate for hormonal stimulation in mating behavior. Komisaruk has provided direct evidence of this in the sensory and hormonal control of the lordosis response in the female rat, and Hinde has proposed a priming effect of light and male song on estrogen stimulation of nestbuilding and nest-related activity in canaries and budgerigars.

There are, in addition, developmental implications of this view. Frank has always been interested in the ontogeny of mating behavior and several of his studies were devoted to the role of experience in the development of mating in male rats. Technical difficulties of rearing rats without mothers prevented Frank from exploring this fully but recent studies that have overcome or circumvented this problem show a strong influence of early experience on the development of mating behavior. Compensating effects of hormones and sensory stimuli suggest that each normally supplements the other in mating behavior: viewed developmentally this implies that the absence of either may impede development of mating behavior. This has certainly proved to be the case in the guinea pig, dog, cow, pig and several nonhuman primates and my own findings in the castrated, sexually inexperienced male cat and Frank's in the male dog also confirm this view. As Hinde has written recently: "Until the mechanisms involved in behavior have been worked out in detail it would be wiser to consider the control of behavior as a network of interactions rather than to assume that the hormones are primary." The same may be said for the development of reproductive behavior.

Frank has always considered individual differences among animals in their display of sexual behavior as an important area of study. This is a natural consequence of his recognition of the multiplicity of factors that govern this behavior. The importance of the social interaction as an essential aspect of mating behavior and situational factors, for instance "anxiety," has important contemporaneous influences on mating. In his studies on the beagle, he has been able to investigate some of the factors governing individual differences, particularly the role of individual preferences for particular sexual partners. The place of these studies in the larger scheme of his thinking will be dealt with below.

Frank's interests as a comparative psychologist ranged over the entire class of vertebrates both in his writings and in several studies he published. His studies on the neural and hormonal basis of maternal behavior in the rat are well known, and recently I came across an early paper by Frank on maternal behavior in the pouchless marsupial *Marmosa cinerea*. The Department of Animal Behavior under G. Kingsley Noble, Frank Beach, and later under T. C. Schneirla and Lester Aronson has been truly a center for the comparative study of reproductive behavior. The Institute of Animal Behavior continued this tradition under Dan Lehrman, who was influenced at an early age by Frank and whose studies on reproductive behavior in the Ring Dove began at the Museum.

In 1947, Frank published a major review (its publication in *Physiological Review* was itself a landmark), which I consider among his principal contributions, in which he proposed a theory of the evolution of sexual behavior in mammals. Frank reviewed the available evidence on the interrelationships between hormonal, neural, and experiential processes governing sexual behavior among the mammals and proposed that there was a gradual shift from primary control by hormonal stimulation to regulation by complex sensory and experiential factors in the evolution of higher mammals. Frank's idea was that in lower mammals (i.e., rodents) hormonal control of sexual behavior is more direct, requiring simpler psychological mechanisms than in higher mammals, (i.e., primates) in which hormonal stimulation is just one of many factors—the others being individual psychological and social factors—regulating sexual behavior. Frank may have underestimated the psychological complexity of sexual behavior among rodents, since only recently has research revealed the subtle behavioral interactions between males and females during mating (Diakow, *Advances in the Study of Behavior*, Volume 5, 1974). He anticipated, however, the field observations of sexual behavior among primates, the recent studies on monkeys by several groups, and the earlier study on the development of sexual behavior in chimpanzees by Nissen. Although Frank proposed this theory thirty years ago, it remains viable today as one of the best evolutionary syntheses of the interrelationship between hormonal, neural, and experiential processes governing reproductive behavior.

Frank's career has been and continues to be richly productive scientifically and in the personal relationships he has established with colleagues throughout the world for several generations. Having been one of the younger people who was strongly influenced by Frank and his work, in part through the tradition he left behind at the Museum, also by the opportunity Frank gave me to meet with the European ethologists at the California conference Robert Hinde describes in his foreword, and in many meetings with Frank, I feel I can express the feeling of gratitude which all of us feel toward him.

Jay S. Rosenblatt

Institute of Animal Behavior
Newark College of Arts and Sciences
Rutgers University, Newark, New Jersey

Preface

The Forewords of Robert Hinde and Jay Rosenblatt are testimony to the affection and respect with which his colleagues hold Frank Beach. The rest of this book is testimony to the affection and respect of Frank's students for him as scientist, teacher, and person.

Beach's research career has spanned about forty years thus far, and he has published actively throughout that time, a tribute to his creativity and stamina. An impression of the long-term influence of Beach's publications may be obtained by consulting the *Science Citation Index*. In the years 1970–1974, Beach's papers were cited 961 times, and only rarely by Beach himself. Almost half the citations were to papers published in 1950 or earlier, including 133 citations of his eleven 1942 papers. These data underestimate Beach's contribution, in part because he has always been generous in deciding questions of authorship. Beach was first author with student collaborators of *his* research, but he rarely allowed his contribution to his *students'* research to be acknowledged by authorship. It is for this reason that Beach is sole or first author on almost all his papers. The *Science Citation Index* data also do not reveal that throughout his career Beach's contributions have included data papers, literature surveys, and theoretical papers that are classics, yet as significant today as when they were written. In fact, many of his students have experienced the frustration of having a good idea, only to discover that Beach published on it 30 years before. So that future scientists may avoid such frustration, we have included in Appendix 1 Frank Beach's bibliography to date.

As a teacher, Frank engendered affection and respect, and he infected us with his contagious enthusiasm for the enterprise of research. We have experienced the delight of being buttonholed by Beach and dragged to the lab to see his newest phenomenon, and he has always been accessible to share our enthusiasm for our newest set of data. His intolerance of our ineptitude seemed tempered by his recall of his early difficulties. For example, a generation of students may be cheered by the revelation (1974, *Autobiography,* Appendix 1) of Beach's first experience with a laboratory rat. Donning the heaviest available gloves, he forced himself to be brave and hold the beast for three minutes. After watching a stopwatch throughout the anxious period, he pridefully released the rat, only to discover that the poor creature had succumbed in his grasp. Frank's students, on the contrary, thrived in his laboratories. For many of us Frank has served as a model to emulate, however unlikely we may be to achieve it. Frank has been known to comment gently on our progress. He once asked one of us "When are you going to stop being a promising young man?" Some of us are no longer quite so young, and for us it might even suffice to have in our 66th year the energy and enthusiasm that Frank is displaying in this, his 66th year.

About a year before Beach's 65th birthday, one of his students noted the approaching milestone. Subsequent discussions with other Beach students revealed the widely held feeling that the event should not go uncelebrated. A plan developed to gather his former doctoral and postdoctoral students for an assessment of the status and prospectus of research in sex and behavior, broadly considered. Beach had organized a similar conference more than 15 years earlier (see Forewords), and it seemed timely to take stock again.

Frank Beach is an unassuming man, and there was some doubt that he would agree to being celebrated in this way. Happily, he was delighted with the idea. For several years he had harbored the fantasy of being wealthy enough to invite all his former students to a reunion in Berkeley. Restrictions of time and money made it impossible for all of us to come, but 24 of his 41 students (see Appendix 2) were able to travel to Berkeley for the symposium in April, 1976.

The meeting was most rewarding for all its participants, and there was no more excited a listener, note taker, and commentator than Frank Beach. (Unfortunately, the discussions were not recorded for possible publication.) The consensus following the symposium was that the proceedings were worth sharing with a larger audience, and this volume is the result.

Our goal in this book, as in the symposium, was a broad consideration of sex and behavior. Additional breadth was gained by adding chapters which had not been given in the symposium. With one exception, all the authors (or first authors of multiauthored chapters) were predoctoral or postdoctoral students of Frank Beach. The diversity of their research is evident in this volume, and reflects the diversity of their mentor's interests.

The volume has been divided into three sections, representing three perspectives of sex and behavior: the evolutionary view, the mechanistic view, and the perspective of human sexuality. The sections also reflect three of the areas in the field of sex and behavior to which Frank Beach has contributed importantly. These viewpoints are not mutually exclusive, and one of the goals of this volume is to emphasize the possibility of cross-fertilization. These reciprocal influences were evident throughout the symposium and are evident in the fact that several chapters in this book could have fit in a different section from the one to which they were assigned.

The first section, Evolution and Natural History, opens with a review by Le Boeuf of recent developments in neo-Darwinian theory, with its emphasis on individual selection and its implications for understanding the social structure of populations, as well as the details of copulatory behavior, parental behavior, and even infanticide. In chapter two, Eaton is concerned with questions of both natural history and physiology in his study of Japanese macaques under quasinatural conditions. Possible differences among congeneric species in the social and endocrine correlates of dominance and breeding success emphasize the importance of avoiding premature judgments about the applicability to humans of conclusions about primate social behavior based only on research on Rhesus macaques. In the next chapter, Harper and Sanders report on sex differences in the naturally occurring play of preschool children. The interpretation, necessarily speculative, is cast in an evolutionary framework, and the chapter closes with a question concerning the justification for social action that bears careful thought. The evolutionary and comparative approach to behavior has been brought into the laboratory by Dewsbury, whose chapter ends this section. His analysis of the copulatory behavior and reproductive physiology of muroid rodents bridges one of the gaps between those who are concerned with ultimate causes and those who are concerned with proximate causes. The availability of such bridges (also evident in the chapters of Eaton and Adler) for those who can see and use them raises doubts about the view that sociobiology and physiology are clearly separable.

The chapters in the next section are concerned with mechanisms controlling reproductive behavior. Adler's chapter explores in depth the reproductive life of *Rattus norvegicus*, tracing many details of the often-neglected role of behavior as a cause, not just an effect, of physiological changes. In analyzing the "species-vaginal code" of *R. norvegicus* (and other species in Dewsbury's chapter), it becomes clear that the behavioral and physiological variables have coevolved to maximize individual fitness by ensuring a fertile, conspecific mating.

In McGill's chapter, the focus is upon the interaction between the genotype of an organism, specifically laboratory mice, and the neuroendocrine control of its behavior. This chapter carries with it important caution flags for methodology and interpretation. Whalen's chapter takes the analysis a step further, considering how hormones act at the cellular level to achieve ultimate translation into behavior.

Sexual behavior includes reflexes that are evident even in males with their spinal cords transected, and a complete understanding of an organism's behavior requires an understanding of how those reflexes become integrated into the behavior pattern. Hart's chapter reviews the evidence on the neural and endocrine control of penile reflexes in a number of vertebrates. He includes evidence that the spinal cord's differential sensitivity to hormones contributes to endocrine control of behavior, just as does the uptake of hormones by the brain. Central to a discussion of spinal reflexes is the role of neural inhibition, especially from supraspinal levels. In his chapter, Clemens dissects the evidence for the long-standing hypothesis that the female's lordotic (receptivity) reflex is potentiated by hormone-activated disinhibition of supraspinal neural systems, and Clemens finds the evidence wanting. Sachs then reviews the status of both conceptual and neural models of the control of masculine copulatory behavior and considers how each may be used in understanding species differences in copulatory behavior. At the close of this section, Sachs and Dewsbury review evidence that some of the mechanisms controlling copulatory behavior in muroid rodents have diverged less than the copulatory patterns themselves or the parameters of copulation.

The final section deals with human sexuality and sex differences. Following a theme developed in the chapter by Harper and Sanders, Kagan's chapter and that of Wilson and coauthors report on social and cognitive sex differences, respectively. Both chapters emphasize that the sex similarities are more substantial and, in many ways, more impressive than the sex differences, but the reliable differences may be revealing of developmental, social, and genetic interactions and influences. Doty then reviews the literature on how the human female's sensation of taste is affected by changes in her endocrine condition. The evidence for what "everyone knows" is skimpy indeed, and one may infer from this chapter that in this instance, as in so many others, "common knowledge" is not to be relied upon. The potential for misinterpretation of studies on sex differences, sexual differentiation, and of the sexual behavior of animals is the topic of Tiefer's critique of current approaches to sex and sexuality. Finally, Bermant considers human sexuality and sexual psychopathology from an ethological-psychoanalytic perspective, which exemplifies the broad concern of those who work in sex and behavior.

Frank A. Beach's reputation as a scientist is so firmly established that his students can do little to enhance it. However, the authors and editors hope that this book will provide some indication of Frank's success in a related and perhaps more difficult area—that of stimulating the men and women who studied with him to pursue further research in Sex and Behavior.

Thomas E. McGill

Donald A. Dewsbury

Benjamin D. Sachs

Contents

PART I: EVOLUTION AND NATURAL HISTORY

PART II: MECHANISMS CONTROLLING REPRODUCTIVE BEHAVIOR

PART I

EVOLUTION AND NATURAL HISTORY

1

Sex and Evolution

Burney J. Le Boeuf

> The courtship of animals is by no means so simple and short an affair as might be thought.
>
> Charles Darwin (1871, p. 577)

When I joined Frank Beach's laboratory as a new graduate student I was anxious to get started on a project in his research specialty, sexual behavior. During one of our initial meetings, he strongly suggested (as he had done with many others before me) that I go down into the basement of the Life Sciences building and watch laboratory rats mate. One thing that immediately caught my interest was the female's habit of running from the male as he approached, stopping abruptly, wriggling her ears, and then darting away again as the male gave chase. This behavior seemed to excite the male and invigorated his pursuit. I saw this episode repeated time and time again and it intrigued me all the more because Lynn Clemens, Bill Westbrook, and Jim Wilson, who were running the "sex tests," assured me that the female was definitely in estrus and receptive to the male's advances. Then why did she run? Why didn't she just stand there and copulate? These simple questions plagued my fledgling graduate student mind and led to my first idea about an experiment on sexual behavior. It was simple enough. To determine the function of the female's puzzling behavior, stop her from performing it, immobilize her and then observe what the male does. Well, for various reasons, the experiment was never done and it is probably just as well. Had the experiment been performed, I'll wager that the male rats would have attempted to copulate with the females even though the latter were completely immobile.

Burney J. Le Boeuf • Crown College, University of California, Santa Cruz, California.

Many vertebrate males can be very indiscriminate in what they mount. Some will even mount dead females as numerous observations on birds, seals, and humans, to mention a few species, have shown. But females do not do this to males! Why the sex difference?

1. Sexual Selection and What Darwin Didn't Know

Darwin (1871) observed striking structural and behavioral differences between males and females, which he was unable to explain with natural selection. The list included the observation that males are almost always the aggressors in courtship and they alone have special weapons for fighting with their rivals. Males are stronger, larger, and more pugnacious than females; males are provided in a much higher degree than females with organs for making sounds or music and with glands for exuding odors; males are more brilliantly and conspicuously colored while females are usually unadorned; and it is the male that is provided with sense organs or structures for finding, getting to, or holding members of the opposite sex.

Darwin concluded that these differences arose because of sexual selection, a type of selection which he distinguished sharply from natural selection. He argued that sexual selection operated whenever certain individuals had a reproductive advantage over others of the same sex and species. Natural selection, on the other hand, depended on the success of both sexes throughout development with respect to the general conditions of life. Thus, natural selection was to him, survival selection, and sexual selection was reproductive selection (Selander, 1972). Darwin believed the sexual struggle to be of two kinds, ". . . in the one it is between individuals of the same sex, generally the males, in order to drive away or kill their rivals, the females remaining passive; whilst in the other, the struggle is likewise between the individuals of the same sex, in order to excite or charm those of the opposite sex, generally the females, which no longer remain passive, but select the more agreeable partners." This dual aspect of sexual selection is often casually (and somewhat erroneously) referred to as male–male competition and female choice.

Fisher (1930) and Huxley (1938) emphasized two basic principles involved in sexual selection. Huxley called behavioral interactions between males and females epigamic selection, and behavioral interactions between members of the same sex, intrasexual selection. Female choice, which occurs when a female selects only one or a few among many male suitors, is an example of epigamic selection and fighting among males to monopolize females, with the result that some males breed more than others, is an example of intrasexual selection. This distinction between the two types of selection is often difficult to make. For example, if a female increases her fitness by choosing among male sexual partners, she outcompetes females that do not choose.

It is worth noting that Darwin was heavily criticized for the distinction he made between natural selection and sexual selection and it is still a matter of debate whether such a sharp distinction can be made between the two modes of selection. Indeed, the separation is less sharp today than it was in Darwin's day because the mathematical geneticists have redefined fitness in terms of single genes as "the contribution to the gene pool of the next generation," rather than the way Darwin meant it in an everyday sense, as well adapted or anything that improved chances for surviving. Under the new definition, sexual selection can be construed simply as one form of natural selection and many students of evolutionary biology considered it so for awhile (Mayr, 1972).

At present, the tendency is to employ the term sexual selection but not to restrict its application, as Darwin did, to display characters or weapons that operate around the time of mating, i.e., characters involved in obtaining mates or matings. It is now understood that "all characters directly affecting reproduction should be attributed to sexual selection, including those that facilitate copulation and fertilization and those involved in parental care" (Selander, 1972, p. 181). From this perspective even events occurring in early development may profoundly affect reproduction in adulthood.

Although Darwin was undoubtedly correct in stating that sexual selection "played an important part in the history of the organic world," he did not know why it affected males and females differently. He did not understand why males were more strongly modified than females by sexual selection. With the limited information available to him, he could not satisfactorily explain why males were so eager to copulate, and so indiscriminate, and why females were so coy. His "explanation," based on the idea that males were more variable than females, is nearly circular and unconvincing (Selander, 1972).

2. *Recent Ideas about How Natural and Sexual Selection Operate*

During the last two decades, the availability of new information and a more detailed understanding of the fundamental manner in which selection operates has produced a novel and far-reaching explanation for the differential action of sexual selection on the sexes and how this situation came about. These ideas have shed new light on this old principle of evolutionary biology. This approach to these problems has stimulated research into new areas, prompted reanalysis and reinterpretation of previous ideas, integrated a vast, scattered, complicated literature, and promises to continue breaking new ground in the future. There is presently a vitality, a level of energy and excitement, surrounding the pursuit of some of these problems that is at once exhilarating and infectious. There is talk of a new synthesis in studying the biological basis of social behavior, a renaissance in behavioral and evolutionary biology. There is also the expected reaction to such a revolution in ideas—opposition and controversy.

It is this new way of thinking about how selection, both natural and sexual, operates—and the central role of sex in this emergent discipline—that I will address in this paper. The history of this topic is lengthy and the literature enormous. Since I must be brief, I will not be able to undertake any in-depth treatment. Rather, my aim is to try to stand back from the action in the trenches and give those who rarely work under the modern evolution-by-natural-selection paradigm, e.g., psychologists studying sexual behavior in the laboratory, a sense of what has been going on at this level of explanation and where it is coming from. Lastly, I intend to suggest how an understanding of this approach can aid those working at reductionistic levels of explanation in asking important questions of nature and comprehending the results of their experiments.

I see this new synthesis, which some call social theory (Trivers, 1974) or sociobiology (Wilson, 1975), as having come about because of the emergence of three basic ideas:

1. Selection operates primarily at the level of the individual.
2. Individuals can increase their fitness through actions affecting their relatives.
3. The operation of sexual selection is governed by the unequal investment of the sexes in offspring.

I will briefly summarize the development of these ideas, especially concentrating on the last one.

2.1. The Level at Which Selection Operates

Theoretically, selection can operate at many levels: the individual, the family, a group of families, the breeding population, or the species. Darwin thought selection operated at the level of the individual. He argued that individuals behave in such a way as to advance themselves because they gain greater representation in the next generation by doing so. Or, to put it bluntly, individuals are selfish because they leave more offspring by behaving this way. So he argued until he encountered a thorny problem he couldn't explain on that level: the evolution of sterile castes in social insects (Darwin, 1859). He introduced the concept of group or kin selection in the *Origin of the Species* to account for this phenomenon. In retrospect, his logic is admirable. If sterile individuals are important to the welfare of fertile relatives in the hive, selection must occur at the family or colony level. Darwin used this explanation only for the social insects. Yet, paradoxically, after publication of the *Origin of the Species,* many students of Darwinian theory adopted the idea that social traits served the good of the group or species. This tradition developed despite the fact that natural selection refers to individual reproductive success. In the early part of this century, several geneticists wrote on the topic of group selection and the extent to which it might be operative. It became fashionable to attribute sex, death, and reproductive cycles to the advantage of the species rather than the individual (Ghiselin, 1974). Even today it remains the dominant interpretation in the social sciences (Trivers, 1974a).

The significance of alternative modes of selection, or the possibility of a dimension of selection, was not fully apparent until Wynne-Edwards brought the argument to the attention of the biological public in 1962 with his controversial book *Animal Dispersion in Relation to Social Behavior.* Wynne-Edwards forced biologists to consider the unit of selection by proclaiming group selection all important in determining social behavior. He proposed that animals avoided overexploiting their food resources by adjusting their own population densities to the fluctuating levels of the resources. He argued that feedback on population size was communicated by social conventions such as territorial behavior and epideictic displays (hypothetically, displays by which animals reveal their presence and density to each other). These modes of communication informed individuals that they should curtail their own individual fitness for the good of group survival, i.e., sacrifice personal fitness for the good of the species. It is these considerations, Wynne-Edwards argued, that shaped the evolution of social behavior. According to Wilson (1975), Wynne-Edwards's contribution carried the "theory of self-control by group selection" to its extreme and forced an evaluation of its strengths and weaknesses.

A large number of biologists took up the challenge against Wynne-Edwards and he became a popular target to shoot at during the 1960s. Critiques, rebuttals, long reviews, and new studies followed and, as is typical of such controversies, the air was charged with excitement, for there was a lot to argue about. A few addressed themselves to the more important issues of group selection and the genetics of social evolution. But Wynne-Edwards's propositions of epideictic displays and social conventions, and his analogy on overfishing, were gradually and convincingly refuted one by one on evidential grounds, or matched by equally plausible explanations of individual selection. The counterattack culminated in G. C. Williams's important book, *Adaptation and Natural Selection* (1966). Williams thoroughly and convincingly documented the argument that the goal of an individual's reproduction is not to perpetuate the population or species, but to maximize the representation of its own germ plasm, relative to that of others in the same population.

The dragon was slain, but according to Wilson (1975, p. 110), ". . . for a long time neither critics nor sympathizers could answer the main theoretical question raised by this controversy: What are the deme sizes, interdemic migration rates, and differential deme survival probabilities necessary to counter the effects of individual selection?" How would populations behave if selection were important at the group level? Population geneticists eventually confronted the problem. Deductions from models attempting to predict the course of evolution by group selection led to the conclusion that selection at this level is feasible but unlikely and improbable. Conclusions from the models suggested further that "self-restraint on behalf of the entire population is *least* likely in the largest, most stable populations, where social behavior is the most highly developed" (Wilson, 1975, p. 113).

News of the demise of group selection, as Wynne-Edwards envisioned it, traveled at a snail's pace. Indeed, it has reached woefully few places as yet. This type of thinking is still evident and all too obvious in many high school and college classrooms and in college introductory biology textbooks (e.g., Ebert *et al.*, 1973, p. 438). It remains the predominant lay interpretation of how natural selection operates as is evident in magazine articles (e.g., *Newsweek,* 12 April 1976) and in movies (e.g., David Wolper's *Birds Do It, Bees Do It*) and this interpretation is so solidly entrenched in the social sciences it is almost ideology. The telltale signs consist of phrases such as "good for the species," "enhances species survival," and "preservation of the population." Obviously, it will take some time before this misinterpretation of Darwinian selection is corrected.

2.2. *Individuals Can Increase Their Fitness through Their Relatives*

In Darwinian terms, every characteristic of an organism is a means and a consequence of reproductive competition among individuals. This idea is so integral to the theory that Darwin issued the bold challenge (1859, p. 201): "If it could be proved that any part of the structure of any one species has been formed for the exclusive good of another species, it would annihilate my theory, for such could not have been produced through natural selection." Similarly, no behavior can benefit another individual at a reproductive cost to the performer (Alexander, 1975; Eberhard, 1975), when reproductive cost and benefit are measured in terms of fitness (see Williams, 1966). Thus, cases of apparent altruism proved difficult to explain until W. D. Hamilton, in a set of seminal papers (1964, 1970, 1971a,b, 1972), extended classical thinking by introducing the concept of *inclusive fitness*. This concept encompasses an individual's personal fitness plus his effects on the related parts of the fitness of all his relatives. It accounts for the individual's total effect on the gene pool of the next generation through the production of his own offspring *plus* his effects on the reproduction of other individuals with whom he shares genes.

This concept explains in a straightforward and simple way, why individuals perform acts of self sacrifice. The answer is because the performer incurs a net benefit. The circumstances under which an "altruistic" act will be performed (or a gene for altruism will evolve) is given by the formula

$$k > \frac{1}{\bar{r}}$$

in which k is the ratio of gain in fitness to loss in fitness and $\bar{r}$ is the average genetic relationship to the relative. Hamilton shows that in order for an individual to risk his life (or his fertility) for his brother (with whom he shares ½ of his genes by common descent, or $\bar{r} = ½$), the brother's fitness must more than double, an uncle's ($\bar{r} = ¼$) would have to be advanced fourfold, a first cousin's ($\bar{r} = ⅛$) eightfold and so on. That is, an act of apparent welfare will be performed if the genetically related recipient is likely to enjoy reproductive success in the future, which more than compensates the performer for his momentary loss. Thus, what appears to be altruism is in reality selfishness or as Ghiselin (1974, p. 247) puts it: "Scratch an 'altruist' and watch a 'hypocrite' bleed."

This type of reasoning enabled Hamilton (1964) to come up with an ingenious solution to Darwin's dilemma about social insects. He hypothesized that the mode of sex discrimination, haplodiploidy, played a central role in the evolution of sociality in these insects. Hamilton pointed out that in haplodiploidy the coefficient of relationship between sisters is ¾, whereas between mothers and daughters, it is ½. This is so because sisters have in common the genes from their father, who is homozygous, and they share on the average ½ of the genes of their mother, who is heterozygous. Since each sister receives ½ of her genes from her father and ½ from her mother, the coefficient of relationship between sisters is

$$(1 \times ½) + (½ \times ½) = ¾$$

In other words, in species with this type of sex determination offspring can increase their inclusive fitness more by caring for younger sisters than by caring for their own offspring with whom they share only ½ of their genes! Species with this type of sex determination should stop having offspring in order to help their sisters, i.e., they should become social. Indeed, Wilson (1975) points out that "true" sociality has evolved at least 11 times in the order Hymenoptera, where sex determination is based on haplodiploidy, and only once in any other Arthropod. Numerous other predictions from the model, based on the genetic relationships between individuals in the hive, are borne out, e.g., hymenopteran males should contribute virtually no labor to the colony.

Besides offering an explanation for altruism in social insects, the concept of inclusive fitness predicts that there are limits to selfish behavior. Selfishness will be selected against when relatives are harmed to such an extent that it lowers the inclusive fitness of the selfish individual. Similarly, spiteful behavior may evolve if it results in an increase in the inclusive fitness of the perpetrator (Wilson, 1975). Hamilton's theory is supported by the finding that most cases of altruism observed in nature occur between relatives, e.g., helpers at the next, cooperative breeding and some cases of adoption (see the review in Wilson, 1975). Trivers (1971) has produced a model showing how altruism between unrelated individuals ("reciprocal altruism") might operate and still be consistent with Hamilton's theory.

The implication of these ideas for the study of social behavior is that "an animal is expected to adjust its behavior (whether altruistic or selfish) toward another individual according to its degree of relatedness to that individual" (Trivers, 1974a, p. 525). It is interesting to note the difference between this prediction and the influential credo of the behaviorist J. B. Watson to the effect that the role of genetics is minimal as a source of human and animal differences (Watson, 1925).

Hamilton's concept of inclusive fitness and the predictions from it have been called the theory of "kin selection" (Maynard Smith, 1964) for good reason. Hamilton clarified the relationship between kin selection and interdemic selection (group selection à la Wynne-Edwards) by demonstrating that kin selection could be explained by selection operating on the individual through his offspring and his relatives. Thus, both kin selection theory and the individual-versus-group-selection controversy emphasized that each individual is selected to behave in such a manner as to maximize his or her reproductive success. This important aspect of the new movement in evolutionary biology represents a return to Darwin and his original interpretation of how selection works. Reinstatement of the individual as the principal unit on which selection operates has important implications for understanding sexual selection and the study of social and sexual behavior.

2.3. *The Operation of Sexual Selection Is Governed by the Unequal Investment of the Sexes in Offspring*

To determine how sexual selection operates, why males compete for females and females choose their male partners, one needs precise data on reproductive success in each sex. Bateman (1948) collected such data in an elegant series of experiments on a single species, *Drosophila melanogaster*. His experiments consisted simply of putting five adult males, each with a different marker gene, in a jar with five adult females. After two days, he removed the females, isolated them until their eggs were hatched and checked the appearance of the progeny to determine male parentage. His results revealed that males and females maximized their reproductive success in quite different ways and that the contributions of males to the next generation was much more variable than that of females. Some males were very successful but almost a quarter of them did not leave any offspring at all despite vigorous attempts to mate. Females simply refused to mate with them. In contrast, almost every female had some offspring but the most successful female had three times fewer progeny than the most successful male. The few females that left no offspring were courted but refused to mate. Each male mated as often as he could and the more females he mated with, the more offspring he produced. Most females mated once and they left no more offspring if they mated more frequently.

Bateman concluded that his results could be explained by the differential energy investment of each sex in their sex cells. Since male flies invest little energy in producing sex cells, which are very small compared to female eggs, a male's reproductive success is not limited by sex cell production but rather by his ability to fertilize eggs, i.e., find willing females with whom to mate. On the other hand a female has no trouble getting her eggs fertilized but is limited by her ability to produce eggs. Because male sex cells are considerably smaller than those of females, in most plants and animals (anisogamy), Bateman interpreted his results as having wide applicability for all but a few primitive species and those with a monogamous breeding system.

More than two decades later, R. L. Trivers (1972) exhumed this long dormant study, perceived the explanatory value of Bateman's insight, and in a landmark paper expanded the argument to apply to all animals independent of their breeding system. He restated Bateman's argument and put it into a more general form with a single variable, parental investment, controlling sexual selection. He defined parental investment as "any investment by the parent in an individual offspring that increases the offspring's chance of surviving (and hence reproductive success) at the cost of the parent's ability to invest in other offspring." Thus defined, parental investment includes all metabolic investment in the sex cells plus any subsequent investment that benefits the young, e.g., feeding or protecting it. Each offspring is viewed as an investment independent of any other offspring produced, and investment in one tends to decrease the potential investment in another. This means that investment is negative and can be considered a cost from the parent's point of view. Trivers's theory assumes that natural selection favors the total parental investment that leads to maximal reproductive success, e.g., in a season or a lifetime.

Trivers's reformulation of Bateman's argument can be summarized as follows: the sex whose parental investment is greater will become a limited resource. Individuals of the sex investing less will compete with each other to mate with members of the sex investing more since they can maximize their reproductive success by producing successive offspring with several members of the limiting sex. This argument leads to the conclusion that the relative parental investment of the sexes in their offspring governs the operation of sexual selection. In other words, the sex with the smaller per-offspring investment will evolve extreme epigamic displays, it will be larger, more aggressive, noisier and more conspicuous. These traits are adaptations useful in intrasexual competition or in attracting the opposite sex. The sex with the larger per-offspring investment will be coy and selective in mating because it has invested more and has more to lose.

In most animals male investment in the offspring is minute, consisting only of the sex cells. The rule is that the female does everything in the way of parental care. Hence, females are a limited resource and males fight for females and females are coy and selective in mating. This explains why males are sexually relatively indiscriminate; they are investing only their sex cells and sperm are cheap relative to eggs. The theory predicts that in species where male investment is equal to that of females such as in many species of monogamous birds, male and female reproductive success should vary in similar ways. Sex differences should be minimal. When male parental investment exceeds that of females, the latter should compete for males and males should be selective in mating. This last possibility provides a good test for the theory. As the theory predicts, sex-role reversal is strong in polyandrous bird species where the male alone broods and cares for the young; in these species, the female is larger and courts the male. Field data for other species are scanty but support the theory (Trivers, 1972; Wilson, 1975).

How did the present parental investment pattern governing sexual selection originate? From the combined writings of Bateman (1948), Trivers (1972), and Parker, Baker, and Smith (1972), one infers that it all goes back long ago to differentiation of the sex cells according to size and mobility. Originally, all sex cells are thought to have been of an equal size, i.e., isogamous. One sex, which we now call males, hit on the strategem of reducing the size of the sex cells, initially producing two small ones instead of one large one. By sacrificing the amount of nutrition in the sex cell package and duplicating the DNA, the total energy expended remained the same but it produced more sex cells. This revolutionary reproductive maneuver must have been much more successful than the previous strategy. Selection for still greater reduction in sex cell size followed. This resulted in an unbalanced situation with many more sperm than ova. Since the small stripped-down-to-the-DNA sperm could not live long, they had to make quick contact with the large sex cells of females. These sexual inequalities set the scene for intense competition among sperm and selected for movement of sperm to the female egg cells. Parker (1970) reasons further than this state of affairs favored males that placed sperm closer to females, the source of eggs, and this culminated in internal fertilization.

The larger egg size relative to that of the sperm, meant a larger investment by the female than by the male and this set the course for things to come. This initial inequality is a key element in Trivers' theory. The female's ". . . initial very great investment commits her to additional investment more than the male's initial slight investment commits him" (Trivers, 1972, p. 144). It is this early imbalance in parental investment that gives rise to different male and female reproductive strategies. It is the origin of what the humorist, James Thurber, called "the war between the sexes." In this conflict, male and female interests rarely coincide. Indeed, as Parker *et al.* (1972) point out, males, in a sense, parasitize females and propagate at their expense.

The elegance of this theory is perhaps best illustrated in Trivers's explanation of desertion and cuckoldry. These reproductive strategies are predicted from the rate at which males and females of a monogamous couple invest in their offspring. At any time, the individual who has invested less is tempted to desert,* and this is particularly true if the disparity in investment is large. Since the individual left with the offspring (usually the female) has more to lose, it will be likely to continue to invest and rear the offspring alone. The deserter benefits from any degree of success. In addition, the deserter may find another mate and may thus be doubly successful. If the female is deserted, it may be to her advantage to cuckold another male into helping to raise young that are not his own. Males are vulnerable to cuckoldry whenever they invest strongly in the young. Vulnerabilities such as desertion and cuckoldry lead to adaptations to reduce their occurrence and counteradaptations to increase them. For example, females should be selected to choose a mate that will invest substantially in the offspring and not desert her. Males should be selected to avoid being cuckolded. A male can avoid this possibility by having a long engagement period before investing or by preventing other males from getting near his mate. The latter is well expressed in a popular song by the male blues singer, B. B. King, "I don't want anybody hanging around my house when I'm not at home." Thus, even when the sexes come together in a seemingly joint effort, there is conflict. Indeed, much of the behaviors we see during courtship are adaptations on the part of one sex to prevent being taken advantage of by the other.

* R. Dawkins and T. R. Carlisle (*Nature* (*London*) 1976, *262*, 131–132) have recently pointed out that this reasoning is fallacious (see also J. Maynard Smith, *Animal Behaviour*, 1977, *25*, 1–9). This flaw does not alter Trivers's theory in a significant way.

In summary, the most important theoretical innovation in our thinking about sex and evolution is a reemphasis on natural selection acting on the individual, and through the individual on his relative. This logic predicts that whenever different genotypes interact, conflict of interest and disagreement can be expected. Thus, conflict lies at the very heart of sexual reproduction (Trivers, 1974a). Much courtship behavior results from a contest between two self-interested, different individuals, a contest between "skilled salesmanship among the males and an equally well-developed sales resistance and discrimination among the females" (Williams, 1966, p 184). In a brief performance, the courting sex, usually the male, is selected to convince the female that he is fit, and if presenting a false image helps, then this deceit (even self-deception) will be selected. In contrast, the courted sex will be strongly selected to distinguish the fit from the merely pretended fit and can be expected to proceed with caution and discrimination. Thus, in terms of this body of theory, behaviors such as selfishness, altruism, spite, deceit, and self-deception are not simply moral issues peculiar to humans and worthy of debate by ethical philosophers, they are universal reproductive strategies that are favored by natural selection because they benefit the performer.

3. *Reproductive Strategies: A Promising Direction for Future Research*

Since the impetus for evolution by natural selection is differential reproduction, and since selection acts primarily on the individual, individual approaches to reproduction warrant careful study. Thus, the body of theory just described placed great emphasis on the study of reproductive strategies. In the remainder of this chapter, I will outline various individual strategies that have been described, others that are predicted from the theory, and the types of studies that are being done from this point of view. I will try to mix judiciously what has gone on recently, what is happening now, and what seems promising for the future. Some of the work I mention may not have been performed with social theory in mind but if it fits, I will use it. My main purpose is to give a sense of the breadth rather than the depth of the approach. Most of the work I refer to is presented in detail elsewhere and is my rationale for treating the material synoptically.

Reproductive strategies of males and females are best considered separately, except for those that are so basic that they are common to both sexes. Moreover, within each sex, it is possible to view reproductive strategies on a temporal dimension ranging from those evident early in life to those operating around the time of mating and even shortly after mating takes place.

3.1. Male Reproductive Strategies

How do males go about maximizing the number of females inseminated while at the same time minimizing their costs or losses? What are some strategies for monopolizing receptive females? How do the reproductive strategies of polygynous males vary with the distribution of females? Do males that conquer other males leave more offspring? It is well known that male reproductive strategies vary considerably with the distribution of females. Among polygynous males, this is most obvious when females are clumped in space and time. In this situation, male competition centers on keeping as many females as possible inaccessible to other males. Some strategies used in this context are commonly employed by males from a wide variety of species. These include territorial defense, dominance relationships, and spacing based on individual distance. These strategies are also commonly referred to as modes of social organization. Despite the enormous amount of time and effort that has been spent studying these behavior patterns, surprisingly few field studies have attempted to correlate the outcome of intermale competition (for territory or for status) with reproductive success. Reasonably complete data are available only for dragonflies, dung flies, a few species of gallinacous fowl that lek, a few frogs, some lizards, the elephant seal, and baboons (see Trivers, 1972, for a review). If one assumes that the form of male competition is a reproductive strategy that enables some males to more readily acquire females, either directly or via possession of a resource, then determining the effectiveness of the strategy should be of paramount importance.

One reason why this question has been hard to answer (or has not been asked in the first place) is because of the difficulty in determining male reproductive success. Since virtually all polygynous males copulate in a brief encounter with the female, and the male is absent at the time of hatching or parturition, it is difficult to determine how many offspring the male sired. Most investigators have had to be content with counting copulations. There are problems with this approach since females of many species mate with more than one male and many copulations occur at night or when the observer is not looking. Indeed, in some species, copulations are rarely observed. A better handle on the male reproductive success variable is needed in order to answer some of these basic questions. Ideally, it should be one that enables the investigator to identify the individual offspring sired by a particular male. Genetic markers, based on color or other morphological variation, although useful in the laboratory (e.g., Bateman, 1948; Guhl and Warren, 1946; De Fries and McClearn, 1970), have not proven feasible in field studies. The technique of starch-gel electrophoretic analysis of polymorphic blood proteins, used so successfully by Selander and coworkers in studying intrapopulation gene flow (e.g., Selander, 1970;

Selander and Johnson, 1972; Selander and Kaufman, 1973), appears to be a promising approach for determining male parentage in future field studies.

How flexible are strategies like defending a territory against other males or achieving a high social rank? How do these strategies vary with changes in the social and physical environment? A short time ago it was thought that having observed a particular social system in a population of animals it could be said that the behavior pattern characterized the species. It was thought that there was a correlation between the intelligence of an animal species and the complexity of its social system (see Crook, 1970). More and more studies are now showing that these behavior patterns vary considerably with the social context, the food supply, or the physical setting in which they occur. This is to be expected if the behavior pattern being observed is a reproductive strategy and the availability of the resource is shifting. Flexible reproductive strategies are apt to be most successful. Males should adjust their behavior so that they maximize mating with the greatest number of females while at the same time minimizing the amount of energy expended to keep females inaccessible to other males. The following studies illustrate this plasticity in male strategies.

Alaska fur seal males, *Callorhinus ursinus*, are rigidly territorial during the breeding season on the Pribilof Islands in the Bering Sea. Each territorial male keeps other males out and herds approximately 10–20 females in the confines of his territory, which is small and fixed. The terrain is rocky and presents numerous cues for setting up boundaries. However, the behavior of males was quite different in a new colony formed in 1968 on San Miguel Island, California (Peterson, Le Boeuf and De Long, 1968). The male in charge of this aberrant harem of 40 females followed the heat-stressed females as they moved to and from the water during the course of the day. Peripheral males also followed the movements of females and occasionally tried to intercept females as they arrived. With so much unoccupied area and so few competitors and females, it would have been a poor strategy, indeed, for the largest, most aggressive bull to remain fixed to one spot and work hard to herd females into this area. Obviously, what was a sound reproductive strategy in a densely populated rookery was less adaptive under the different conditions in the new colony. In the latter, the best strategy was to remain with the female group wherever it went and keep other males away. At what point does this local strategy become ineffective as the number of females increase? Will the fact that the new colony is located on shifting sands make it difficult for males to set up rigid territories with fixed boundaries in between?

The northern elephant seal provides an example that is almost the reverse of the above. In most colonies, females in a harem may number up to 300, and competing males exhibit a clear-cut social hierarchy. The alpha male tries to prevent all males from encroaching into the harem of females.

Males immediately below him in rank keep others out while at the same time avoiding the number one male. The number of males that enter the harem is a function of its size. In a few rookeries, the density of females is so great that harems become confluent and a breeding beach appears to be one large blanket of females and pups interspersed with competing males. In these super rookeries, the most dominant males—those that would be the alpha males in smaller, discrete harems—stop trying to keep all other males out of the harem and away from females. This is futile and to attempt it would mean that they would have time for nothing else, i.e., more important matters like mating. In effect, the most dominant males become more territorial. They situate in an area dense with females, mate with them, and only work to keep other males out of this immediate area. As the season progresses and the density of females declines and harems become discrete once again, alpha males switch back to their former strategy of trying to keep all males out of the harem.

Lastly, an excellent study by Campanella and Wolf (1974) illustrates the flexibility of male strategies in the mating of the dragonfly, *Plathemis lydia.* Females of this species oviposit on restricted, predicted areas of ponds during the middle of the day. Males arrive early in the day and compete aggressively for the highest dominance status in a particular area. Since females only appear on the leks around midday, males compete most vigorously for this time of day and the winner does most of the mating. In this breeding situation, the most dominant male commands the optimal time in a particular place.

If the dominant adult male employs a strategy that enables him to monopolize females and keep other males away from them, what do the other males do? What kinds of strategies do young, smaller, or less aggressive males employ? What are some alternative or minor reproductive strategies and how effective are they? In some species, peripheral males wait for their opportunity to depose the top male and take his place or they adopt a completely different strategy which gives some small measure of success. Increasing attention is being paid to these alternative reproductive strategies and an excellent example occurs in the bluehead wrasse, *Thalassoma bifasciatum,* a coral reef fish found throughout the Western Atlantic (Warner, Robertson, and Leigh, 1975). During spawning, females visit a restricted reef area near midday where several dozen large males hold their spawning territories. They choose the largest and most brightly colored males for mating. Large males spawn more than 40 times a day and small ones less than twice a day. The preference of females for large males prompts smaller males to tamper with females waiting to be served. Small males often achieve some success by rushing into a large male's territory and releasing their sperm just as the spawning pair are shedding their gametes into the water. This strategy is called "streaking." Another alternative strategy, "sneaking," takes this form: the small male may unobtrusively follow a female into a large bluehead's territory and attempt to stimulate a spawning rush with a female as she rises to the actively courting territory owner. Small males can fertilize eggs in still one other way. They intercept females en route to the reef edge or within a large bluehead's territory. As a group, they pursue her and subject her to repeated tactile stimulation until she releases her eggs. As Warner *et al.* (1975) point out, it is external fertilization that makes these alternative strategies of streaking, sneaking, and group spawning possible. An important part of the strategy of the dominant large bluehead is to vigorously defend his territory, and the females within it, to prevent her eggs from being fertilized by other males' sperm.

A male reproductive strategy that has been given very little attention is rape, i.e., copulation when the female appears unreceptive but the male appears to overcome her rejecting behavior. Some basic questions are: How forcefully can a male pursue copulation with a female and still fertilize her eggs? How effective is rape as a reproductive strategy and in what animals? When is the optimal time in a female's estrous cycle for a male to deposit his sperm, particularly when other males may mate before or after him?

A major obstacle to studying rape as a male reproductive strategy has become a semantic one. Estrus is usually defined as that time interval between a female's first and last copulation during an estrous cycle or breeding season (Beach and Le Boeuf, 1967). In other words, a female is receptive if she copulates. This definition of estrus precludes the possibility of rape. A female mates or she doesn't and female animals cannot tell the investigator whether or not they were willing, so the issue of rape in animals is often avoided. Yet few would deny that in many species males try to induce females to mate before they are ready. If so, how forceful must a male be in order to call a copulation a rape? At what point do we infer that copulation occurred despite the female's objections, particularly when the "objections" are subject to interpretation (e.g., Cox and Le Boeuf, 1977). The fact is that "the boundaries between seduction and rape are often far from clear" (Ghiselin, 1974, p. 136).

Despite these semantic and methodological difficulties, the phenomenon deserves further study. This kind of strategy is predicted by sexual selection theory; individuals of one sex can be expected to abuse individuals of the other sex if there is a reproductive benefit to be gained. Thus, males will rape or desert the opposite sex and females will cuckold males.

In species where male reproductive success is limited solely by the number of females fertilized, there may be an obvious selective advantage to being the first male to mate with a female if the first male to mate is more likely to inseminate her. Pressing the female to copulate early in her estrous period would appear to be a fruitful strategy. In sexually dimorphic species, in particular, one might expect males to try to overpower the smaller females simply because it may be to their advantage and they are capable of doing it. This strategy should be most apparent in pro-estrus.

Rape or attempted rape appears to be a predominant strategy among male northern elephant seals who are two to three times larger than females (Cox and Le Boeuf, 1977). Males attempt to copulate with pregnant females as soon as they arrive on the rookery. They continue to mount forcefully until copulation occurs on approximately day 19 after parturition. Throughout this period, and even during estrus, i.e., when copulation occurs, the female attempts to wriggle free when mounted and she "protests" mounts by uttering threatening calls and biting the male on the neck.

Other recent studies suggest that the occurrence of rape in nature may be far more common than has been suspected. Parker (1970) thinks that rape may account for some cases of observed sperm competition in Drosophila. De Long (1975) witnessed interspecific rape on an island where several pinniped species breed at the same time; Steller sea lion males raped and often killed smaller California sea lion females and California sea lion males overpowered smaller Alaska fur seal females. According to casual reference in the literature, rape also occurs in dung flies and some frogs, where amplexus is so hard that the male does not let go until the female releases her eggs.

Certainly the possibility of rape is high in the standard laboratory sex test where the female has no avenue of escape and the male can repeatedly try to force copulation with her. Might this unnatural situation elicit attempts to rape females? In one species on record, captive female white-tailed deer have been killed by males repeatedly trying to mate with them *shortly before* they were ready to ovulate (Severinghaus, 1955). A great deal of work needs to be done on this neglected aspect of male behavior. The basic questions are whether this is a favored reproductive strategy, to what extent it is effective, and in what species? We know that males will mount females before they are receptive, but can a male force a female to copulate before she is ready and is this a good reproductive strategy? How far in advance of ovulation is a mating apt to lead to fertilization? When a female copulates with several different males, which one fertilizes her egg(s) and why? Is it likely to be the first male, the last one, or the one that mates nearest the time when ovulation occurs? It is well known that the second male to mate has the advantage in some insects (Parker, 1970). How does this male strategy affect the female's reproductive success?

Although female choice of males at mating time is far more common than the reverse, especially in polygynous species, mate selection among males might be beneficial to males of some species under certain conditions. Males might be expected to prefer females that will produce the greatest number of progeny. In several fishes, males prefer the largest females, e.g., the swordtail, *Xiphorus helleri* (J. Nelson, personal communication). Large females carry more eggs. When a male pairs for a season or a lifetime, what is the basis for his choice of mate? Is it his choice? Much work remains to be done on male choice.

Many other male strategies occur around the time of mating, but space does not permit me to enumerate. For example, what makes the behavior of some males more attractive to females than that of others? How do males maximize their chances of being chosen? Instead, I will describe a few strategies that are employed before mating is imminent. One reasonable male strategy is to speed up the onset of estrus and ovulation in the female encountered. When nomadic male langurs, *Presbytis entelus*, usurp a troop, they drive off all resident males and routinely kill all of the infants (Sugiyama, 1967; Mohnot, 1971). Besides avoiding investing in another's offspring, killing the infants causes females to come into estrus faster than if they continued to nurse. Thus, the usurpers can more quickly inseminate the females. A similar form of infanticide occurs in male African lions of the Serengeti plains (Schaller, 1972; Bertram, 1975). The Bruce effect in mice does not kill the infants directly. When a strange male mouse is introduced to a pregnant female, his odor is sufficient to cause her to abort and she quickly becomes available for reinsemination (Bruce, 1967, 1969). Obviously, this type of strategy is in conflict with the reproductive success of females and the latter will develop counter-measures to prevent being exploited in this way.

Some species in which females are induced ovulators, i.e., copulation induces ovulation, might be put on this same dimension. It is well known that copulation is violent in most mustelids because sensory stimulation during copulation (primarily to the vagina and cervix, but generalized stimuli will also suffice) triggers ovulation. In these species, it is reasonable that selection would favor males that are rough and aggressive during mating. Such a male would benefit by causing ovulation when his sperm is present in the female's reproductive tract. Females might benefit from these matings because only adult males are large and tough enough to provide the necessary stimulation for ovulation, these males are theoretically more fit than younger males and, hence, will provide her with more viable offspring. Courtship synchrony in colonial sea birds might also be viewed from a similar perspective.

In reproductive competition, the jockeying for position may begin early in life and battles won long before maturity may influence mating and have important reproductive consequences. Consider the case of the wild turkey, *Meleagris gallopavo*, in southeastern Texas (Watts and Stokes, 1970). Young males break away from their brood flock in late fall at six to seven months of age. During the first winter, pairs of brothers struggle with each other until one individual becomes dominant. Next, brotherhoods meet in contests with other such brotherhoods until one, usually the largest, is dominant to all others in the winter flock. Finally, different flocks compete and one emerges dominant. The result of this elaborate tournament is that one male holds the dominant position in the entire local population of turkeys. In February, all brotherhoods gather on the mating grounds where each tries to outstrut and outfantail the other. Brothers display in synchrony with each other before passing females. But when a female is ready to mate and chooses a brotherhood, the subordinate brother yields to his dominant brother and subordinate brotherhoods yield to more dominant ones. Consequently, only a fraction of the males present do all of the mating. Obviously, the outcome of early sibling fights has a great bearing on reproductive success in adulthood. As for the subordinate brother in a pair, his chances of ever mating are slim. His best strategy for leaving genes in the next generation is to help his dominant brother mate since any success of his close relative increases his inclusive fitness.

The competition for the alpha male position and the enormous reproductive advantage it yields begins shortly after birth in northern elephant seals. Among males, great size helps in fighting and fighting determines social rank. It is apparent from the following observations that natural selection has favored a growth strategy. Males are on the average 4.5 kg heavier and 63 mm longer at birth than females (Le Boeuf and Briggs, 1977). The male mortality rate is greater than that of females during the four-week suckling period, in large part because males will more readily try to suckle alien females in addition to their own mothers. This is a gamble with the payoff being a greater fat store at weaning time if the pup wins. The risk involved is being injured and perhaps killed by an alien female. At four weeks of age, weaning occurs when the mother simply leaves the rookery and goes to sea. Males manage to suckle their mothers, on the average, one full day longer than females (perhaps to the mutual advantage of parent and young). At this point male "weaners," much more so than females, attempt to reenter the harem and steal milk from alien females. Twenty-six males compared to four females were observed stealing milk in one harem at Año Nuevo Island in 1975. Not only were males more successful but they tried harder when they failed. On 49 occasions, males were thwarted by females in their attempts to suckle nursing females as compared to five thwarted attempts for female pups (J. Reiter and N. Stinson, personal communication). Once again, the gamble is that the winner may be bitten and seriously injured, but if successful in stealing milk, it achieves a size that is markedly greater than its peers. If one assumes that size at weaning is correlated with size in adulthood (see Trivers and Willard, 1973), then the selective advantage is obvious. The fact that the canine teeth of males do not erupt until they are about seven to eight weeks of age whereas the teeth in females erupt at three weeks of age (A. Hoover, personal communication) may be an adaptation that better enables males to steal milk without being detected.

Why don't females gamble as readily as males to get extra milk? The wide gap in reproductive potential between males and females in this species helps to explain the appearance of sex differences early in life. A successful male in a colony may sire 250 offspring in a lifetime but the very best that a female can do is to produce 10 to 12 pups in her lifetime (Le Boeuf, 1974). The catch is that nearly all females are somewhat successful but only a few males enjoy any success at all and these males are supersuccessful. Slightly greater size among females would not lead to significantly greater reproductive success as it does in males and the young female who attempts to steal milk incurs the very same risks. For a male to be successful, he must have everything going for him. Thus, a high-risk, high-gain gambling strategy should be favored by natural selection and this strategy should be apparent early in life. A great deal more work needs to be done on lifetime strategies such as these.

Male reproductive strategies do not end with mating. In some species males have evolved numerous behavioral and morphological traits to prevent additional males from introducing their sperm into a recently mated female. This is an important strategy in many animals, such as certain insects, where sperm competition within the female reproductive tract is intense and the temporal order of sperm introduction is important (Parker, 1970). Postcopulatory strategies may be as diverse as sequestering the female, remaining *in copula* several hours or depositing a vaginal plug. In the ceratopogonid fly, *Johanseniella nitida,* the male blocks the entrance to other males with his own body which the female eats, leaving his genitalia attached to her (Wilson, 1975). The importance of sperm competition in vertebrates is largely unknown and is just beginning to be investigated. This unexplored battleground for male–male competition, and female choice, offers many intriguing puzzles and great promise for future research. Is it possible that males produce a varied assortment of sperm as an adaptation to fluctuations in the female reproductive tract, e.g., some that swim fast, others than live long, some specialized to compete with the sperm of other males that might be present?

3.2. Female Reproductive Strategies

Besides being aimed at producing the greatest quantity of offspring, like the predominant strategy of most males, female reproductive strategies stress the quality of the offspring and survival during the period of maternal care. One approach to implementing these strategies is for females to choose among male suitors. There are two basic reasons why a female might be selected to discriminate among males: (1) to get the best genotype for her offspring, and (2) to get the male that will provide the best resources for feeding and protecting the young.

Selection should favor females that discriminate between potential mates because one half of the genetic complement to their offspring comes from the father. This should hold true whether the father of a female's offspring helps to rear them or not. In species where males invest little or nothing in the offspring, females should be selected to choose on the basis of genotype alone (Trivers, 1972). How does a female determine which male will provide the best genotype for her offspring? How does she maximize the probability of mating with the most fit male available? Trivers (1972) and Selander (1972) argue that females can enhance the quality of their offspring by choosing males that have demonstrated their fitness. This seems to be the basis for female choice in birds that lek, such as grouse and the ruff (Wiley, 1973). The attributes of the preferred males are that they are older, more aggressive, and more sexually experienced than other males present. If mortality is partially a function of genotype then males that survive to maturity have demonstrated an important aspect of their fitness. Similarly, highly aggressive males that achieve high social rank or the best territories must be in good physiological condition, and this characteristic will be more likely to be evident in their progeny. In many animals, especially in birds that lek, female choice is influenced by a male's "courting technique," a behavior pattern that develops with experience. Thus, it seems that females can enhance the quality or probable success of their offspring by mating preferentially with older, aggressive males. A similar basis for female choice is found in a few ungulates where males compete for mating territories, e.g., the Uganda kob, *Adenota thomasi* (Buechner and Schloeth, 1965; Leuthold, 1960). Female choice, based on more subtle discriminations among many appropriate males, has proven more difficult to document.

Instead of passively allowing males to compete and then choosing to mate with the winner, females of some species play a more active role. They incite competition among males (Cox and Le Boeuf, 1977). Theoretically, the benefit of this strategy is that the female can better survey the field, so to speak, and screen males according to fitness. This strategy is most apparent in species where several males compete to mate with females, where a social hierarchy exists, and where interference with mating is common. These conditions exist in many species of fishes, lizards, toads, frogs, gallinaceous birds, carnivores, ungulates, and primates. The method of inciting competition varies among species. For example, I pointed out earlier that female elephant seals respond to attempted copulations with loud vocalizations and vigorous escape movements. This blatant response catches the attention of all males in the vicinity and, if any onlooker is dominant to the mounting male, he will prevent copulation from occurring by threatening and chasing off the mounter. Thus, the probability that the mounting male will be interrupted is a function of his social rank. The net effect of the "signaling" behavior of the female is to increase the probability that mating will occur with a high ranking male, the type of male that will presumably provide the best genotype for the female's offspring. It is interesting to note in connection with the earlier discussion of rape as a male strategy that the female elephant seal's response to mounts selects for males that attempt to rape her. Only the most dominant male is unaffected by a female's vocal display when she is mounted because all males in the vicinity are subordinate to him. The result is that he can continue to try to force the female to copulate without being interrupted. An important consequence of this is that the initial copulations of most females occur with high ranking males at the end of a mount of long duration in which the female resisted and protested throughout. Earlier, I mentioned the advantage of this strategy for the male. Paradoxically, it may also be to a female's advantage to be raped, or more precisely, to be inseminated by a male who is most capable of raping her because he is probably the most fit of all males present. The highest ranking male is mature and since his position is gained and maintained by fighting this would indicate that he is in good health.

Female incitement of male competition is only one way in which a female can influence the genotype of her offspring. No doubt, selection has favored various other approaches to this problem. The manner in which females go about solving this problem has just begun to be studied.

In species in which males engage in parental care, female choice should also involve the male's willingness and ability to be a good parent (Orians, 1969; Trivers, 1972). How does a female determine which male will make the best parent? What cues does she use? There is much evidence in several bird species that females choose males on the basis of how rich in food are their territories (e.g., Hinde, 1956; Verner, 1964). The resources the territorial male can supply to the female and her offspring are so important that she may choose to mate with a male with a superior territory who already has one or more females in it rather than with a single male in a poorer territory. Females might also be expected to look for information in the courtship behavior of their potential mates that predicts their promise as parents (Trivers, 1972). In a study of a colony of common terns, *Sterna hirundo,* Nisbet (1973) observed that the amount of food the male feeds the female during courtship predicted the degree of his investment in the offspring. He recorded great variation among pairs of terns in total weight of the clutch, the amount of food fed to the chicks by the male, the weight of the brood and the number of chicks fledged. All of these variables were positively correlated with the amount of food fed to the female by the male during courtship. Erickson and Martinez-Vargas (1975) also considered this problem in their studies of the ring dove, *Streptopelia risoria*, a species in which the male and the female share in nest building, incubation and squab care. They point out that it would be advantageous for each animal, particularly the female, to know at an early stage whether the mate being selected is physiologically prepared to carry through with the entire breeding sequence. Since physiological readiness of males may vary with age and time of the season, a female's choice of mate could be critical in determining her reproductive success. In this species, physiological readiness to court, nest build, and incubate are all influenced by secretions of gonadal steroids. This coincidence led to the hypothesis that sex-steroid levels, being common to all these reproductive behaviors, may provide the important cue to male investment. If the sensitivity of the system mediating the courtship displays is lower than that of the later occurring parental behaviors, then the emergence of courtship displays guarantees that hormone levels are adequate for nest building and incubation. This is an intriguing idea, and illustrates the power of combining sociobiological thinking with physiological studies.

How does a female maximize the quantity of progeny produced in one season? Bateman (1948) demonstrated that the number of eggs a female fruit fly produced was limited to her body size. This constraint seems to hold for most arthropods. In vertebrates the limiting factor is the number of offspring a female can feed and wean or fledge in a healthy condition without endangering her own survival. This point was well documented in Lack's classic studies on nidicolous birds (Lack, 1954), in which he showed that clutch size was adapted through natural selection to correspond with the maximum number of offspring for which the parents could find food without seriously harming themselves.

The number of offspring a female can fledge or wean in healthy condition is in large part a function of the amount of help she gets in the task of rearing the young. Thus, selection should favor those females that reduce the burden of investing in the young as much as possible. Conceivably, help could come from a conspecific male or female, a relative or stranger, or even a member of another species. A cursory inspection of the literature on animal behavior shows that all of these possibilities occur. Most frequently, the female is helped by her mate, a conspecific male, and we call this monogamy. Monogamy is thought to have evolved in species where both parents could optimize their reproductive success by combining their efforts in rearing the young (Lack, 1968). This system is most widespread in birds and two factors were instrumental in developing it. A variable food supply provided strong selection pressure for the parents to cooperate in feeding the young, and the fact that birds do not nurse made it possible for the male to play a more active role. Young adults commonly assist their parents in rearing the young, particularly among birds and higher primates. "Helpers" at the nest are common in moorhens, thornbills, anis, and several other birds (Skutch, 1961). Other relatives commonly give aide among many primates and in chimpanzees, sisters may often adopt orphaned siblings (Lawick-Goodall, 1971). Under certain circumstances, unrelated females may adopt orphans in primates (Itani, 1959; Rowell, 1963) and in the elephant seal two apparently unrelated females may nurse the same pup until it is weaned. Among wild dogs and wolves, all members of a pack may help the female to feed the developing pups (Estes and Goddard, 1967; Mech, 1970). One of the most unusual solutions to the problem of enlisting help in rearing the young is the strategy adopted by the brood parasites, many species of birds, including cuckoos and cowbirds, which exploit another species and cuckold them into rearing their young (Lack, 1968). The adaptations that have arisen in the 80 or so species that employ this strategy are remarkable. For example, some parasitic females are able to lay their eggs quickly during the short time that the host female leaves her nest. The parasitic male partner may help by serving as a decoy leading the host away from her nest. The eggs mimic those of the host and require less incubation time so that they hatch first. Parasitic young of some species push the host's eggs out of the nest before they hatch whereas others are equipped with mandibular hooks which are used to kill the host's young. Even the gaping mouths of the parasitic young mimic those of the host's young. The advantages of these strategies are obvious. A female who cuckolds individuals of another species into rearing her young need not spend time and energy building a nest, incubating the eggs and caring for the young. She can use the time and energy to produce more eggs. The entire area of parasite–host relations is a fertile ground for future research.

Females may employ a wide variety of strategies to insure the survival of their young once they are born or hatched. This is a broad topic and there is insufficient space to review the multiple strategies employed. I will mention one aspect of sexual selection that has been neglected and deserves much more intensive study. This is the matter of competition among females related to survival of the young. Competition among females is usually not as physical and, hence, as obvious as male–male competition but the outcome of the competition may have great significance for a female's reproductive success. Competition among females is perhaps most obvious in colonial breeding species of birds, ungulates, and pinnipeds. Females may compete for a variety of resources such as the parturition or nest site that is most favorable or a food resource for herself and her offspring. In the highly colonial northern elephant seal, female competition is one of the main causes of pup mortality on the rookery (Le Boeuf and Briggs, 1977). Conflicts between females, resulting in mother–pup separation and pup biting, increase with density and with physical agents such as storms that cause disturbance among females. Pups that lose their mothers are forced to attempt to suckle alien females and the latter, protecting their milk for their own young, bite orphans and drive them away. These interfemale conflicts of self interest occur constantly and vary greatly in intensity. Females threaten other females that approach their pups and threat vocalizations set the predominant tone in the rookery. Interfemale rivalry often culminates in one or more females killing the newborn of a neighboring female. The circumstances and cause of this apparently spiteful behavior warrant further study. High infant mortality in several other species can be attributed to interfemale rivalries, e.g., Herring and Western gulls (Paynter, 1949; Hunt and Hunt, 1976) and Belding ground squirrels (P. Sherman, personal communication). A systematic appraisal of female–female competition is long overdue.

The topic of adoption also deserves special attention because it would not be predicted by social theory unless relatives are involved. Yet it is known that females adopt apparently unrelated young in a variety of animals as diverse as birds (Brown, 1974; Power, 1975), seals (Le Boeuf and Briggs, 1977), and primates (Itani, 1959; Rowell, 1963). Williams (1966) attributes many cases of adoption to "reproductive error." A female who has just lost her newborn is physiologically primed and behaviorally set to respond maternally. When her own pup gets lost or dies, her response generalizes to a similar stimulus. This is a plausible explanation. I have seen female elephant seals give birth to a stillborn pup, or lose their pup shortly after parturition, then immediately set out roaming about the harem issuing pup attraction calls. This behavior often ends with the female literally stealing another female's pup which she nurses as she would her own until weaning. But obviously, this behavior should be subject to strong negative selection pressure unless it enhances the inclusive fitness of the performer. Future work may shed more light on this puzzle.

Finally, females may compete with each other by employing sex to obtain something that insures their own survivial or that of their offspring. In some birds, a deserted female may copulate with a potential cuckold as an enticement to pair-bond quickly and get him to help rear the young. Similarly, Wolf (1975) observed female hummingbirds, *Eulampis jugularis*, mate outside the breeding season in order to gain access to rich food sources being defended by males. He called this strategy "prostitution" behavior, because the female employed sexual behavior to obtain an energetic benefit (see also Zahavi, 1971). These examples stand in contradiction to the idea that sexual receptivity of females is completely regulated by hormones.

3.3. Strategies Common to Both Sexes

The sex ratio of an animal's progeny has an important influence on the reproductive success of both males and females. R. A. Fisher (1930) showed that natural selection favored those parents that produced offspring in a 50/50 sex ratio. Others (e.g., Hamilton, 1967; Leigh, 1970) refined the solution and showed that each parent is selected to make equal investments in sons and daughters. But deviations from parity occur occasionally and for a long time these aberrancies defied explanation. Recently, Trivers and Willard (1973) came up with an amazingly simple and far-reaching explanation for these deviations and when to expect them. They argue as follows: In most mammals, the majority of whom are polygynous, a male in good condition at the end of the parental investment period will out-produce a sister in similar condition while the reverse is true if both are in poor condition. This is because there is greater variation in reproductive success among males than among females. As I pointed out earlier for elephant seals, a few males will be very successful, far more successful than the most successful female, but to be in this category a male must have everything going for him from the beginning. Thus, as a female deviates more and more from an optimal nutritional condition immediately before and during pregnancy, she should show an increasing bias to produce females because male progeny will be more harmed by an initial disadvantage than female progeny. Conversely, if conditions during the period of parental investment are ideal, females should show an increasing bias towards producing males because of the higher potential for perpetuating her genes.

Females that could influence the sex of their offspring in response to available nutrients would obviously enjoy greater reproductive success than those who could not do so. Control of sex by the female is presumed to be by differential mortality either of the sex cells or of the developing young. The Trivers and Willard model makes three assumptions for which data are presented:

1. The condition of the young at the end of the period of parental investment is correlated with the condition of the mother during parental investment.
2. Differences in the condition of the young at the end of parental investment will still be apparent in adulthood.
3. Adult males will be helped more than females by slight advantages in condition.

The beauty of this model is its simplicity and breadth. It provides a novel explanation for many previously unexplained data on mink, pigs, sheep, deer, seals, and man (see Trivers and Willard, 1973). For example, laboratory rats fed a suboptimal diet during pregnancy should produce more females in their litters and indeed this is the case (Rivers and Crawford, 1974). The model generates numerous questions and easily testable hypotheses. What is the ideal time in a female's life to have males, or females? A female mammal's best strategy might be to give birth to females early in life, when she is not quite fully grown or proficient at getting food and protecting her young and then give birth to males in her prime. Do females care for male and female progeny differently as food resources fluctuate? Are weak, sickly male progeny discriminated against more readily than female progeny in a similar condition? Hypotheses such as these are currently being tested. I expect that much interesting work will flow from this novel theory.

Finally, the evolution of sexual reproduction, the most basic reproductive strategy of them all, is currently a topic of great interest and debate. Asexual reproduction is a simple, fast, and efficient way of reproducing. Why isn't it universal? Why and how did such an elaborate and in a sense, wasteful, process such as sexual reproduction evolve? The transition from asexual to sexual reproduction has long confronted biologists with a seemingly insolvable paradox: a female that reproduces sexually cuts her genetic contribution to each gamete, or her average genetic relationship to her offspring, by one-half. This is done despite the fact that in most species the female receives no help from the male in rearing the young. In asexual reproduction, all of the genes in the offspring are identical to those of the mother. Therefore, by reproducing sexually, a female throws away half of her investment! Obviously, the advantage of sex must be great to overcome this 50% initial deficit. Williams (1975) has recently presented a convincing solution to the dilemma. He makes a case for sexual reproduction conferring an immediate advantage to the individuals involved, a solution that eluded earlier writers (e.g., Crow and Kimura, 1965; 1969). Williams argues that sexual reproduction evolved because it allows each parent to diversify its offspring, and this variable strategy is most effective in dealing with unpredictable changes in the environment that occur from one generation to the next. A key point in his argument is that sexual progeny vary more in fitness than asexual progeny and that selection is far more intense than we previously thought. Only a few individuals, those possessing extreme traits or the very highest "fitness dosage," survive during each cycle. Thus, genes for sexual reproduction are consistently favored over those for asexual reproduction. The latter process persists and is favored under conditions when the environment is extremely stable over generations. These ideas on this basic subject will undoubtedly stimulate further thinking and research in the future.

4. Concluding Remarks

I have no doubt that the neo-Darwinian approach to the study of the biological basis of social behavior that I have described briefly will become firmly established, stimulate research efforts in many new directions, and attract many new adherents. Studies in the near future will undoubtedly focus on various aspects of sexual selection and predictions from sex-ratio theory and kinship theory will be put to the test. Since much of the viability of social theory stems from the fact that it is basically testable, because most of its predictions are formulated in genetic terms, this will no doubt create a priority for studies in which the genetic relationships between individuals is known. Field studies of animals that can be marked, identified individually, and followed for generations will assume greater importance and will serve as the basis for testing predictions. The pattern, amount, and duration of parental investment will be studied in many species, and the generality of this variable as a predictor of behavior central to sexual selection will be further evaluated both in the field and in the laboratory.

This approach to the study of social behavior will influence the social sciences and eventually the study of man. The applicability of some of the predictions from social theory to man has already been argued (Trivers, 1974b; Alexander, 1974, 1975; Wilson, 1975). Some statements about man have already provoked a spirited controversy (e.g., the *New York Review of Books,* November 13, 1975; *Bioscience,* March 1976; *Time,* August 1, 1977).

One moral from this brief outline of sociobiology needs mention and it is this. To interpret function satisfactorily, whether in the laboratory or the field, whether on the population level or the suborganismic, the data must be consistent with the reproductive strategy of the individual. For example, the Coolidge effect—the effect of a novel sexual stimulus in reducing the latency with which a sexually sated animal resumes copulation—makes sense as a male reproductive strategy. But if the phenomenon were reported for females, or as being stronger in females than males, it would be suspicious. It would not agree with the theory I have presented or with what we expect from our understanding of female reproductive strategies. It would suggest that the theory or the facts are false. Except for Bateman's experiments, I know of no others that have been done to determine whether the sexes differ in their response to novel sexual partners.

Since this chapter has a title in common with G. C. Williams's recent book (Williams, 1975, p 169), it seem appropriate that it have the same ending, too: "Clearly the contest of ideas on these fundamental problems has only just begun. History has afforded a rare opportunity to ardent participants and alert spectators in the years ahead."

References

Alexander, R. D. The evolution of social behavior. *Annual Review of Ecology and Systematics,* 1974, *5,* 325–383.

Alexander, R. D. The search for a general theory of behavior. *Behavioral Science,* 1975, *20,* 77–100.
Bateman, A. J. Intrasexual selection in Drosophila. *Heredity,* 1948, *2,* 349–368.
Beach, F. A., and Le Boeuf, B. J. Coital behaviors in dogs, I. Preferential mating in the bitch. *Animal Behavior,* 1967, *15,* 546–558.
Bertram, B. C. R. The social system of lions. *Scientific American,* 1975, *232,* 54–65.
Brown, J. L. Alternate routes to sociality in jays—with a theory for the evolution of altruism and communal breeding. *American Zoologist,* 1974, *14,* 63–80.
Bruce, H. Effects of olfactory stimuli on reproduction in mammals. In G. Wolstenhome (Ed.), *Effects of external stimuli on reproduction* (CIBA Foundation Study Group No. 26), Boston: Little and Brown, 1967.
Bruce, H. M. Pheromones and behavior in mice. *Acta Neurologica Belgica,* 1969, *69,* 529–538.
Buechner, H. K., and Schloeth, R. Ceremonial mating behavior in Uganda kob (*Adenota kob thomasi* Neuman). *Zhurnal fuer Tierpsychologie,* 1965, *22,* 209–225.
Campanella, P. J., and Wolf, L. L. Temporal leks as a mating system in a temperate zone dragonfly (Odonata: Anisoptera), I: *Plathemia lydia* (Drury). *Behavior,* 1974, **51,** 49–87.
Cox, C. R., and Le Boeuf, B. J. Female incitation of male competition: A mechanism in sexual selection. *American Naturalist,* 1977, *111,* 317–335.
Crook, J. H. Introduction—social behavior and ethology. In Crook, J. H. (Ed.), *Social behavior in birds and mammals,* London: Academic Press, 1970.
Crow, J. F., and Kimura, M. Evolution in sexual and asexual populations. *American Naturalist,* 1965, *99,* 439–450.
Crow, J. F., and Kimura, M. Evolution in sexual and asexual populations. *American Naturalist,* 1969, *103,* 89–91.
Darwin, C. *On the origin of species by means of natural selection, or the preservation of favored races in the struggle for life.* London: John Murray, 1859.
Darwin, C. *The descent of man and selection in relation to sex.* London: John Murray, 1871.
DeFries, J. C., and McClearn, G. E. Social dominance and Darwinian fitness in the laboratory mouse. *American Naturalist,* 1970, *104,* 408–411.
De Long, R. L. Interspecific reproductive behavior among four otariids at San Miguel Island, California. Talk given at the Conference on the Biology and Conservation of Marine Mammals, 4–7 December 1975, University of California, Santa Cruz, 1975.
Eberhard, M. J. W. The evolution of social behavior by kin selection. *Quarterly Review of Biology,* 1975, *50,* 1–33.
Ebert, J. D., Loewy, A. G., Miller, R. S., and Schneiderman, H. A. *Biology.* New York: Holt, Reinhart, and Winston, 1973.
Erickson, C. J., and Martinez-Vargas, M. C. The hormonal basis of cooperative nest-building. In P. Wright, P. G. Caryl, and D. M. Vowles (Eds.), *Neural and endocrine aspects of behaviour in birds.* Amsterdam: Elsevier, 1975.
Estes, R. D., and Goddard, J. Prey selection and hunting behavior of the African wild dog. *Journal of Wildlife Management,* 1967, *31,* 52–70.
Fisher, R. A. *The genetical theory of natural selection.* Oxford: Clarendon Press, 1930.
Ghiselin, M. T. *The economy of nature and the evolution of sex.* Berkeley: University of California Press, 1974.
Guhl, A. M., and Warren, D. C. Number of offspring sired by cockerels related to social dominance in chickens. *Poultry Science,* 1946, *25,* 460–472.
Hamilton, W. D. The genetical theory of social behaviour, I, II. *Journal of Theoretical Biology,* 1964, *7,* 1–52.
Hamilton, W. D. Extraordinary sex rations. *Science,* 1967, *156,* 477–488.
Hamilton, W. D. Selfish and spiteful behaviour in an evolutionary model. *Nature* (*London*), 1970, *228,* 1218–1220.

Hamilton, W. D. Geometry for the selfish herd. *Journal of Theoretical Biology*, 1971a, *31*, 295–311.

Hamilton, W. D. Selection of selfish and altruistic behavior in some extreme models. In J. F. Eisenberg and W. S. Dillon (Eds.), *Man and beast: Comparative social behavior*. Washington: Smithsonian Institution Press, 1971b.

Hamilton, W. D. Altruism and related phenomena, mainly in social insects. *Annual Review of Ecology and Systematics*, 1972, *3*, 193–232.

Hinde, R. A. The biological significance of territories in birds. *Ibis*, 1956, *19*, 340–369.

Hunt, G. L., Jr., and Hunt, M. W. Gull chick survival: The significance of growth rates, timing of breeding and territory size. *Ecology*, 1976, *57*, 62–75.

Huxley, J. S. The present standing of the theory of sexual selection. In G. DeBeer (Ed.), *Evolution*. New York: Oxford University Press, 1938.

Itani, J. Paternal care in the wild Japanese monkey, *Macaca fuscata fuscata*. *Primates*, 1959, *2*, 61–93.

Lack, D. The evolution of reproductive rates. In J. Huxley, A. C. Hardy, and E. B. Ford (Eds.), *Evolution as a process*. New York: Collier Books, 1954.

Lack, D. *Ecological adaptations for breeding in birds*. London: Chapman and Hall, 1968.

Lawick-Goodall, Jane van. *In the shadow of man*. Boston: Houghton–Mifflin, 1971.

Le Boeuf, B. J. Male–male competition and reproductive success in elephant seals. *American Zoology*, 1974, *14*, 163–176.

Le Boeuf, B. J., and Briggs, K. T. The cost of living in a seal harem. *Mammalia*, 1977 (in press).

Leigh, E. G. Sex ratio and differential mortality between the sexes. *American Naturalist*, 1970, *104*, 205–210.

Leuthold, W. Variations in territorial behavior of Uganda kob, *Adenota kob thomasi* (Newmann, 1896). *Behaviour*, 1966, *27*, 215–258.

Maynard-Smith, J. Kin selection and group selection. *Nature*, 1964, *201*, 1145–1147.

Mayr, E. Sexual selection and natural selection. In B. Campbell (Ed.), *Sexual selection and the descent of man, 1871–1971*. Chicago: Aldine, 1972.

Mech, L. D. *The wolf: The ecology and behavior of an endangered species*. Garden City, New York: Natural History Press, 1970.

Mohnot, S. M. Some aspects of social changes and infant-killing in the Hanuman langur, *Presbytis entellus* (Primates: Cercopithecidae), in western India. *Mammalia*, 1971, **35**, 175–198.

Nisbet, I. C. T. Courtship feeding, egg-size, and breeding success in common terns. *Nature* (*London*), 1973, *241*, 141–142.

Orians, G. H. On the evolution of mating systems in birds and mammals. *American Naturalist*, 1969, *103*, 589–604.

Parker, G. A. Sperm competition and its evolutionary consequences in insects. *Biological Reviews of the Cambridge Philosophical Society*, 1970, *45*, 525–567.

Parker, G. A., Baker, R. R., and Smith, V. G. F. The origin and evolution of gamete dimorphism and the male–female phenomenon. *Journal of Theoretical Biology*, 1972, *36*, 529–533.

Paynter, R. Clutch size and the egg and chick mortality of Kent Island Herring Gulls. *Ecology*, 1949, *30*, 146–166.

Peterson, R. S., Le Boeuf, B. J., and DeLong, R. L. Fur seals from the Bering Sea breeding in California. *Nature* (*London*), 1968, *219*, 899–901.

Power, H. W. Mountain bluebirds: Experimental evidence against altruism. *Science*, 1975, *189*, 142–143.

Rivers, J. P. W., and Crawford, M. A. Maternal nutrition and the sex ratio at birth. *Nature* (*London*), 1974, *242*, 297–298.

Rowell, T. E. Behaviour and female reproductive cycles of rhesus macaques. *Journal of Reproduction and Fertility*, 1963, *6*, 193–203.
Schaller, G. B. *The Serengeti lion: A study of predator-prey relations*. Chicago: University of Chicago Press, 1972.
Selander, R. K. Behavior and genetic variation in natural populations. *American Zoologist*, 1970, *10*, 53–66.
Selander, R. K. Sexual selection and dimorphism in birds. In B. Campbell (Ed.), *Sexual selection and the descent of man, 1871–1971*. Chicago: Aldine, 1972.
Selander, R. K., and Johnson, W. E. Genetic variation among vertebrate species. *Proceedings of the XVII International Congress of Zoology*, 1972, 1–31.
Selander, R. K., and Kaufman, D. W. Genic variability and strategies of adaptation in animals. *Proceedings of the National Academy of Sciences, USA*, 1973, *70*, 1875–1877.
Severinghaus, C. W. Some observations on the breeding behavior of deer. *New York Fish and Game Journal*, 1955, *2*, 239–241.
Skutch, A. F. Helpers among birds. *Condor*, 1961, *63*, 198–226.
Sugiyama, Y. Social organization of hanuman langurs. In S. A. Altmann (Ed.), *Social communication among primates*. Chicago: University of Chicago Press, 1967.
Trivers, R. L. The evolution of reciprocal altruism. *Quarterly Review of Biology*, 1971, *46*, 35–37.
Trivers, R. L. Parental investment and sexual selection. In B. L. Campbell (Ed.), *Sexual selection and the descent of man, 1871–1971*. Chicago: Aldine, 1972.
Trivers, R. L. Book review of *The Economy of Nature and the Evolution of Sex* by M. T. Ghiselin. *Science*, 1974a, *186*, 525–526.
Trivers, R. L. Parent-offspring conflict. *American Zoologist*, 1974b, *14*, 249–264.
Trivers, R. L., and Willard, D. E. Natural selection of parental ability to vary the sex ratio of offspring. *Science*, 1973, *179*, 90–91.
Verner, J. Evolution of polygamy in the long-billed marsh wren. *Evolution*, 1964, *18*, 252–261.
Warner, R. R., Robertson, D. R., and Leigh, E. G., Jr. Sex change and sexual selection. *Science*, 1975, *190*, 633–638.
Watson, J. B., *Behaviorism*. New York: Norton, 1925.
Watts, C. R., and Stokes, A. W. The social order of turkeys, *Scientific American*, 1971, *224*, 112–118.
Williams, G. C. *Adaptation and natural selection*. Princeton: Princeton University Press, 1966.
Williams, G. C. *Sex and evolution*. Princeton: Princeton University Press, 1975.
Wiley, R. H. Territoriality and non-random mating in Sage Grouse, *Centrocercus urophasianus*. *Animal Behavior Monographs*, 1973, *6*, 85–169.
Wilson, E. O. *Sociobiology*. Cambridge, Massachusetts: The Belknap Press, 1975.
Wolf, L. L. "Prostitution" behavior in a tropical hummingbird. *Condor*, 1975, *77*, 140–144.
Wynne-Edwards, V. C. *Animal dispersion in relation to social behavior*. London: Oliver and Boyde, 1962.
Zahavi, A. The social behavior of the White Wagtail *Motacilla alba alba* wintering in Israel. *Ibis*, 1971, *113*, 203–211.

2

Longitudinal Studies of Sexual Behavior in the Oregon Troop of Japanese Macaques

G. Gray Eaton

1. Introduction

1.1. Aims and Objectives

For the past six years my colleagues and I have spent over 8,000 man-hours studying the Oregon troop of Japanese macaques (*Macaca fuscata*). The aim of this research has been to understand the mechanisms that regulate sexual and social behavior in these nonhuman primates. The investigations have followed Beach's (1960) strategy for investigating species-specific behavior:

> One merely attempts to answer three questions: How did the behavior get to be what it is? What are the external causes of the behavior? What internal mechanisms mediate it? Phrased in slightly more sophisticated terms each of these questions can be said to deal with a different class of *determinants* of behavior. The determinants of behavior can be subdivided into three broad categories: *Historical* determinants, *environmental* determinants, and *organismic* determinants. (p. 5)

G. Gray Eaton • Unit of Primate Behavior, Oregon Regional Primate Research Center, Beaverton, and the Department of Medical Psychology, University of Oregon Health Sciences Center, Portland, Oregon. Publication No. 880 of the Oregon Regional Primate Research Center, supported in part by grants HD-05969 and RR-00163 from the National Institutes of Health, U.S. Public Health Service.

We have, therefore, through a combination of longitudinal and cross-sectional studies, investigated social, seasonal, and hormonal influences on the behavior of the macaques. In addition, questions of evolutionary and adaptive significance have been considered in interpreting the results of these studies.

Members of the genus *Macaca* are of scientific interest because they are the most widely distributed of all nonhuman primates and are, therefore, relatively successful from an evolutionary point of view. Moreover, their subfamily, the Cercopithecinae, tend to have social orders more complex than most other nonhuman primates (Nagel and Kummer, 1974). Macaques have also exhibited social learning of food preference and acquisition that appear to depend on tradition (Itani, 1965; Kawai, 1963; Kawamura, 1963). This social evolution has been termed "precultural" behavior (Kawai, 1965a). Furthermore, their behavior can be remarkably similar to that of people. For example, the reproductive behavior of the females of some macaque species appears to be closer in some respects to that of women than that of other nonhuman primates. Like women, female pigtailed macaques (*Macaca nemestrina*) and stumptailed macaques (*M. arctoides*) can be sexually receptive throughout the entire ovarian cycle when studied under laboratory conditions (Eaton, 1973; Eaton and Resko, 1974; Goldfoot *et al.*, 1975), whereas female chimpanzees (Young and Orbison, 1944) and gorillas (Nadler, 1975) are receptive principally during the follicular phase of the cycle. Moreover, the sexual behavior of macaques is of intrinsic interest because it constitutes a potential ethological barrier to random mating, since "ethological barriers to random mating constitute the largest and most important class of isolating mechanisms in animals." And, "the mechanisms that isolate one species reproductively from others are perhaps the most important set of attributes a species has, because they are by definition the species criteria." (Mayr, 1963, pp. 95 and 89).

1.2. Troop History

Genetic relationships, critical factors in primate social and sexual behavior, have been preserved in the Oregon troop of Japanese macaques, which is of natural origin. The troop has occupied an 8,058-m^2 (2-acre) grassy enclosure at the Oregon Regional Primate Research Center since October, 1966 (Alexander, Hall, and Bowers, 1969). The troop was captured in February, 1964 near Mihara City in the Hiroshima prefecture of Japan after the monkeys had been damaging crops. Forty-six of the 49 troop members were trapped and housed at the Japan Monkey Center until they were moved to Oregon in May, 1965. They were then housed for 18 months in an indoor–outdoor run of 134 m^2 until their move to the 2-acre enclosure (Alexander and Bowers, 1967). Confinement does not appear to have qualitatively altered troop patterns of social (Alexander and Bowers, 1967, 1969; Alexander, 1970), aggressive (Alexander and Roth, 1971; Alexander and Hughes, 1971; Eaton, 1974), or sexual behavior (Hanby, Robertson, and Phoenix, 1971; Eaton, 1972, 1974). A central–peripheral macaque troop structure has been described by Imanishi (1963) and Sugiyama (1960): an alpha (leader) or dominant male occupies the center of the troop with four or five other high-ranking (subleader) males, the adult females, and their offspring, while low-ranking males remain on the periphery. This structure does not occur in the Oregon troop except during controlled feeding conditions (Alexander and Bowers, 1967). However, this spatial arrangement is not known to occur in unprovisioned feral troops, and varies widely in provisioned feral troops (Yamada, 1971).

2. *Previous Studies*

2.1. *Seasonal Mating*

The seasonality of the macaques' reproductive behavior is remarkable. Beginning in the fall, feral troops have a five- to six-month mating season followed by a two- to four-month birth season (Tokuda, 1961–1962). Lancaster and Lee (1965) have pointed out that although decreasing day length, falling temperature, and the end of the rainy season are common to the initiation of mating, local differences in these factors do not necessarily correlate with onset of mating. In the Oregon troop, mating usually starts in September and continues until February so that while day length is declining, temperature may not be, and fall marks the end of our dry season.

We hypothesized that an evolutionary trend toward a fall mating season (so that the young would be born in the spring to maximize their chance of survival) may be influenced by environmental conditions but intensified by endogenous endocrine events. To test this hypothesis we planted modified intrauterine contraceptive devices (IUDs) in nine females of the Oregon troop (Eaton, 1972). We theorized that endogenous endocrine conditions, for example, those associated with the cessation of lactation, the onset of puberty, or a response to male courtship behavior, interact with other conditions to initiate female receptivity in the fall, and that endocrine conditions related to pregnancy also interact with other conditions to end the receptive period. We predicted that females with IUDs would not become pregnant and, therefore, would have a longer mating season than pregnant females. This hypothesis was supported. The normal six-month mating season was extended by one month in the females with IUDs. Failure to conceive increased the duration of the female's season unless she had a recent infant (less than a year old); in this case, the presence of an infant had an inhibiting effect on the duration of the mother's mating season and nullified the differences between pregnant and nonpregnant females. Females without infants received more ejaculations than those with infants, and nonpregnant females without infants received more ejaculations than those that became pregnant (Figure 2.1). However, when we computed the mean frequency of ejaculations per day per female, we found no significant differences between the groups of females. Some frequency totals were higher because of the longer seasons. Therefore, the factors that timed the breeding season did not affect the intensity of female activity within the season.

Since the females with IUDs eventually stopped mating, the factors associated with pregnancy (presumably hormonal) cannot be the only ones involved in the termination of the mating season. Like male rhesus monkeys (Gordon, Rose, and Bernstein, 1976) Japanese macaque males probably have a seasonal testosterone cycle and thus contribute to the seasonal nature of the reproductive behavior of this species.

Many factors could control the timing of the breeding season (Rowlands, 1966). The low correlation between environmental conditions and the onset of the mating season in feral troops of Japanese macaques (Lancaster and Lee, 1965) suggests that circannual endocrine and behavioral mechanisms are involved. These controls could function by: (1) inhibiting reproduction in potentially continuous breeders, or (2) initiating reproduction during critical periods in a fundamentally cyclic system (Conaway,

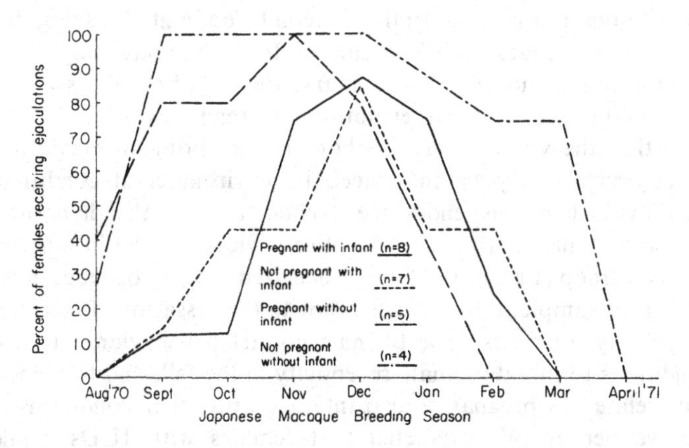

Figure 2.1. Percentage of Japanese macaque females which received at least one ejaculation during each calendar month of the mating season. From Eaton, G. Gray. Seasonal sexual behavior: Intrauterine contraceptive devices in a confined troop of Japanese macaques. *Hormones and behavior*, 1972, *3*, 133–142; reproduced by permission of Academic Press, New York.

1971). The inhibitory effects of the infant and pregnancy found in our study support the first of these hypotheses (Eaton, 1972).

2.2. *Dominance and Mating*

Dominant males have been reported to mate more frequently than lower ranking males in multimale troops of baboons (DeVore, 1965), rhesus monkeys (Carpenter, 1942; Kaufmann, 1965), and Japanese macaques (Tokuda, 1961–1962; Hanby *et al.*, 1971). However, none of the authors gave the criteria for dominance or the data; instead most authors simply stated that dominance was determined by displacement at food or water sources or by the exchange of aggressive and submissive postures. These analyses apparently ignored reversals or deviations from linearity, and ties in the hierarchy.

In studies of the relation between dominance rank and social organization in the Oregon troop of Japanese macaques, we have demonstrated that male rank, determined by aggressive interactions, is not correlated with mating success, determined by ejaculation frequency (Eaton, 1974). We recorded four patterns of aggressive behavior (cf. Alexander and Roth, 1971) in terms of increasing severity: (1) threat—lunging toward another animal, gaping or ear-flattening, or emitting a "woof" vocalization; (2) chase—pursuing with threats; (3) punish—noninjurious biting of brief duration, and/or pinning to the ground, striking, leaping on, or pulling fur or body parts; and (4) assault—biting that injures or is prolonged, accompanied by vigorous head movement while the teeth grip the victim. If more than one type of behavior occurred during a single sequence, only the most severe was scored, e.g., if one animal chased, caught, and punished another, only a punish was scored. With each incident of aggression, one of three forms of response was also recorded: (1) submission—moving away, grimacing, cringing, or crouching; (2) stand-off—submission not apparent or submissive gestures interspersed with aggressive actions; (3) ignore—no aggressive or submissive behavior in response to the attack. The criteria used to construct the males' dominance ranking involved only aggressive interactions that were unambiguously followed by submission (cf. Sade, 1967). Cases in which more than one attacker was involved were not included in this analysis.

We delineated the dominance hierarchy by constructing a matrix of all the adult males of the troop ($N = 21$) and entering the frequency of each animal's attacks that were met by submission by each of the others (Table 2.1). We then renumbered ranks in order to obtain a minimum number of reversals of attack and submission behaviors between individuals in the linear order. Ties in position in the new hierarchy were broken by giving precedence first to the male in the tied pair that had more frequently

Table 2.1. Dominance Rank of Adult Male Japanese Macaques Compiled from 646 Attacks in Which Submissive Behavior Was Observed[a]

Dominance rank	Male number	Field mark	Attacked males[b] (frequency)																					Number of males attacked[c]
			Δ	>	∇	B	3	E	α	ſ	L	—	h	ʅ	ꟻ	F	φ	L^N	⅄	7	9	V	X	
1	1723	Δ		8	2	0	10	3	0	1	0	0	2	0	8	0	1	1	1	0	1	1	2	13
2	1751	>			7	5	3	2	2	0	2	3	15	5	6	0	4	2	4	1	3	3	1	17
3	1728	∇				9	9	1	1	0	4	0	16	10	5	1	20	3	5	0	1	12	3	15
4	1766	B			2		5	2	6	1	3	0	6	4	5	3	5	3	3	2	2	13	2	16
5	1734	3				1		3	9	2	1	0	12	5	6	2	11	0	3	1	4	6	0	13
6	2378	E							1	4	1	0	1	4	1	0	0	1	0	1	1	0	1	9
7	1725	α						1		0	0	0	0	1	1	3	3	0	1	1	1	6	0	8
8	3223	ſ									1	21	2	0	0	0	0	0	0	0	0	1	0	4
9	1719	L										1	1	1	0	3	1	1	9	3	1	13	2	9
10	3302	—											2	1	3	0	2	0	0	0	0	1	0	5
11	2191	h					2		1					3	6	1	10	1	1	3	1	3	1	10
12	1768	ʅ									1				6	0	2	0	8	3	2	9	1	7
13	1756	ꟻ														1	4	2	1	0	0	3	0	5
14	1736	F															0	0	0	5	3	3	1	4
15	1741	φ									5							8	5	1	0	6	5	5
16	1735	L^N								1									1	0	0	0	0	1
17	1716	⅄									1			1				1		6	3	33	6	4
18	1733	7													1		1				1	1	1	4
19	1753	9																				4	3	2
20	1721	V											3	1	2			2	1				4	2
21	1764	X								2		1									1	1		2
Number of males[d] attacking			0	1	2	2	4	5	5	6	7	4	9	8	11	7	10	10	11	10	13	16	14	

[a] From Eaton, G. Gray, Male dominance and aggression in Japanese macaque reproduction. In W. Montagna and W. A. Sadler (Eds.), *Reproductive behavior.* New York: Plenum Press, 1974, p. 291; reproduced by permission of Plenum Publishing Corporation. [b] Entries below diagonal line indicate reversals of dominance rank. [c] Minus those males that had more frequently attacked the male in question. [d] Minus those males that had been more frequently attacked by the male in

attacked the other; then, if a pair was still tied, to the one that had been attacked by a fewer number of other males; and finally, if the tie persisted, to the one that had attacked the greater number of other males.

The basic units of sexual behavior recorded were adult heterosexual mount series, which were defined as two or more mounts in less than five minutes by the same pair of animals. Mounts were single or double foot clasps with thrusts. The occurrence of intromission could not be ascertained. Each series was assigned at the time of the first mount to one of four categories of initation: (1) the male approached from more than 10–15 m or had followed the female; (2) the female approached from more than 10–15 m or had followed the male; (3) both approached or had walked together or were seated less than 5 m apart and away from other animals; or (4) initiation unknown. After a series was initiated, one of six terminations was recorded: (1) ejaculation, recognized by a pause in thrusting and the rigid posture of the male, or by semen on the female's perineum; (2) male movement at least 10–15 m away and failure to return to mount within five minutes; (3) female movement at least 10–15 m away and failure to return for mounting within five minutes; (4) separation of both partners of more than 10–15 m and failure to return to mount within five minutes; (5) failure of a mount to occur within five minutes even though the pair remained less than 10–15 m apart; or (6) termination unknown (either observations ceased while a pair was still mounting, or a pair that had been mounting was observed to have separated).

The dominance rank of the males was then correlated with their rank in each of the sexual behavior categories. All correlations were low and not significant, e.g., the Spearman rho coefficient between dominance rank and rank in frequency of ejaculation was 0.10, $n = 21$, $p > .05$.

We concluded, therefore, that there was little or no relation between male dominance rank and mating behavior in the Oregon troop of Japanese macaques. That such a relation has been observed in some troops of baboons and macaques under some conditions serves to emphasize the plasticity of individual and group behavior that characterizes all primates (Eaton, 1974).

2.3. *Testosterone and Dominance*

In another study designed to investigate the determinants of dominance hierarchy in the Oregon troop of Japanese macaques, we measured plasma testosterone levels in the 21 adult males (Table 2.2) and attempted without success to correlate these levels with dominance rank or frequency of aggressive behavior (Eaton and Resko, 1974). Plasma testosterone titers had been reported to correlate with dominance rank and aggressive behavior in a large all-male group of rhesus monkeys (Rose, Holaday, and

Table 2.2. Male Dominance Rank, Plasma Testosterone Titers, and Number of Aggressive Attacks in a Confined Troop of Japanese Macaques[a]

Male number	Dominance rank		Testosterone (ng/ml)		Number of attacks	
	1971	1972	1971	1972	1971	1972
1723	1	1	8.3	2.3	41	29
1751	2	2	6.9	2.1	68	31
1728	3	3	16.4	2.3	99	62
1766	4	4	8.3	1.3	67	44
1734	5	5	11.8	1.3	66	21
2378	6	6	1.6	7.7	16	13
1725	7	9	2.1	1.5	18	11
3223	8	7	0.5	5.9	25	3
1719	9	12	8.3	1.2	36	38
3302	10	Dead[b]	0.2	Dead[b]	9	Dead[b]
2191	11	8	1.0	6.8	33	60
1768	12	11	0.3	2.4	32	33
1756	13	13	4.0	14.6	11	3
1736	14	14	18.1	7.0	12	17
1741	15	15	1.8	1.7	30	35
1735	16	16	4.9	3.1	2	2
1716	17	17	3.7	8.2	51	39
1733	18	18	14.3	2.7	5	4
1753	19	19	15.9	19.8	7	4
1721	20	20	3.1	2.9	13	14
1764	21	21	1.0	1.8	5	0

[a] From Eaton, G. Gray, and Resko, J. A. Plasma testosterone and male dominance in a Japanese macaque (*Macaca fuscata*) troop compared with repeated measures of testosterone in laboratory males. *Hormones and Behavior*, 1974, *5*, 251–259; reproduced by permission of Academic Press, New York.

[b] Arbitrarily assigned the rank of 10.

Bernstein, 1971). However, in subsequent experimental studies wherein individual rhesus males were introduced to established groups, repeated physical assaults and threats by the group were correlated with depressed levels of testosterone in the attacked individuals or an increase in the individual's sexual behavior was correlated with an increase of testosterone (Rose, Gordon, and Bernstein, 1972). Therefore, the earlier correlation between dominance rank and testosterone quantities in rhesus males (Rose *et al.*, 1971) may have been related to the establishment of a dominance hierarchy in the all-male group structure. Thus, the lack of correlation between dominance rank and testosterone levels in the Oregon troop of Japanese macaques may be related to the feral origin of the troop. Because these males matured together, those lower in rank may not receive the same stress as lower ranking males in groups of strange males that have established their dominance through fighting.

In feral troops of Japanese macaques and rhesus monkeys, an individual's rank correlates with that of its mother's; this correlation decreases, however, as males reach maturity (Kawai, 1965b; Missakian, 1972). Furthermore, the canine teeth of the adult males are unnecessary to attain or maintain high rank, which indicates that more than fighting is involved in the dominance relationships of a natural troop (Alexander and Hughes, 1971).

2.4. *Displays and Mating*

Many descriptions of nonhuman primate behavior include references to stereotyped locomotor activities performed in various social contexts. These activities which have been labeled "displays" presumably serve to direct attention to the displayer and to transmit information to other individuals. Some of these, such as branch shaking, generate noise and motion as part of the performance. Displays and sexual behavior have been tentatively correlated within the genus *Macaca,* but quantitative data to support this relation are lacking.

To investigate this relationship in the Oregon troop of Japanese macaques, Kurt B. Modahl and I defined and recorded the frequency of five types of display—shaking, kicking, leaping, tossing, and swinging—during a complete annual cycle of breeding and nonbreeding seasons (Modahl and Eaton, in press).

The results revealed a significant sex difference. Males displayed much more frequently than females, and males that displayed most frequently also mated most frequently. There was also a striking increase in the frequency of male (but not female) displays during the mating season (Fig. 2.2). However, the male display frequency rank did not correlate with individual dominance rank or attack frequency rank.

Because of the significant correlation between displays and mating we concluded that these displays may serve as a mechanism of sexual selection and influence the females' choice of males. However, the sequence of the behaviors and other contextual information must be analyzed before the extent of this function can be determined (Modahl and Eaton, in press).

3. *Longitudinal Studies*

Our longitudinal investigations into the determinants of Japanese macaque sexual behavior comprise over 7,000 man-hours of observation from September 1970 to February 1975. The size and composition of the troop during each mating season is given in Table 2.3. Animals that died or were removed from the corral because of severe injury or disease are not included. General methods and procedures have been previously defined (Eaton, 1972, 1974). The data were collected by two or three observers during the first three hours of daylight, five days a week, 52 weeks a year. However, only data from September to March (the mating season) of each year are presented here.

Hanby (1972) reported that in this same troop over 50% of all mount

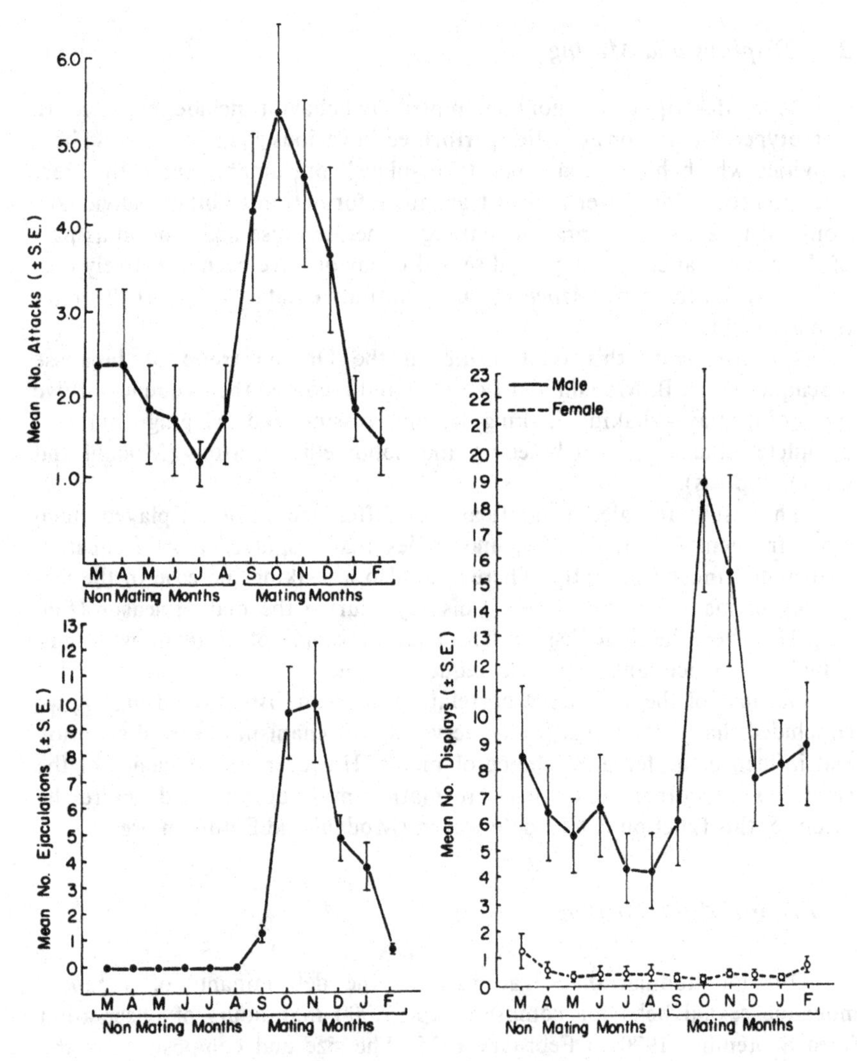

Figure 2.2. Temporal patterning of adult Japanese macaque male aggression, mating and displays during the 1972–1973 seasons.

Table 2.3. Numerical Composition of the Oregon Troop of Japanese Macaques[a]

	Mating season				
Measure	1970–1971	1971–1972	1972–1973	1973–1974	1974–1975
Total	98	109	124	150	162
Adult males	22	21	21	30	33
Adult females	34	37	44	52	58
Juvenile males	22	26	30	34	30
Juvenile females	20	25	29	34	41

[a] Juvenile males were three years old and younger. Juvenile females were two years old and younger.

series and ejaculations occurred during the first three hours after dawn; that no pairs mounted exclusively in the mornings or afternoons; and that no change occurred in the diurnal pattern of mounting (after the large morning peak a small secondary peak occurred just before dusk) when the feeding time was varied.

3.1. Heterosexual Behavior

Macaques, like people, do not simply walk up to each other and start to copulate no matter how well acquainted. Their courtship and processes of pair-bond formation are complex and, again like those of people, not always obvious. For example, if a female approaches a male and sits down beside or behind him about one meter away (the typical female "beginning" of a mount series) we score a female approach. But that is probably only a small part of the sequence. He may have sent roses or winked at her; or more likely, he may have smacked his lips, or raised his tail or fur or eyebrows, or flipped his ears at her a few seconds, hours, or days before. Or he may have attacked her by sneaking up behind her, pouncing like a cat and then biting her on the back or rump hard enough to draw blood, and then run off while she chased him screaming as if in pain, anger, and frustration. We are only beginning to recognize these acts of "courtship" and it will be some time before they can be reduced to a series of formulas encompassing primate pair-bond formation. However, once the male mounts the female (from the rear by grasping her ankles with his feet and resting his hands on her rump while she stands quadrapedally) certain patterns and sequences of behavior tend to reappear. For example, once a consort pair has mounted twice in less than 5 min (an arbitrary definition of a sexual mount series to differentiate them from singleton "social" or "courtesy" mounts) the male mounts the female an average of once every 1.5 min for 12.5 min until the series is terminated in one of six ways: ejaculation, pause, male leave, female leave, both leave, or unknown (as defined in Section 2.2).

We could not always determine the occurrence of intromission, but we believe that it occurred during almost every mount. The males thrust 0–8 times during each mount but we have not systematically recorded these frequencies. The number of mounts varied with the termination type of the series (Table 2.4), and the intermount interval of 1.5 min remained relatively constant (recorded only during the 1970–1971 season). The 1970–1971 data in Table 2.4 ae presented only for illustrative purposes and were not used to compute the grand means because some of the females were implanted with IUDs, as previously discussed. It is apparent from Table 2.4 that about twice as many mounts occurred during an ejaculatory series as during any other type of series. This is consistent with the definitions of the terminations in that the animals either leave or stop mounting (pause) before an ejaculation occurs.

It is also apparent from Table 2.4 that once a series started it was most likely to end with an ejaculation, second most likely to end with the pair remaining together but not mounting (pause), and third most likely to end with the male leaving. On the other hand, the female leaving or both partners leaving at once occurred relatively infrequently, as did the "unknown" termination. Inasmuch as "female leave" and "both leave" were usually scored when a more dominant male broke up the pair (almost invariably by chasing the female away), this suggests that these interruptions were rarely successful in keeping the mounting pair apart.

Some of these data are remarkably similar to those reported by Hanby *et al.* (1971) in their study of the 1968–1969 breeding season in this troop. For example, they reported that 34.3% of the mount series ended in ejaculation; the overall mean of our study was 38.6% with a range of 33.1%–42.7%. Furthermore, they reported that a pause in mounting occurred in 24.2% of the mount series and our overall mean for this termination was 32.0% with a range of 28.0%–34.5%. The other types of terminations are not directly comparable, but Hanby *et al.* also noted that high-ranking males tended to chase and attack females and again were relatively unsuccessful in keeping consort pairs apart. On the other hand, Wolfe (1976) in her study of the Arashiyama West troop (the Arashiyama A Troop transplanted from Japan to a 107-acre enclosure near Laredo, Texas in 1972) reported that 65% of the mount series she observed during the 1973–1974 season ended with ejaculation. The reason for this descrepancy is not readily apparent, although an unusually large ratio of females to males existed in that troop. This will be discussed in more detail below.

About 40% of the females mounted about 50% of the males in the Oregon troop (Table 2.5) with either a male-like double-foot-clasp mount or more frequently by sitting and rubbing the perineum on the male's back. These mounts characteristically preceded male mounts and have been previously described as female courtship (Hanby *et al.*, 1971; Stephenson, 1973). They have also been classified as a type of precultural behavior by

Table 2.4. Mean Frequency of Mounts per Type of Mount Series Termination and Percent of Total Mount Series

	Mating season					
Termination	1970–1971 f(%)	1971–1972 f(%)	1972–1973 f(%)	1973–1974 f(%)	1974–1975 f(%)	Grand means[a] f(%)
Ejaculation	9.3 (45.4)	8.6 (37.0)	8.1 (42.7)	7.7 (33.1)	9.1 (34.8)	8.4 (36.9)
Pause	4.5 (16.6)	4.0 (31.2)	3.6 (28.0)	4.0 (34.5)	4.6 (34.2)	4.1 (32.0)
Male leave	3.6 (24.1)	3.4 (17.5)	3.5 (15.4)	3.4 (11.3)	3.7 (10.1)	3.5 (13.6)
Female leave	3.6 (6.9)	3.0 (5.8)	3.5 (.55)	2.6 (3.3)	5.5 (3.9)	3.7 (4.6)
Both leave	3.2 (2.5)	4.5 (1.7)	3.2 (2.0)	4.5 (1.5)	2.7 (0.4)	3.7 (1.4)
Unknown	10.1 (4.5)	5.4 (6.7)	5.8 (6.5)	4.5 (16.2)	5.4 (16.5)	5.3 (11.5)

[a] Grand means exclude 1970–1971 data because some of the troop females were implanted with IUDs during that season.

Table 2.5. Measures of Females Mounting Males

Measure		Mating season				Mean
		1971–1972	1972–1973	1973–1974	1974–1975	
Females	(*n*)	16	18	16	24	
	(%)	43.2	40.9	30.8	41.4	39.1
Males	(*n*)	11	11	14	17	
	(%)	52.3	52.4	46.7	51.5	50.1
Total series	(*n*)	141	63	55	68	
Mounts/series		2.9	2.2	3.1	3.8	3.0

Stephenson (1973), who observed provisioned feral troops; only females in the lower half of the social hierarchy mounted males of the Miyajima troop, whereas females of all ranks mounted males in the Arashiyama troop, and no females mounted in the Koshima troop. Wolfe (1976) has reported that 56.1% of the females in the Arashiyama West troop mounted males and that rank was not a factor. These females averaged 3.1 mounts per series, which is remarkably similar to the overall mean of 3.0 for the Oregon females (Table 2.5), where rank is also unrelated to this behavior.

3.2. Homosexual Behavior

The data for female homsexual behavior are given in Table 2.6, which starts with the 1971–1972 season because this type of behavior was not systematically recorded during the previous year. Females tended to mount other females in the same manner as they mounted males, that is, by sitting on the mountee's back and rubbing the perineum along the base of her spine. The females did not appear to adopt a consistent sex role during these series but mounted each other alternatively. The percentage of females that were involved in homosexual mount series has slightly, but not significantly, increased across the last four mating seasons, while the mean number of mounts per series has fluctuated widely. Hanby *et al.* (1971) reported only one pair of adult females (Big T and Infinity) mounting during the 1968–1969 season in the Oregon troop. On the other hand, Wolfe (1976) reported that 34 of 46 (73.9%) adult females were involved in homo- or bisexual behavior in the Arashiyama West troop during the 1973–1974 season, and 21 of 57 (36.8%) during the 1974–1975 season. Over half the adult males were killed or lost shortly after this troop was released in Texas, and Wolfe (1976) hypothesized that female homosexual behavior may have been an adaptation by the females to the extraordinarily high female to male ratio (1973–1974: 3.87 females per male; 1974–1975: 3.0 females per male).

Males also exhibited homosexual behavior in the Oregon troop by pairing with other males and mounting them in series (Table 2.7). Male mount series were defined in the same manner as female mount series (two or more mounts in less than five minutes) to differentiate them from social mounts between males, which occur frequently throughout the year. With rare exceptions females do not indulge in such social mounting and, like the males, only mount in series during the mating season.

Approximately 15% of the males mounted other males each season with a mean of 7.4 mounts per series (Table 2.7). This is higher than that reported by Hanby *et al.* (1971), but this data for the 1968–1969 season correspond to those of 1971–1972 season in Table 2.7, and our data are somewhat variable across the years. Furthermore, Wolfe (1976) saw no male homosexual behavior during the two seasons that she studied the Arashiyama West troop. Like the high incidence of female homosexual behavior, this may have been due to the unusual sex ratio.

3.3. *Age and Female Sexual Behavior*

In order to analyze the relation between advancing age and sexual behavior we selected the eight oldest females of the Oregon troop for comparison with a group of nine young females who had reached puberty in 1970. The exact ages of the old females are unknown, but female Japanese macaques reach full body weight at 6–8 years, so we know that these females were at least six upon arrival in Oregon in 1965, or at least 11 years old at the start of this study in 1970. However, I believe 11 to be a conservative estimate; from casual comparison of tooth wear and facial wrinkles, I have estimated that in 1970 two of them Red Witch and Mrs. Perfect) were at least 20; four were at least 15 (Lucretia, Squaw, Infinity and Big Y), and only two were 11 (Split and Big W). These estimates give a mean age of 15 years in 1970. The nine pubescent females were all born in the spring and early summer of 1967 and were thus about 3.5 years old when the mating season started in the fall of 1970. Four of these pubescent females were

Table 2.6. Measures of Female Homosexual Behavior

Measure		Mating season				
		1971–1972	1972–1973	1973–1974	1974–1975	Mean
Mounters	(*n*)	2	4	6	7	
	(%)	5.9	9.1	11.5	12.1	9.7
Mountees	(*n*)	2	3	6	6	
	(%)	5.9	6.8	11.5	10.3	8.6
Total series	(*n*)	5	21	38	44	
Mounts/series		3.4	10.1	11.2	6.4	7.8

Table 2.7. Measures of Male Homosexual Behavior

Measure		Mating season				
		1971–1972	1972–1973	1973–1974	1974–1975	Mean
Mounters	(*n*)	2	10	4	13	
	(%)	9.1	21.7	8	22	15.2
Mountees	(*n*)	2	7	6	11	
	(%)	9.1	15.2	12.0	18.6	13.7
Total series	(*n*)	4	127	35	66	
Mounts/series		7.5	10.1	6.6	5.5	7.4

daughters of the old females discussed above (Little x—Mrs. Perfect; Little s—Squaw; Little t—Big W; and Little r—Infinity); the remaining five were not known to be related to the other females.

The mean frequencies of mount series terminations per female are given in Table 2.8. A two-way analysis of variance with repeated measures (Winer, 1962) was computed to analyze the effect of infants on their mother's behavior and to assess changes in behavior with increasing age within groups. The old females were considered to range in age from a mean of 11.5–16.5 years at the beginning of the 1970 season to a mean of 16.5–21.5 years at the beginning of the 1974 season; the young females ranged from 3.5–8.5 years during the same period.

As expected, the effect of the infants was highly significant in reducing the frequency of almost all types of mount series for both groups of females. The only behaviors that were not significantly reduced ($p > .05$) were the three categories of "leave" for the old females. However, no significant effects were related to increasing age in either group. The categories of "female leave," "both leave," and "unknown" termination differed significantly across the years but not systematically with increasing age.

The two groups did not differ with respect to the number of infants they bore after the 1970–1971 season. During the next four years the eight old females had a total of 18 infants and the nine young females had a total of 19. Hence, the mean frequencies of their terminations under the infant and noninfant categories in Table 2.8 were combined and a two-way analysis of variance with repeated measures (Winer, 1962) was computed to compare the behavior of the young females with that of the old (as with all the statistical analyses in this report, the data from 1970–1971 were not used because of the IUD experiment conducted that season).

The two groups of females did not differ significantly in any category except "female leave," and the reason for this difference was not apparent. This type of termination did have the lowest frequency of occurrence. For example, each young female without an infant was involved in a mean of 43.0 mount series per season for each of the four years; across the same period "female leave" occurred 0.8 times per season. The single significant

Table 2.8. Mean Frequency per Female per Mating Season of Mount Series Terminations for Old and Young Females With and Without Infants

		Mating season											
		1970–1971		1971–1972		1972–1973		1973–1974		1974–1975		Grand means[c]	
Termination	Females	Inf.[b]	Number	Inf.	Number	Inf.	Number	Inf.	Number	Inf.	Number	Inf.	Number
Total series	Old	23.8	91.0	23.0	47.8	21.2	25.5	16.2	29.7	7.3	40.5	16.9	35.9
	Young	—	18.7	4.3	37.0	15.0	44.8	8.6	62.3	9.3	28.0	9.3	43.0
Ejaculation	Old	10.8	41.5	8.3	21.6	9.7	10.5	4.2	9.0	2.5	18.5	6.2	14.9
	Young	—	8.3	0.7	14.0	4.3	19.4	2.4	14.7	3.5	7.0	2.7	13.8
Male leave	Old	6.8	20.0	5.0	7.4	2.8	2.5	2.4	3.3	0.5	3.5	2.7	4.2
	Young	—	4.2	1.0	4.8	2.7	5.6	0.6	4.7	1.0	2.7	1.3	4.5
Female leave	Old	0.2	1.5	0.3	0.4	0.3	0.5	0.6	0	0	0	0.3	0.2
	Young	—	0.3	0	0.8	1.0	1.2	0	1.3	0	0	0.3	0.8
Both leave	Old	1.7	4.0	1.0	2.4	1.2	1.5	0.2	1.0	0.5	0.3	0.7	1.3
	Young	—	0.4	1.5	4.0	3.5	2.6	0	1.5	0.2	2.0	1.3	2.5
Pause	Old	3.8	20.5	7.3	13.6	6.0	9.5	7.8	9.3	2.3	13.3	5.9	11.4
	Young	—	3.8	1.0	11.6	3.0	13.8	4.2	23.0	2.8	11.3	2.7	14.9
Unknown	Old	0.5	3.5	1.0	2.4	1.2	1.0	1.0	7.0	1.5	5.0	1.2	3.9
	Young	—	1.5	0	1.8	0.5	4.2	1.4	17.0	1.8	5.0	0.9	7.0

[a] The number of old females was 8, of young females, 9. The young females reached puberty during the 1970–1971 season.
[b] Inf. = live offspring less than one year old.
[c] Grand means exclude 1970–71 data because some of the old females were implanted with IUDs during that season.

Table 2.9. Mean Frequency per Female per Mating Season of Ejaculations Received by Different Groups of Females

	Mating season				
Females	1971–1972 $f(n)$	1972–1973 $f(n)$	1973–1974 $f(n)$	1974–1975 $f(n)$	Grand mean
Without infant	23.9 (16)	39.7 (18)	34.5 (18)	16.9 (31)	28.7
With infant	7.5 (17)	6.6 (18)	3.6 (26)	3.7 (19)	5.3
Pubescent	7.7 (4)	14.6 (8)	12.0 (8)	3.7 (8)	9.5

difference between young and old females in this category appears, therefore, to be a spurious effect.

Unfortunately, 1970–1971 was the pubescent season for the young females and it appeared (Table 2.8) that they were less active than the old females. Therefore, an additional analysis was undertaken to compare the number of ejaculations received by all adult females (4.5 years old and up) of the troop with those females that reached puberty (3.5 years) during each of four mating seasons (Table 2.9). Frequency of ejaculation was chosen as the critical measure because it not only correlates with total mount series but is also the ultimate criterion of mating success.

The pubescent females received significantly fewer ejaculations than adult females without infants, but they did not differ significantly from those with infants (Table 2.9). Therefore, the effects of puberty and of infants are similar even though the underlying mechanisms are apparently different. The effect of the infant is to inhibit a developed system whereas at the time of puberty the system is presumably still developing. It remains for future research to determine the extent to which hormonal–behavioral interactions produce these effects. For example, both sets of females may be less attractive to the males; the old females with infants may be less receptive because the hormones of lactation inhibit those of ovulation; the pubescent females may have to learn how to complete a mating sequence, or their own endocrine systems may not be completely developed—they are, after all, still several years away from adult body size.

3.4. Age and Male Sexual Behavior

The eight oldest males in the troop were selected for study through a procedure similar to that used for selecting the oldest females. However, the males do not reach full body growth until they are 8–10 years old. Therefore, four males were estimated to be at least eight, one to be nine, and three to be 13 years old in 1970. As with the females, these estimates were believed to be conservative and the ages of three males were probably much greater (Arrowhead and Bruno 20, and Perfect 25 years old in 1970).

The mean age of these old males ranged, therefore, from at least 10–13.3 years in 1970. There were only one or two pubescent males (4.5 years) each year until 1969. Therefore, the data for the old males range from the 1970–1971 to 1974–1975 season, whereas those for the young males cover only the 1973–1974 and 1974–1975 seasons (Table 2.10).

We computed a one-way analysis of variance with repeated measures for each of the mount series terminations to analyze age changes in the old males, and *t* tests for independent groups were used to compare the behavior of the young and old males during the 1973–1974 and 1974–1975 seasons (Winer, 1962). As with the analysis of the female data, the 1970–1971 season was included only for illustrative purposes and not used in the statistical analysis of age changes.

Unlike that of the old females, the sexual behavior of the old males declined significantly across the mating seasons on all measures except "female-leave," "both leave," and "pause." The behavior of the young males did not differ, however, from that of the old males during their pubescent season (1973–1974) or the following season when they were 5.5 years old. These two sets of data suggest, therefore, that sexual vigor (defined by the frequency of mounting and ejaculation) first increases and then declines with advancing age in these Japanese macaques. However, because of the different histories of the young (born in the corral in Oregon) and old males (born in the wild in Japan) only the decline in vigor with

Table 2.10. Mean Frequency per Male per Mating Season of Mount Series Terminations for Old and Young Males

		Mating season				
Termination	Males[a]	1970–1971	1971–1972	1972–1973	1973–1974	1974–1975
Total series	Old	80.0	82.3	75.1	59.1	52.3
	Young	—	—	—	28.7	34.3
Ejaculation	Old	38.5	35.3	37.9	22.6	28.1
	Young	—	—	—	7.1	10.0
Male leave	Old	20.0	16.0	16.0	7.6	5.6
	Young	—	—	—	3.9	4.6
Female leave	Old	6.4	5.3	4.4	1.9	2.3
	Young	—	—	—	1.4	1.9
Both leave	Old	1.7	1.1	0.9	0.5	0.4
	Young	—	—	—	0.7	0.3
Pause	Old	10.6	19.4	13.3	17.6	10.9
	Young	—	—	—	9.0	11.6
Unknown	Old	2.7	5.3	4.0	8.9	5.0
	Young	—	—	—	5.0	6.0

[a] The number of old males was 8, of young males, 7. The young males reached puberty during the 1973–1974 season.

advancing age should be considered a reliable outcome. Continuing studies of the young males are necessary in order to assess any increase in their sexual behavior.

4. *Discussion and Summary*

4.1. *Effects of Confinement*

It is difficult to assess the effects of confinement because the macaques were not studied before their capture. The problem is confounded by intertroop differences in sexual behavior such as the failure of females to mount males in some troops and the occurrence of this behavior among only certain classes of females in other troops (Stephenson, 1973). It does seem probable, however, that provisioning plus confinement increase the frequency of all types of social interaction. It may also be that the macaques are not crowded in the sense of Stokols' (1972) definition of crowding as a motivational state aroused through the interaction of spatial, social, and individual factors. For example, in informal observations I found the Arashiyama West troop to be usually grouped into a space of less than 2 acres while they moved through 8–10 acres a day and through the entire 107 acres every few days. It may be, therefore, that the movement of the Oregon troop is restricted by the fence, but the condition of crowding does not exist (Eaton, 1976). Some support for this hypothesis comes from our observations of adult aggressive behavior, which has not increased parallel to the increase in population density. In fact, there has been a significant decline in aggressive behavior during the past few years (Eaton, Modahl, and Johnson, in preparation).

4.2. *Macaque Sexual Behavior*

These studies have described and analyzed some of the relationships between biological and social forces that influence the sexual behavior of a confined troop of Japanese macaques. Confinement appears not to have qualitatively affected the macaques' behavior but it may have increased the frequency of all types of social interaction. Confinement has not affected the seasonal nature of their sexual behavior. However, the mechanism that controls this seasonal cycle is unknown. Pregnancy has been shown to have a significant inhibitory effect on the duration of the mating season (Eaton, 1972). On the other hand, reports of seasonal endocrine changes in laboratory rhesus macaques living under constant environmental conditions (Plant *et al.*, 1974; Robinson *et al.*, 1975) suggest that circannual rhythms are set early in life. These rhythms may interact with other endocrine changes, as suggested by our studies to be associated with puberty and pregnancy and with behavioral–endocrine stimuli (Vandenbergh, 1969; Vandenbergh and Drickamer, 1974), to time the periodicity of seasonal reproduction.

Our studies have also shown that male dominance and seasonal mating success are not correlated (Eaton, 1974). One of the values of a confined troop is in the ease of accurate observation and identification of all troop members, which allows unbiased frequency counts of behavior. Drickamer (1974) has reported that high-ranking rhesus males at La Parguera were more conspicuous than low-ranking males, and has suggested that such a bias in field studies may account for many of the reports that dominant males mate more frequently than low-ranking males. Furthermore, in a recent paternity study of a group of rhesus monkeys, Duvall, Bernstein, and Gordon (1976) have reported that high-ranking males did not father more infants than low-ranking males during any one year, although by way of increased longevity high-ranking males may father more offspring over a lifetime.

We have also recorded seasonal changes in the frequency of male displays and computed a significant correlation between them and the frequency of mating, which suggests that they serve a courtship function (Modahl and Eaton, in press). However, a more refined analysis of sequential behaviors is necessary to demonstrate a casual relation between displays and mating success.

The macaques' sexual behavior includes both hetero- and homosexual aspects as part of the "normal" pattern. Protocultural variations of some of these patterns have already been discussed but it is well to remember the extreme variation in behavior that characterizes individuals and groups of primates. This plasticity of behavior has apparently played a major role in the evolutionary success of primates by allowing them to adapt to a variety of social and environmental conditions.

After a female Japanese macaque has passed puberty the major influence on her sexual behavior is the profound inhibitory effect of an infant conceived during the previous mating season. This effect may operate in feral troops as a method of population limitation but it appears to be less effective in provisioned troops that are assured of an adequate food supply.

The fact that a mount series is likely to be completed once it has been initiated suggests that dominant males are not successful in keeping consort pairs apart. This observation is supported by the lack of correlation between male dominance rank and mating rank. It also emphasizes the fact that although these macaques are promiscuous and mate with several different partners, these matings are not random couplings and require emotional bonds of some substance between the consort pairs.

The data on mount series terminations also suggest that while an infant has an inhibitory effect on its mother's sexual behavior, increasing age has no effect on sexual behavior or fertility in female Japanese macaques. Longevity is unknown, however, and it may be another 10 or 15 years before the "old" females can be considered really old. On the other hand, increasing age did reduce the frequency of sexual behavior in the group of "old" males. Robinson *et al.* (1975) have also recently reported that the sexual activity of male rhesus monkeys older than 20 years was significantly less than that of younger males. The testosterone and dihydrotestosterone levels of these males did not differ from those of the young males, so the mechanisms involved remain unknown.

5. Prospectus

Our prospectus for studies of macaque sexual behavior includes continuing the longitudinal studies while conducting a detailed analysis of male and female courtship behavior. We plan to investigate the relation of courtship to the initiation of the mating season and to the development of partner preference in consort relationships. The relation between male aggressive behavior and courtship will also be analyzed. A newly constructed second corral will allow us to observe the formation of a new social order (if the troop divides); then relations between two troops may be studied, such as the exchange of males during mating seasons. With these and other studies we hope not to "prove" anything about human sexual behavior but to provide a comparative frame of reference from which we may more clearly view some aspects of our own humanity. For example, the variability and plasticity of the behavior displayed by macaques suggests an optimistic or "*maximal* view of human potentialities and limitations" (Matson, 1976) rather than a pessimistic or *minimal* view of man as a biological machine functioning on the basis of instinct. This minimal view based on the fang-and-claw school of Darwinism finds little support in the evidence of protocultural evolution in nonhuman primates. Future studies of primate behavior should provide additional resolution to humanity's image of itself.

ACKNOWLEDGMENTS

I wish to thank Kurt B. Modahl, Deanne F. Johnson, and Jens Jensen for their technical assistance.

References

Alexander, B. K. Parental behavior of adult male Japanese monkeys. *Behaviour*, 1970, *36*, 270–285.

Alexander, B. K., and Bowers, J. M. Social organization of a troop of Japanese monkeys in a two-acre enclosure. *Folia Primatologia*, 1969, *10*, 230–242.
Alexander, B. K., and Bowers, J. M. The social structure of the Oregon troop of Japanese macaques. *Primates*, 1967, *8*, 333–340.
Alexander, B. K., and Hughes, J. Canine teeth and rank in Japanese monkeys (*Macaca fuscata*). *Primates*, 1971, *12*, 91–93.
Alexander, B. K., and Roth, E. M. The effects of acute crowding on aggressive behavior of Japanese monkeys. *Behaviour*, 1971, *39*, 73–90.
Alexander, B. K., Hall, A. S., and Bowers, J. M. A primate corral. *Journal of the American Veterinary Medical Association*, 1969, *155*, 1144–1150.
Beach, F. A. Experimental investigations of species specific behavior. *American Psychologist*, 1960, *15*, 1–18.
Carpenter, C. R. Sexual behavior of free ranging rhesus monkeys (*Macaca mulatta*). I. Specimens, procedures and behavioral characteristics of estrus. *Journal of Comparative Psychology*, 1942, *33*, 113–142.
Conaway, C. H. Ecological adaptation and mammalian reproduction. *Biology of Reproduction*, 1971, *4*, 239–247.
DeVore, I. Male dominance and mating behavior in baboons. In F. A. Beach (Ed.), *Sex and behavior*. New York: John Wiley, 1965, pp. 266–289.
Drickamer, L. C. Social rank, observability, and sexual behavior of rhesus monkeys (*Macaca mulatta*). *Journal of Reproduction and Fertility*, 1974. *37*, 117–120.
Duvall, S. W., Bernstein, I. S., and Gordon, T. P. Paternity and status in a rhesus monkey group. *Journal of Reproduction and Fertility*, 1976, *47*, 25–31.
Eaton, G. G. The social order of Japanese Macaques. *Scientific American*, 1976, *235*, 96–106.
Eaton, G. G. Male dominance and aggression in Japanese macaque reproduction. In W. Montagna and W. A. Sadler (Eds.), *Reproductive behavior*. New York: Plenum Press, 1974, pp. 287–297.
Eaton, G. G. Social and endocrine determinants of sexual behavior in simian and prosimian females. In C. H. Phoenix (Ed.) *Primate reproductive behavior. Symposium of the IVth international congress of primatology*, Vol. 2. Basel: Karger, 1973, pp. 20–35.
Eaton, G. G. Seasonal sexual behavior and intrauterine contraceptive devices in a confined troop of Japanese macaques. *Hormones and Behavior*, 1972, *3*, 133–142.
Eaton, G. G., and Resko, J. A. Plasma testosterone and male dominance in a Japanese macaque (*Macaca fuscata*) troop compared with repeated measures of testosterone in laboratory males. *Hormones and Behavior*, 1974, *5*, 251–259.
Eaton, G. G., Modahl, K. B., and Johnson, D. F. Aggressive behavior: A longitudinal study in a confined troop of Japanese macaques. In preparation.
Goldfoot, D. A., Slob, A. K., Scheffler, G., Robinson, J. A., Wiegand, S. J., and Cords, J. Multiple ejaculations during prolonged sexual tests and lack of resultant serum testosterone increases in male stumptail macaques (*M. arctoides*). *Archives of Sexual Behavior*, 1975, *4*, 547–560.
Gordon, T. P., Rose, R. M., and Bernstein, I. S. Seasonal rhythm in plasma testosterone levels in the rhesus monkey (*Macaca mulatta*): A three year study. *Hormones and Behavior*, 1976, *7*, 229–244.
Hanby, J. P. The sociosexual nature of mounting and related behaviors in a confined troop of Japanese macaques (*Macaca fuscata*). Doctoral dissertation, University of Oregon, 1972.
Hanby, J. P., Robertson, L. T., and Phoenix, C. H. The sexual behavior of a confined troop of Japanese macaques. *Folia Primatologica*, 1971, *16*, 123–143.
Imanishi, K. Social behavior in Japanese monkeys, *Macaca fuscata*. In C. H. Southwick (Ed.), *Primate social behavior*, Princeton: Van Nostrand, 1963, pp. 68–81.
Itani, J. On the acquisition and propagation of a new food habit in the troop of Japanese

monkeys at Takasakiyama. In S. A. Altmann (Ed.), *Japanese monkeys: A collection of translations*, Edmonton, Canada: published by the editor, 1965, pp. 52–65.
Kaufmann, J. H. A three-year study of mating behavior in a free-ranging band of rehsus monkeys. *Ecology*, 1965, *46*, 500–512.
Kawai, M. On the newly acquired behaviors of the natural troop of Japanese monkeys on Koshima Island. *Primates*, 1963, *4*, 113–115.
Kawai, M. Newly acquired precultural behavior of the natural troop of Japanese monkeys on Koshima Island. *Primates*, 1965a, *6*, 1–30.
Kawai, M. On the system of social ranks in a natural troop of Japanese monkeys: I. Basic rank and dependent rank. In S. A. Altmann (Ed.), *Japanese monkeys: A collection of translations*. Edmonton, Canada: published by the editor, 1965b, pp. 66–86.
Kawamura, S. The process of subculture propagation among Japanese macaques. In C. H. Southwick (Ed.), *Primate social behavior*. Princeton: Van Nostrand, 1963, pp. 82–90.
Lancaster, J. B., and Lee, R. B. The annual reproductive cycle in monkeys and apes. In I. Devore (Ed.), *Primate behavior*. New York: Holt, Rinehart, and Winston, 1965, pp. 486–513.
Matson, F. W. *The idea of man*. New York: Delacorte, 1976.
Mayr, E. *Animal species and evolution*. Cambridge: Harvard University Press, 1963.
Missakian, E. A. Genealogical and cross-genealogical dominance relations in a group of free-ranging rhesus monkeys (*Macaca mulatta*) on Cayo Santiago. *Primates*, 1972, *13*, 169–180.
Modahl, K. B., and Eaton, G. G. Display behaviour in a confined troop of Japanese macaques. *Animal Behavior*. In Press.
Nadler, R. D. Sexual cyclicity in captive lowland gorillas. *Science*, 1975, *189*, 813–814.
Nagel, U., and Kummer, H. Variation in cercopithecoid aggressive behavior. In R. L. Holloway (Ed.), *Primate aggression, territoriality, and xenophobia*. New York: Academic Press, 1974, pp. 159–184.
Plant, T. M., Zumpe, D., Sauls, M., and Michael, R. P. Annual rhythm in plasma testosterone of adult male rhesus monkeys maintained in laboratory. *Journal of Endocrinology*, 1974, *62*, 403–404.
Robinson, J. A., Scheffler, G., Eisele, S. G., and Goy, R. W. Effects of age and season on sexual behavior and plasma testosterone and dihydrotestosterone concentrations of laboratory-housed male rhesus monkeys (*Macaca mulatta*). *Biology of Reproduction*, 1975, *13*, 203–210.
Rose, R. M., Gordon, T. P., and Bernstein, I. S. Plasma testosterone levels in the male rhesus: influences of sexual and social stimuli. *Science*, 1972, *178*, 643–645.
Rose, R. M., Holaday, J. W., and Bernstein, I. S. Plasma testosterone dominance rank and aggressive behaviour in male rhesus monkeys. *Nature*, (*London*), 1971, *231*, 366–368.
Rowlands, I. W. *Comparative biology of reproduction in mammals*. New York: Academic Press, 1966.
Sade, D. S. Determinants of dominance in a group of free-ranging rhesus monkeys. In S. A. Altmann (Ed.), *Social communication among primates*. Chicago: University of Chicago Press, 1967, pp. 99–114.
Stephenson, G. R. Testing for group specific communication patterns in Japanese macaques. In E. W. Menzel, Jr. (Ed.), *Precultural primate behavior. Symposium of the IVth International Congress of Primatology*, Vol. I. Basel: Karger, 1973, pp. 51–75.
Stokols, D. On the distinction between density and crowding: some implications for future research. *Psychological Review*, 1972, *79*, 275–277.
Sugiyama, Y. On the division of a natural troop of Japanese monkeys at Takasakiyama. *Primates*, 1960, *2*, 109–149.

Tokuda, K. A study on the sexual behavior in the Japanese monkey troop. *Primates*, 1961–1962, *3*, 1–40.

Vandenbergh, J. G. Endocrine coordination in monkeys: Male sexual responses to the female. *Physiology and Behavior*, 1969, *4*, 261–264.

Vandenbergh, J. G., and Drickamer, L. C. Reproductive coordination among free-ranging rhesus monkeys. *Physiology and Behavior*, 1974, *13*, 373–376.

Winer, B. J. *Statistical principles in experimental design*. New York: McGraw-Hill, 1962.

Wolfe, L. Sexual behavior of the Arashiyama West troop of Japanese macaques (*Macaca fuscata*). Doctoral dissertation, University of Oregon, 1976.

Yamada, M. Five natural troops of Japanese monkeys on Shodoshima Island: II. A comparison of social structure. *Primates*, 1971, *12*, 125–150.

Young, W. C., and Orbison, W. D. Changes in selected features of behavior in pairs of oppositely sexed chimpanzees during the sexual cycle and after ovariectomy. *Journal of Comparative Psychology*, 1944, *37*, 107–143.

3

Sex Differences in Preschool Children's Social Interactions and Use of Space: An Evolutionary Perspective

Lawrence V. Harper and Karen M. Sanders

1. Introduction

Although the bases for anatomical and physiological gender dimorphisms in man are fairly well understood (Tanner, 1970), the origins of behavioral sex differences are the subject of vigorous debate (cf. Birns, 1976; Hutt, 1972). Many of the disagreements appear to revolve about the relative contributions of organismic as opposed to environmental determinants of sex-typed activities. It seems to be all too easy to fall into the old trap of characterizing sex differences as due to either "nature" or "nurture," rather than as the coaction of both. In these debates, there is a tendency to lose sight of the facts that the differences under consideration are matters of degree rather than kind, and that there is no *a priori* reason to expect that all of them result from the same causes. Analysis of the problem is further hampered by uncertainties concerning the empirical demonstrability of a number of alleged sex differences. In some cases, the dimensions of comparison, such as "dependence–independence" or "person versus object orientation" seem to be too broad and ill-defined to permit unequivocal demonstrations of gender differences (cf. Maccoby and Jacklin, 1974). In

---

Lawrence V. Harper and Karen M. Sanders • Department of Applied Behavioral Sciences, University of California, Davis, California.

other comparisons, as with "timidity" or "aggressiveness," it is also unclear whether the research findings reflect the methods used to obtain them rather than actual differences in behavior. For example, ratings may reflect observers' implicit expectations, instead of the subjects' activities (Birns, 1976). Even when objective measures of behaviors are available, as in the case of the amount and extensiveness of physical activity or time spent caring for small children, the meaning of the data is clouded by the fact that behavioral dimorphisms may result from culturally-imposed roles rather than individual choice (cf. Whiting and Edwards, 1973).

1.1. Use of Space and Social Interactions

In most human societies there are stereotyped expectations concerning how the sexes should differ behaviorally. In Western cultures the degree of physical "vigor," "dependence," and "sociability" are among the many characteristics that are supposed to distinguish males from females. According to prevailing stereotypes, males should be physically more active than females and expand more energy in commerce with their surroundings. Females are supposed to rely more upon others for emotional support and help in dealing with practical problems, and they are expected to be more interested in intimate personal relationships than males (cf. Maccoby and Jacklin, 1974).

Across both Western and non-Western societies, there is sexual division of labor in subsistence-related activities that seems to correspond to several of these stereotyped expectations. The tasks allotted to men are more strenuous and involve more extensive travel and cooperation; women's activities usually involve less physical effort and mobility, and are more solitary—although females are responsible for child care (D'Andrade, 1966). Comparable sex differences appear quite early in the play of young children: by the third year young boys expend more energy in outdoor play than do girls (Bell, Weller, and Waldrop, 1971). This sex difference in energy expenditure is not a product of boys' greater susceptibility to perinatal insult (Pedersen and Bell, 1970), and it is even displayed during solitary play (Moore, Evertson, and Brophy, 1974). Between three and five years of age boys spend more time playing outdoors and use greater amounts of space than do girls (Harper and Sanders, 1975). As early as the first year, under certain conditions, boys play farther away from their mothers than girls (Lewis and Weinraub, 1974), and from preschool through grade school, boys consistently play at a greater distance from their teachers than do girls (Omark and Edelman, 1973). This phenomenon is not limited to American youngsters. British preschool boys spend less time than girls near teachers (Blurton-Jones and Konner, 1973; Waterhouse and

Waterhouse, 1973), and children from an African hunting–gathering culture (Blurton-Jones and Konner, 1973; Draper, 1973) and from five additional non-Western, agricultural societies (Whiting and Edwards, 1973) also show comparable sex differences. It is possible that the greater centrifugal tendency of boys is related to their higher activity level: 12- to 36-month-old American boys more frequently moved away from their mothers than did girls of the same age while playing in a public park. Although there were no sex differences in the duration or distance traveled per sortie, the net result of the boys' more frequent departures was they they spent more time away from their mothers and covered more ground (Ley and Koepke, 1975). However, this finding still does not account for the boys' greater activity level or its taxic component.

Although differences in young children's proximity seeking are sometimes explained as reflecting culturally mediated responses in the service of "dependency motives" (cf. Mussen, Conger, and Kagan, 1974), this hypothesis is contradicted by the evidence. In Maccoby and Jacklin's (1974) exhaustive review, their Tables 6.1 and 6.2 list 31 observational studies of one- to five-year-old children's tendencies to hover near adults and to protest at separation from parents. When these studies are recategorized according to whether or not the situation could be considered stressful (an unfamiliar setting or parental departure), in five of the six stressful situations in which differences were found, boys displayed greater proximity seeking; in seven of the eight more neutral settings (free play or exploration) girls spent more time near to adults if the sexes differed at all (Maccoby and Jacklin, 1974, pp. 194–198). Thus the relatively greater proximity seeking of girls seems to be confined to nonstressful situations—the reverse of social learning predictions based upon cultural stereotypes of the "brave, independent" male.

The allegedly greater "sociability" of females also cannot explain these phenomena; the situation is much more complex. Observational studies of young children's free play indicate that boys may engage in more peer-related social contacts than girls (Blurton-Jones and Konner, 1973; Omark and Edelman, 1973; Maccoby and Jacklin, 1974; Waterhouse and Waterhouse, 1973). However, this difference may also depend upon the situation. Among preschoolers confined to a playroom, girls engaged in more "cooperative" (including "helpful") interactions than boys and more peer-related activity in general (Brindley, Clarke, Hutt, Robinson, and Wethli, 1973). In contrast, whereas no sex difference appeared indoors, boys engaged in more interactive fantasy play than girls when outdoors in a very spacious preschool setting (Sanders and Harper, 1976).

In summary, the evidence indicates that: (1) boys, more than girls, engage in vigorous, outdoor activity involving the extensive use of space; (2)

during free play, but not in stressful situations, girls spend more time near adults than do boys; and (3) the amount of peer interaction shown by either sex may depend upon the quality of the activity or its location.

1.2. An Evolutionary Hypothesis

In our work we have found that a comparative-evolutionary approach provides a framework within which the foregoing observations can be ordered, and is a source of testable hypotheses. These phenomena can be explained as the result of natural selection if we accept the view that hunting and gathering represented a basic human adaptation. According to current anthropological theories, man subsisted in this way during the better part of his evolutionary history (Lee and DeVore, 1968). Studies of contemporary hunter-gatherers suggest that gathering *per se* does not require a gender-related division of labor or physiological specialization. Hunting, however, does seem to impose such demands; among the majority of these societies it is predominantly a male activity (Hamburg, 1974; Watanabe, 1968). This fact can be understood as a biological adaptation if we assume that the burdens imposed by gestation, postnatal care, and the infant's early dependence upon mother's milk reduced the net advantage from females' joining the hunt. In addition, given an about equal sex ratio, the female is the limiting factor in the group's potential reproductive rate; the "cost" of losing a female during dangerous hunts would be much greater than that of losing a male (Dornbusch, 1966). Thus, very little gain and even a possible loss of fitness would result from female hunting. Finally, insofar as most terrestrial primates are physically dimorphic with the male being the larger and stronger (Crook, 1972), it is likely that humans evolved from a stock in which the male was already larger, stronger, and probably given to greater violent activity. Certainly, among most contemporary primates, the male is more likely than the female to spend time away from the main body of the group and is generally more "aggressive" (Crook, 1966; Izawa, 1970). Given the foregoing, one would expect selection to enhance mechanisms favoring development of the male-as-hunter and the female-as-gatherer. To the extent that large or potentially dangerous game was hunted, the male's greater strength and larger size would be maintained and his greater geographic mobility would continue to be favored whenever prey were scarce or timid.

Cooperative behavior among hunting species also appears to be advantageous. Students of wolves (Mech, 1966) and wild dogs (Estes and Goddard, 1967) have concluded that, compared to solitary stalking, group hunting is generally more efficient in obtaining large game. In his study of the spotted hyena (a species more closely related to cats than dogs), Kruuk (1972) observed a tendency for pack size to increase with the size and potential danger posed by the prey. If we may extrapolate, and given that early humans did engage in the pursuit of large and potentially dangerous animals (Speth and Davis, 1976), fitness would have been enhanced whenever males hunted in closely coordinated groups.

From the foregoing arguments, the males' greater tendency to wander afield and to engage in mutual give-and-take with peers would be interpreted as biological predispositions to behave in ways adaptive for group hunting.

The early appearance of this sex difference during children's play can be explained by the finding that mammalian play involves behavioral patterns "basic" to species survival (Loizos, 1967). Selection would favor such early play because young males' forays away from camp should enhance fitness by familiarizing them with the group's home range, hunting grounds, and the habits of available prey (Harper and Sanders, 1977). Interactive play at such times might also facilitate the development of abilities to coordinate individual actions to those of the group. The latter would be particularly advantageous where early playmates were likely to hunt together later in life. Moreover, the male's greater activity afield can also benefit the group and—to the extent that group members are relatives—increase fitness. More frequent departures from the main body, combined with the tendency to return to the group when confronted by strangers or other sources of danger, would also provide an "early warning" or buffer zone around the periphery of the group. The more centripetal, adult-oriented tendencies of girls would leave them relatively less vulnerable; while their more expendable brothers help to further shield them from danger (Harper and Sanders, 1977).

In short, we have reasoned that, if hunting, as opposed to gathering, usually involves greater geographic mobility (Watanabe, 1968) and group cooperation (Etkin, 1964), young males will spend more time away from the main group and in interaction with one another. Furthermore, insofar as males are adapted to pursue game, we also predicted that boys would display more interactive play than girls while moving about vigorously.

2. Some Empirical Findings

Over the last four years we have gathered data that appear to be consistent with the position stated in Section 1.2.

2.1. Subjects and Procedures

Since the fall of 1971 we have been studying the free-play activities of three- to five-year-old children who were enrolled at the laboratory nursery school at the University of California at Davis. During the winter and spring of 1973–1974 we also observed the behavior of a smaller group of children who were enrolled in a private, parent cooperative preschool in the nearby community of Dixon. As of spring, 1975 we have data on 51 girls and 56 boys from Davis and 6 girls and 8 boys from Dixon. For all but the Dixon group, average observation time per child exceeded 4 hr per season; the data discussed here represent over 1,000 hours of free-play behavior.

Our basic procedure was to record the activities of individual children every 15 sec for 35–50 consecutive minutes once each week throughout the 30-week school year. (At Dixon, our observation periods were limited to 20–35 min.) Our observations were confined to time periods in which the children were active and free to engage in the occupations of their choice; they could change from one to another without adult interference. Thus, all of their spontaneous behavior was considered "play," including such creative pursuits as painting or such adult-directed activities as tie-dyeing or dancing. In each time interval, we recorded the child's location, whether he or she was playing alone, with peers, and/or interacting with adults. We also obtained a measure of the child's geographic mobility (time "in transit") by noting every time that he or she entered 3 or more (previously defined) areas in a single 15-sec interval. Play with peers was categorized as follows: (a) parallel: when the subject was within 2 m of and engaging in a similar activity to that of, another child without influencing the other child; (b) interactive: when the child's activity was in response to another child's behavior or elicited a reciprocal response from the other; and (c) cooperative: when the child and his or her playmates were engaging in an activity that centered about a shared theme or goal or if the child complied with a peer request. These categories were mutually exclusive. From the fall of 1972, we also noted whether each level of play occurred while the children were "in transit." In addition, from the fall of 1973, all levels of play were qualified according to whether they involved a "fantasy" theme (see Sanders and Harper, 1976) and whether or not interaction involved "helping" another child; that is, spontaneously providing assistance or materials to a peer. For the purposes of this report, adult contacts were simply categorized as to whether or not they were "disciplinary." Nondisciplinary adult contacts were also scored as child initiated or adult initiated whenever it was obvious who originated a bout of interaction. Interobserver agreement was measured by number of agreements divided by number of agreements plus disagreements on the presence of a behavior (agreements on absence were not counted). With this conservative measure, mean agreement on all categories mentioned here ranged from 70%–98%.

The Davis facility is very large, having over 186 m² indoor space and 1,780 m² outdoor area available for free play. The Dixon school was housed in a community church with 141 m² indoors and 375 m² outdoors. Both facilities had the usual range of toys and equipment, but there were more playthings available at Davis. The programs were more or less traditionally child centered, although more emphasis was placed upon preacademic training at Dixon.

Since enrollment at Davis varied somewhat from quarter to quarter (generally ranging between 25 and 30 children), and the weather conditions during winter discouraged outdoor play, we analyzed each year's data separately according to season. "Winter" constituted the rainy season when skies were usually overcast and temperatures remained below 60° F. "Fall" and "spring" were typified by clear, sunny days with temperatures consistently above 60° F.

The staff at Davis changed every year (the student teachers changed each quarter), and several extensive modifications to both indoor and outdoor facilities occurred during the course of our study. Nevertheless, the behavior of the children across years was remarkably stable.

2.2. *Results*

Like the investigators mentioned in the introduction, we found a weak but consistent trend for girls to spend more time than boys in nondisciplinary contact with their teachers. This held true for both the Davis and Dixon schools (Harper and Huie, 1978). In accord with our speculations concerning the origins of these sex differences, the evidence indicates that, while the girls remained closer to the teachers, the boys more often wandered afield. On the basis of data collected during the first two years at Davis we reported that boys exceeded girls in outdoor play, number of play areas visited, and, less clearly, in the amount of time spent in transit (Harper and Sanders, 1975). These findings were replicated in the 1973–1975 observations at Davis. More importantly, a similar trend for boys to engage in more outdoor play than girls was observed at Dixon as well. (See Table 3.1)

An analysis of the location in which each category of play occurred indicated a stable pattern in which all classes of activity showed a sex difference. Consistent with the view that the male hunter should display more interaction while afield, the boys regularly engaged in significantly greater proportions of their interactive and cooperative play than girls while outdoors in both the Davis and Dixon samples. At Davis only, the boys also engaged in significantly more total interactive play than girls and girls regularly spent significantly more time in parallel play than boys. The observation that the boys at Davis engaged in more cooperative play outdoors also is in accord with our theory.

In addition, we analyzed the relationships between the amount of time spent playing outdoors and each category of play. For each child the pro-

Table 3.1. Percent of Observation Time Spent Playing Outdoors, in Transit, and Number of Different Areas Used

				Outdoors	Transit	Number of areas
		University of California Laboratory Nursery School				
1971–1972[c]	Fall	Boys,	$n = 15$	56.8	11.2	35.7
		Girls,	$n = 12$	39.2[a]	8.9[a]	31.3
	Winter	Boys,	$n = 13$	30.8	6.7	39.5
		Girls,	$n = 14$	11.5[b]	5.4	31.1[a]
	Spring	Boys,	$n = 13$	64.0	7.5	53.8
		Girls,	$n = 15$	35.0[d]	6.3	44.2[b]
1972–1973[c]	Fall	Boys,	$n = 12$	63.2	5.0	38.1
		Girls,	$n = 11$	44.3[b]	5.0	37.2
	Winter	Boys,	$n = 13$	46.8	5.8	49.8
		Girls,	$n = 12$	22.9[e]	3.9[a]	37.6[b]
	Spring	Boys,	$n = 14$	70.9	4.5	43.4
		Girls,	$n = 12$	55.3[b]	4.2[a]	39.1
1973–1974	Fall	Boys,	$n = 12$	52.8	2.8	29.0
		Girls,	$n = 12$	34.4[a]	3.2	27.7
	Winter	Boys,	$n = 13$	31.1	2.8	39.4
		Girls,	$n = 14$	15.6[a]	2.6	31.0[a]
	Spring	Boys,	$n = 16$	64.5	4.9	40.1
		Girls,	$n = 14$	48.7[a]	3.5[b]	36.6
1974–1975	Fall	Boys,	$n = 11$	69.6	3.5	32.2
		Girls,	$n = 12$	37.7[b]	3.3	26.7[a]
	Winter	Boys,	$n = 12$	39.8	4.7	44.7
		Girls,	$n = 12$	18.9[b]	3.0[b]	33.7[b]
	Spring	Boys,	$n = 11$	61.6	5.7	47.7
		Girls,	$n = 12$	47.6[a]	3.9[a]	36.2[b]
		Dixon Community Preschool				
1972–1973	Winter	Boys,	$n = 8$	23.5	4.0	23.7
		Girls,	$n = 6$	9.6[a]	5.4	22.7
	Spring	Boys,	$n = 8$	47.5	4.4	23.4
		Girls,	$n = 6$	30.7	4.9	18.8

[a] $p < .05$ (Mann–Whitney test).
[b] $p < .01$ (Mann–Whitney test).
[c] Figures differ slightly from Harper and Sanders (1975, 1977) due to the inclusion of "time in transit."
[d] $p < .0001$ (Mann–Whitney test).
[e] $p < .001$ (Mann–Whitney test).

portion of the total observation time he or she was outdoors was compared with the percent of total observation time during which he or she engaged in each category of play, regardless of its location. Despite the fact that boys spent proportionately more time engaging in all categories of play outdoors, the rank-order correlation coefficients presented in Table 3.2 indicate a preponderance of positive relationships between the amount of time boys spent outdoors and the total amounts of time in interactive and cooperative play, but not in solitary or parallel play. For girls, the data are less consistent. These findings are also in accord with the view that interaction and cooperation would be most valuable to the male (hunter) while in the field.

The foregoing interpretations were largely *post hoc*. A "test" of a

Table 3.2. Rank-Order Correlations between Proportion of Time Spent Outdoors and Total Amount of Time Spent in Each Type of Peer Play

				Solitary	Parallel	Inter-active	Coopera-tive
			University of California Davis Laboratory Nursery School				
1971–1972	Fall	Boys,	$n = 15$	−.157	−.600[a]	+.564[b]	+.293
		Girls,	$n = 12$	−.007	−.252	−.056	+.070
	Winter	Boys,	$n = 13$	−.159	−.416	+.687[b]	+.214
		Girls,	$n = 14$	+.100	−.509	+.526	+.085
	Spring	Boys,	$n = 13$	−.275	−.148	+.110	+.027
		Girls,	$n = 15$	−.082	+.093	+.050	+.061
1971–1973	Fall	Boys,	$n = 12$	−.643[a]	+.430	+.329	+.301
		Girls,	$n = 11$	+.164	−.600	−.054	+.252
	Winter	Boys,	$n = 13$	−.407	−.637[a]	+.797[b]	+.692[b]
		Girls,	$n = 12$	+.329	+.007	−.308	+.026
	Spring	Boys,	$n = 14$	−.335	−.447	+.381	−.053
		Girls,	$n = 12$	+.035	−.133	−.042	−.140
1973–1974	Fall	Boys,	$n = 12$	+.035	−.533	+.482	+.615[a]
		Girls,	$n = 12$	−.217	−.624[a]	+.482	+.343
	Winter	Boys,	$n = 13$	−.154	−.165	+.088	+.258
		Girls,	$n = 14$	−.200	−.279	+.200	+.200
	Spring	Boys,	$n = 16$	−.259	+.144	−.053	+.509[a]
		Girls,	$n = 14$	−.007	+.257	−.284	−.103
1974–1975	Fall	Boys,	$n = 11$	−.064	−.636	+.291	+.209
		Girls,	$n = 12$	+.154	−.608[a]	+.657[a]	+.545
	Winter	Boys,	$n = 12$	−.189	−.406	+.427	+.098
		Girls,	$n = 12$	+.636[a]	−.378	−.468	−.343
	Spring	Boys,	$n = 11$	−.473	−.464	+.582	+.300
		Girls,	$n = 12$	+.420	+.496	−.587[a]	−.497
			Dixon Community Preschool				
1973–1974	Fall	Boys,	$n = 8$	+.119	.000	+.524	.000
		Girls,	$n = 6$	−.600	+.257	−.829	−.200
	Winter	Boys,	$n = 8$	−.286	−.595	+.571	+.714[a]
		Girls,	$n = 6$	+.600	−.771	+.257	−.543

[a] $p < .05$ (two-tailed test).
[b] $p < .01$ (two-tailed test).

theory involves prediction. Before analyzing our data, we reasoned that, if group hunting requires rapid, joint movement, then boys should display more interactive play while in transit than girls, and interactive play should correlate positively with time in transit for boys. The data presented in Tables 3.3 and 3.4 generally support these contentions when the Davis group is considered. Although the differences are not consistent with expectations in the Dixon group, the reversals are not statistically significant. These data are presented at this time, not because they are sufficient to "confirm" our position but rather as examples of the testability of hypotheses derived from an evolutionary orientation. Clearly, more data are necessary for adequate evaluation of our reasoning. However, in view of the facts that the children were in transit for not much more than five per cent of their time and that sample sizes were uniformly small (especially for the Dixon group), we find the consistencies more impressive than the variability. For example, in Table 3.3, in 9 of the 11 comparisons of interaction in transit, the differences were in the predicted direction and two of them were statistically significant, whereas neither of the reversals were. Likewise, in Table 3.4, for boys, 10 of 13 correlations were as predicted and 4 of them reached statistical significance. In contrast, for the girls, the consistency was somewhat less and none of the coefficients reached conventional levels of statistical significance. Thus we feel that we are dealing with

Table 3.3. Percentage of Interactive Play Occurring While "in Transit"

		Boys % (*n*)	Girls % (*n*)
	University of California Davis Laboratory Nursery School		
1972–1973	Fall	2.3 (12)	1.1 (11)
	Winter	3.4 (13)	1.3 (12)[a]
	Spring	1.8 (14)	1.7 (12)
1973–1974	Fall	0.6 (12)	1.7 (12)
	Winter	1.5 (13)	1.0 (14)
	Spring	1.8 (16)	1.7 (14)
1974–1975	Fall	1.8 (11)	1.5 (12)
	Winter	3.8 (12)	1.3 (12)[b]
	Spring	4.6 (11)	2.9 (12)
	Dixon Community Preschool		
1973–1974	Winter	2.7 (8)	2.1 (6)
	Spring	2.6 (8)	4.8 (6)

[a] $p < .05$.
[b] $p < .001$.

Table 3.4. Rank-Order Correlations between Time Spent in Transit and Interactive Play

		Boys r (n)	Girls r (n)
University of California Davis Laboratory Nursery School			
1971–1972	Fall	+.475 (15)	+.468 (12)
	Winter	+.824 (13)[a]	+.314 (14)
	Spring	+.407 (13)	−.211 (15)
1972–1973	Fall	+.287	.564
	Winter	+.121	+.461
	Spring	−.196	+.203
1973–1974	Fall	+.580 (12)[a]	+.287 (11)
	Winter	−.286 (13)	+.037 (12)
	Spring	+.244 (14)	−.218 (12)
1974–1975	Fall	+.682 (12)[a]	−.021 (12)
	Winter	+.503 (13)	−.294 (14)
	Spring	+.645 (16)[a]	+.323 (14)
Dixon Community Preschool			
1973–1974	Winter	+.595 (8)	−.600 (6)
	Spring	−.167 (8)	+.600 (6)

[a] $p < .05$ (two-tailed test).
[b] $p < .01$ (two-tailed test).

a real sex difference—although not a large one—in the social behavior of preschool children.

So far, the data are generally in accord with our position. But Brindley *et al.* (1973), also reasoning from an evolutionary perspective, suggested that girls should engage in more cooperative behavior than boys, and they obtained data consistent with that position. As they defined it, "cooperation" also involved what we classified as "helping" behavior; however, in our study there were no significant tendencies for girls to do more helping than boys at either Davis or Dixon. Since there was no evidence that the boys engaged in more helping at either facility, our failure to replicate their findings cannot be taken as a clear refutation of their hypothesis. Rather, we are intrigued by the possibility that the difference may be accounted for by the fact that the Brindley *et al.* studies were conducted entirely indoors, whereas the children we observed engaged in substantial amonts of outdoor play.

Brindley *et al.* (1973) also reported that when they analyzed their data according to who "cooperated" with whom, they found that the older girls were helping the younger ones significantly more often than the reverse. In contrast, when such behavior occurred among boys, the younger ones were more likely to join in the ongoing activities of their "elders." To the extent that Brindley *et al.* are correct in concluding that the older girls, but not the younger boys, were engaging in caregiving-like activities, the contradictions between their data and ours may be more apparent than real.

3. Discussion

In summary, we have found sex differences in the free play behavior of young children that (1) correspond to sex differences in adult subsistence activities common to many human cultures and (2) are generally in agreement with the hypothesis that they may represent sexually dimorphic predispositions which evolved when hunting and gathering typified most, if not all, human groups. Our clearest data relate to sex differences in vigorous far-ranging outdoor activity. The correlations among outdoor activity and interactive and cooperative play are generally consistent in sign but not large. Similarly, although both predictions yielded only a few significant differences and correlations, they all were in the predicted direction.

The facts that the magnitudes of the latter differences were not uniformly large and that several reversals occurred cannot be construed as evidence against an evolutionary interpretation. In the first place, we are dealing with quantitative, not qualitative, differences within the species. It can be adaptive for females to hunt occasionally just as it is advantageous for males to care for children; selection should be for the degree, not the kind of activity. Thus, given the small sizes of our groups and considerable overlap in the population from which we sampled, even if all the variability were endogenously determined, occasional reversals should be expected. In the second place, we have no intention of arguing that environmental conditions are unimportant or that a single factor suffices to explain these differences. Indeed, we feel that the physical setting may be a critical factor in accounting for some of the differences between the Davis and Dixon groups, and we are fully cognizant of the fact that rearing environments affect the degree to which biological predispositions will be manifested. Thus, in view of the many sources of variance, we feel that the data are reasonably consistent with our theoretical position.

3.1. Evidence for Social Learning

However, since plausible arguments can be offered for social determination of behavioral sex differences (cf. Chapter 15), we shall briefly review some of the empirical findings in this area.

With respect to the tendency for girls to have more contact with adults, the data are mixed. At both Davis and Dixon, boys tended to spend more time than girls with male teachers. Hence, the fact that girls in both preschools spent more time in contact with adults could simply reflect the fact that female teachers predominated in each setting. However, the averages of the ratios of child- to adult-initiated interactions indicated that adults were responsible for the greater proportion of time in contact with children of either sex (Harper and Huie, 1978).

Such evidence as is available also does not clearly support the possibility that girls might be receiving more "reinforcement" for maintaining proximity. Although we were not specifically looking for it, we noticed no differential responsiveness of the teachers who held egalitarian, if not feminist, views. Indeed, when we discussed our preliminary findings with the incoming staffs in 1973 and 1974, we were confidently told that "it will be different this year." The data, however, suggest that their tactics were not entirely successful. The only study known to us that directly examined (female) teachers' responses to preschool children yielded ambiguous results. The girls did stay closer to the teachers than the boys, and teachers did respond somewhat differently according to child gender; however, no clear relationship could be shown between the girls' proximity-maintenance and the teachers' behavior (Serbin, O'Leary, Kent, and Tonick, 1973).

If we turn now to possible explanations for the boys' greater outdoor behavior, we find that frequently invoked hypotheses involve the transmission of cultural stereotypes through the media or differential parental reinforcement. Although television programs directed toward children do portray females as more passive and less often rewarded for activity than males (Sternglanz and Serbin, 1974), we know of no investigation that has unambiguously related such input to overt behavior. Moreover, in one study in which preschoolers were asked to recall and state preferences for activities mentioned in stories that had been read to them, the girls more readily voiced acceptance of opposite-sexed role behaviors than boys (Jennings, 1975). Similarly, despite the fact that parents do "equip" their children differently with playthings according to their gender (Rheingold and Cook, 1975), neither the children's knowledge of the gender appropriateness of toys nor their preferences for them could be related to parental attitudes (Thompson, 1975). Furthermore, four- to five-year-old children spontaneously chose to play with gender-appropriate toys in a test situation regardless of whether their own toys were sex stereotyped—according to maternal reports (Fein, Johnson, Kosson, Stork, and Wasserman, 1975).

Fagot (1974) did observe that, in the home, 18- to 24-month-old boys were more often left to play on their own than were girls of the same ages. However, she also suggested that the gender-specific responsiveness shown by parents was as likely to be a reaction to existing sex differences in offspring behavior as a cause of them. Our own less extensive attempts to assess parental attitudes and practices indicated no relationship to outdoor behavior in the 1972–1973 sample (Harper and Sanders, 1977). The sex difference in outdoor play at Davis was one of our most stable findings despite the fact that the boys received disproportionately more discipline outdoors relative to the time they spent outdoors, whereas this disproportion seldom occurred among the girls. On balance, we must agree with Maccoby and Jacklin's (1974) conclusion that, if anything, boys' rather than girls' independent exploratory behavior receives more restrictive responses. No clear evidence currently exists indicating that the boys' preference for vigorous, outdoor activity is the simple product of reinforcement contingencies or other cultural pressures.

This brings us to the sex difference in peer contacts. Traditional stereotypes of sex-typed behavior hold that sociability is a "feminine" characteristic (cf. Maccoby and Jacklin, 1974). If parents were actually shaping their children's behavior according to cultural expectations, then the girls should have displayed significantly more interactive play at both Davis and Dixon. They did not. However, we did find that, unlike the Davis groups, the boys in the Dixon facility did not engage in more interactive play than did the girls (Harper and Huie, 1978). In view of the fact that the arrangement of the physical setting, the subject population, and the supervisory personnel changed regularly at Davis, the stability of sex differences in parallel and interactive play is impressive. If we assume that an apparent, although not significant, reversal of these trends at Dixon is not due to the smaller sample and more limited period of observation, it is possible that the amount of play area may account for part of the difference. Since we found that the Dixon boys still engaged in more outdoor play (even including fantasy) than girls, we are intrigued by the possibility that ample outdoor space might be a condition favoring interactive play among boys. Unfortunately, however, that is just speculation at this point.

In summary, we have surveyed evidence for parental or other cultural antecedents of the sex differences we have observed. From the data known to us we conclude that the case for "socialization" in the traditional sense is not proven; there are some data consistent with that hypothesis, but none of them are unequivocal. There is no reason to accept social learning as a sufficient condition for the development of gender-related differences in behavior. Thus, it seems reasonable to pursue our alternative hypothesis further.

3.2. *A Possible Hormonal Basis*

We have already outlined a plausible account of how natural selection might have favored the development of predispositions for these behavioral dimorphisms. Now we need to consider the biological mechanisms that could give rise to them.

The broad outlines of the organismic bases for a variety of behavioral sex differences are becoming clear. The sexes clearly differ chromosomally. Although it is still an open question whether structural–genetic information for the Y chromosome is necessary for early sexual differentiation of the gonads (Mittwoch, 1973), there is general agreement that most sexually dimorphic characteristics in mammals are regulated by hormones (Caspari, 1972). Many sex differences in behavior appear to be potentiated early in development by gonadal secretions during or just after the period at which they lead to embryonic differentiation of the reproductive tract. The effects of such early hormones are frequently manifest in later life—even in the absence of further secretions. Current evidence suggests that the effective substances in early development are usually androgenic hormones; in their absence, development (of behavior) proceeds in a typically female direction (cf. Friedman, Richart, and Vande Wiele, 1974). These secretions do not create absolute differences; behaviorally, they act upon essentially the same genetic substrate to alter relative thresholds or probabilities of response (cf. Beach, 1971), and the actual manifestation of such "biases" still requires a facilitative environment (Goy and Goldfoot, 1974).

The biological bases for the human male's greater strength and activity level are well documented (e.g., Tanner, 1970). Furthermore, there are observations consistent with the view that the male's greater preference for outdoor activity may have a hormonal basis. Girls who were exposed to androgenic hormones while *in utero* either because of a congential metablic defect or due to drugs given to their mothers were reported to be "tomboys" and to prefer vigorous outdoor activies more often than unaffected children. Similarly, a male infant who was accidentally penectomized at birth and subsequently castrated and raised as a girl nonetheless showed this preference for outdoor activity (Money and Ehrhardt, 1972). The foregoing data, which were obtained from patients seen at Johns–Hopkins, have been replicated with a different sample from upstate New York (Ehrhardt and Baker, 1974). Unfortunately, the children's behavior was not directly observed in either study; behavioral evaluations were based upon maternal report. Nevertheless, these data are consistent with experimental studies of animals, and, although not conclusive, they provide strong support for the view that the development of at least the sex differences in activity level and outdoor play involve biases established by early hormonal events.

To our knowledge, there exist no data that bear directly upon the organismic determinants of the sex difference in peer interaction. This variable was not discussed in either study of fetally androgenized girls. However, it would be of considerable interest to do so, since, according to our hypothesis, the difference should be most apparent during vigorous outdoor activity—which is affected by early hormones. That the boys' greater peer interaction may depend upon setting variables for its appearance does not contradict the possibility that it is based at least in part upon an endocrine mechanism. Early hormonal influences usually alter the probabilities of behaviors under certain conditions—not "across the board." Thus, our contention is still compatible with Brindley *et al.*'s (1973) hypothesis that a predisposition toward cooperative or "helping" behavior in females may be determined by similar processes.

4. Concluding Remarks

The mechanisms we suggest to account for these phenomena involve nature and nurture; their coaction can be observed, although indirectly, and hypotheses concerning the ways in which hormonally induced behavioral biases are activated by setting are potentially testable. Furthermore, this mechanism can be understood in an evolutionary framework. We may conjecture that, where selection favored the development of sexually dimorphic patterns, the frequency of genotypes compatible with the development of these attributes would have increased in the population as a whole. At the same time, sensitivity to sex hormones would have evolved, which limited the maximal expression of these behavioral tendencies to members of one sex (cf. Caspari, 1972). In short, an evolutionary hypothesis can be invoked to explain sex differences in young children's use of space. The same hypothesis has proven adequate to account for subsequent findings of a related gender dimorphism in social play, and has generated testable predictions. Moreover, this explanation is compatible with clinical and comparative data which suggest that the development of at least some of these differences depends upon early hormonal secretions.

4.1. The Applicability and Value of a Comparative-Evolutionary Approach

At this point, it may be worthwhile to pause long enough to consider the applicability and value of evolutionary theorizing in the study of human behavior. Having no stomach for debates concerning what constitutes an "adequate" explanation, at the outset we will stipulate that adequacy depends upon the kind of question one seeks to answer. For example, if one is interested in the immediate sources of a response, acute subjective experience may be as appropriate as a description of environmental releasers. For a developmentalist, however, the question usually relates to origins, the historical antecedents of responses. Given these concerns, an obvious source of answers is prior experience—including subjective responses. Such data clearly contribute to our understanding of development. The question of "how it works" is no more—or less—relevant; given a developmental phenomenon and some notion of its historical antecedents, some may wish to ask questions about the organismic mechanisms that mediate between past experience or the immediate situation and present activity. For those so inclined the origin of mechanism also poses a question, and, for us at least, evolutionary theory appears to provide the most promising source of answers at this level. Obviously, each step in this seemingly infinite regression depends upon the existence of solid foundations provided by preceding steps. All are valid contributions to science.

Our findings demonstrate that a comparative-evolutionary orientation provides a framework within which we can understand human behavioral development and predict hitherto unexplored phenomena. Although it is quite clear that evolutionary accounts are not in themselves sufficient to explain the development of behavior, this approach can be particularly helpful for students of child development for a number of additional reasons. First of all, in its modern form (e.g., Dobzhansky 1951, 1970), the theory of natural selection emphasizes the interplay between organism and environment. It thus helps us to avoid the temptation to categorize development as due exclusively to either heredity or environment because it emphasizes that all growth results from the coaction of both. Second, the model of gene–environment coaction indicates that similar phenotypic outcomes may result from a variety of antecedent factors. Thus we are dissuaded from reasoning that because one combination of conditions gives rise to a particular syndrome, all instances of that syndrome must have originated in the same way. Third, the fact of opportunism in natural selection shows that components of complex patterns can evolve from a variety of sources, thereby alerting us to the possibility that apparently homogeneous, global behaviors may represent the outcome of a number of different mechanisms. Fourth, the fact that selection operates throughout the life cycle encourages us to consider carefully the organismic and environmental contributors to behavior characteristic of every stage of development and emphasizes that there is no *a priori* reason to assume that later-appearing behaviors (or differences) are due exclusively to cultural influences. Finally, a comparative–evolutionary orientation is particularly helpful to students of sex differences by focusing upon the species as the unit of comparison. We are thus reminded that gender dimorphisms represent variations on a theme rather than absolutes.

Briefly stated then, a comparative-evolutionary approach not only serves as a source of testable hypotheses, but it also helps us to avoid many of the conceptual pitfalls that have hampered progress in the study of human behavioral development.

4.2. Summary and a Plea for Caution

Returning to our findings, we feel that preschool boys', as compared with girls', greater outdoor activity and peer interaction and the girls' greater time spent with adults can be explained as endogenous predispositions, which evolved as animal food became a significant component of the hominid diet. We believe that our hypotheses are at least as plausible as more traditional, social-learning accounts, but we are not suggesting that these sex differences are inevitable. As the comparative data show, early hormonal stimulation does not create qualitatively different organisms; rather, it affects the relative thresholds of response systems common to both males and females (cf. Beach, 1971). We argue only that early modifiability is probably biologically conditioned and that the environmental conditions facilitating the manifestation of these sex differences are more subtle and indirect than those posited by social learning accounts. Essentially, we are affirming that both biological and experiential conditions are necessary.

However, we must also emphasize that the fact of early modifiability should not, in itself, be taken as a go-ahead for (re)programming life histories. There now exists an extensive literature that demonstrates that events occurring early in the life cycle may affect not only the individual's immediate responses, but also behavioral patterns that will not be manifest until later (Thompson and Grusec, 1970). Moreover, the sexes sometimes differ in their subsequent responses to what appear to be the same early experiences. For example, in a longitudinal study of the effects of four different preschool enrichment programs on later I.Q. and grammar-school achievement, it was found that a Montessori approach in the prekindergarten years resulted in the best performance by second grade. However, this outcome was almost entirely the result of I.Q. gains and academic achievement by boys, not girls. In short, the Montessori preschool experience seemed to facilitate boys' later achievement; no such enhancement was apparent for the girls. Furthermore, of the several preschool programs studied, the Montessori method was not clearly the most effective in increasing preschool measures of school readiness for either sex (Miller and Dyer, 1975).

Thus, insofar as the later consequences of early treatments cannot always be predicted from the child's immediate reactions, it would seem inadvisable to undertake any extensive programs designed to modify sex-differentiated behavior without a great deal more longitudinal research. Even if there were a consensus that currently existing behavioral sex differences should be changed or reduced, there is not enough solid information about the long-term consequences of different childrearing patterns to be sure that the desired results would be achieved. Moreover, should any particular goals be attained, there still exists no guarantee that they will not be achieved at the expense of other, equally desired, attributes. One needs only to ponder Bettleheim's (1969) report of the consequences of group rearing to realize that in order to obtain one end, another equally valued characteristic may have to be sacrificed. Just as the rodent's central nervous system has the potential to be biased in either a male or female direction according to early hormonal milieu, there may exist mutually exclusive alternate routes in human development. Therefore, we urge that no large-scale attempts be made to influence the development of sex differences—or any other aspect of human behavior—until we are in a position to predict at least the major consequences of our actions.

ACKNOWLEDGMENTS

This research was supported by the Agriculture Experimental Station, University of California.

References

Beach, F. A. Hormonal factors controlling the differentiation, development and display of copulatory behavior in the ramstergig and related species. In E. Tobach, L. R. Aronson, and E. Shaw (Eds.), *The biopsychology of development.* New York: Academic Press, 1971.

Bell, R. Q., Weller, G. M., and Waldrop, M. F. Newborn and preschooler: Organization of behavior and relations between periods. *Monographs of the Society for Research in Child Development,* 1971, *36,* (1) and (2).

Bettleheim, B. *The children of the dream.* New York: Macmillan, 1969.

Birns, B. The emergence and socialization of sex differences in the earliest years. *Merrill-Palmer Quarterly,* 1976, *22,* 229–254.

Blurton-Jones, N. G., and Konner, M. J. Sex differences in behaviour of London and Bushman children. In R. P. Michael and J. H. Crook (Eds.), *Comparative ecology and behaviour of primates.* New York: Academic Press, 1973.

Brindley, C., Clarke, P., Hutt, C., Robinson, I., and Wethli, E. Sex differences in the activities and social interactions of nursery school children. In R. P. Michael and J. H. Crook (Eds.), *Comparative Ecology and behaviour of primates.* New York: Academic Press, 1973.

Caspari, E. Sexual selection and human evolution. In B. G. Campbell (Ed.), *Sexual selection and the descent of man.* Chicago: Aldine, 1972.

Crook, J. H. Gelada baboon herd structure and movement. A comparative report. In P. A. Jewell and C. Loizos (Eds.), *play, exploration and territory in mammals. Symposia of the Zoological Society of London,* Vol. 18. New York: Academic Press, 1966.

Crook, J. H. Sexual selection, dimorphism, and social organization in the primates. In B. G. Campbell, (Ed.), *Sexual selection and the descent of man.* Chicago: Aldine, 1972.

D'Andrade, R. G. Sex differences and cultural institutions. In E. E. Maccoby (Ed.), *The development of sex differences.* Stanford: Stanford University Press, 1966.

Dobzhansky, T. *Genetics and the origin of species,* 3rd ed., revised. New York: Columbia University Press, 1951.

Dobzhansky, T. *Genetics of the evolutionary process.* New York: Columbia University Press, 1970.
Dornbusch, S. M. Afterword. In E. E. Maccoby (Ed.), *The development of sex differences.* Stanford, California: Stanford University Press, 1966.
Draper, P. Crowding among hunter–gatherers: the !Kung bushmen. *Science,* 1973, *182,* 301–303.
Ehrhardt, A., and Baker, S. W. Fetal androgens, human central nervous system differentiation, and behavioral sex differences. In R. C. Friedman R. M. Richart, and L. R. Vande Wiele (Eds.), *Sex differences in behavior.* New York: Wiley, 1974.
Estes, R. D., and Goddard, J. Prey selection and hunting behavior of the African wild dog. *Journal of Wildlife Management,* 1967, *31,* 52–70.
Etkin, W. Types of social organization in birds and mammals. In W. Etkin (Eds.), *Social behavior and organization among vertebrates.* Chicago: University of Chicago Press, 1964.
Fagot, B. I. Sex differences in toddlers' behavior and parental reaction. *Developmental Psychology,* 1974, *10,* 554–558.
Fein, G., Johnson, D., Kosson, N., Stork, L., and Wasserman, L. Sex-stereotypes and preferences in the toy-choices of 20-month-old boys and girls. *Developmental Psychology,* 1975, *11,* 527–528.
Friedman, R. C., Richart, R. M., and Vande Wiele, R. L. (Eds.). *Sex differences in behavior.* New York: Wiley, 1974.
Goy, R. W., and Goldfoot, D. A. Experiential and hormonal factors influencing development of sexual behavior in the male rhesus monkey. In F. O. Schmitt, and F. G. Worden (Eds.), *The neurosciences. Third study program.* Cambridge, Massachusetts: MIT Press, 1974, pp. 571–581.
Hamburg, B. A. The psychobiology of sex differences: An evolutionary perspective. In R. C. Friedman, R. M. Richart, and R. L. Vande Wiele (Eds.), *Sex differences in behavior.* New York: Wiley, 1974.
Harper, L. V., and Sanders, K. M. Preschool children's use of space: Sex differences in outdoor play. *Developmental Psychology,* 1975, *11,* 119.
Harper, L. V., and Sanders, K. M. Preschool children's use of space: Sex differences in outdoor play. Extended report of the above reference published in R. C. Smart and M. S. Smart (Eds.), *Readings in child development and relationships,* 2nd ed. New York: Macmillan, 1977.
Harper, L. V., and Huie, K. S. The development of sex differences in human behavior: Cultural impositions, or a convergence of evolved response-tendencies and cultural adaptations? In M. Bekoff and G. Burghardt (Eds.), *Ontogeny of behavior.* New York: Garland Publishing Co., 1978.
Hutt, C. *Male and female.* New York: Penguin, 1972.
Izawa, K. Unit groups of chimpanzees and their nomadism in the savannah woodland. *Primates,* 1970, *11,* 1–46.
Jennings, S. A. Effects of sex typing in children's stories on preference and recall. *Child Development,* 1975, *46,* 220–223.
Kruuk, H. *The spotted hyena.* Chicago: University of Chicago Press, 1972.
Lee, R. B., and DeVore, I. (Eds.). *Man the Hunter.* Chicago: Aldine, 1968.
Lewis, M., and Weinraub, M. Sex of parent × sex of child. Socioemotional development. In R. C. Friedman, R. M. Richart, and R. L. Vande Weile (Eds.), *Sex differences in behavior.* New York: Wiley, 1974.
Ley, R. G., and Koepke, J. E. Sex and age differences in the departures of young children from their mothers. Paper presented at the biennial meeting of the Society for Research in Child Development, Denver, Colorado, April 1975.

Loizos, C. Play behavior in the higher primates: A review. In D. Morris (Ed.), *Primate ethology*. Chicago: Aldine, 1967.
Maccoby, E. E., and Jacklin, C. N. *The psychology of sex differences*. Stanford, California: Stanford University Press, 1974.
Mech, L. D. *The wolves of Isle Royale. U.S. National Parks fauna series*, No. 7. Washington, D.C.: US Government Printing Office, 1966.
Miller, L. B., and Dyer, J. L. Four preschool programs: Their dimensions and effects. *Monographs of the Society for Research in Child Development*, 1975, *40*(5–6).
Mittwoch, U. *Genetics of sex differentiation*. New York: Academic Press, 1973.
Money, J., and Ehrhardt, A. A. *Man and woman, boy and girl*. Baltimore: Johns-Hopkins University Press, 1972.
Moore, N. V., Evertson, C. M., and Brophy, J. E. Solitary play: Some functional considerations. *Developmental Psychology*, 1974, *10*, 830–834.
Mussen, P. H., Conger, J. J., and Kagan, J. *Child development and personality*, 4th ed. New York: Harper & Row, 1974.
Omark, D. R., and Edelman, M. Peer group social interactions from an evolutionary perspective. Paper presented at the biennial meeting of the Society for Research in Child Development, Philadelphia, March 1973.
Pedersen, F. A., and Bell, R. Q. Sex differences in preschool children without histories of complications of pregnancy and delivery. *Developmental Psychology*, 1970, *3*, 10–15.
Rheingold, H. L., and Cook, K. V. The contents of boys' and girls' rooms as an index of parents' behavior. *Child Development*, 1975, *46*, 459–463.
Sanders, K. M., and Harper, L. V. Free-play fantasy behavior in preschool children: Relations among gender, age, season, and location. *Child Development*, 1976, *47*, 1182–1185.
Serbin, L. A., O'Leary, D., Kent, R. N., and Tonick, I. J. A comparison of teacher response to preacademic and problem behavior of boys and girls. *Child Development*, 1973, *44*, 796–804.
Speth, J. D., and Davis, D. D. Seasonal variability in early hominid predation. *Science*, 1976, *192*, 441–445.
Sternglanz, S., and Serbin, L. A. Sex role stereotyping in children's television programs. *Developmental Psychology*, 1974, *10*, 710–715.
Tanner, J. M. Physical growth. In P. H. Mussen, (Ed.), *Carmichael's Manual of Child Psychology*, 3rd ed., Vol. 1. New York: Wiley, 1970.
Thompson, S. K. Gender models and early sex role development. *Child Development*, 1975, *46*, 339–347.
Thompson, W. R., and Grusec, J. Studies of early experience. In P. H. Mussen (Ed.), *Carmichael's Manual of Child Psychology*, 3rd ed., Vol. 1. New York: Wiley, 1970.
Watanabe, H. Subsistence and ecology of northern food gatherers with special reference to the Ainu. In R. B. Lee and I. DeVore (Eds.), *Man the hunter*. Chicago: Aldine, 1968.
Waterhouse, M. J., and Waterhouse, H. B. Primate ethology and human social behaviour. In R. P. Michael and J. H. Crook (Eds.), *Comparative ecology and behaviour of primates*. New York: Academic Press, 1973.
Whiting, B., and Edwards, C. P. A cross cultural analysis of sex differences in the behavior of children aged three through eleven. *Journal of Social Psychology*, 1973, *91*, 171–188.

4

The Comparative Method in Studies of Reproductive Behavior

Donald A. Dewsbury

> Thirty years ago in this country a small group of scientists went Snark hunting. It is convenient to personify them collectively in one imaginary individual who shall be called the Comparative Psychologist. The Comparative Psychologist was hunting a Snark known as Animal Behavior. His techniques were different from those used by the Baker, but he came to the same unhappy end, for his Snark also proved to be a Boojum. Instead of animals in the generic sense he found one animal, the albino rat, and thereupon the Comparative Psychologist suddenly and softly vanished away.
>
> F. A. Beach (1950, p. 115)

In "The Snark was a Boojum," Frank Beach did a masterful job of providing a retrospective view of the utilization of the comparative method in psychology. Over the course of this century, psychologists had come to concentrate their efforts on a few species of mammals, especially albino rats. Beach uttered a plea for increased utilization of the comparative method in the study of an increased variety of behavioral patterns. Since publication of the "Snark" paper in 1950, albino rats have remained the animals of choice for the vast majority of experiments. However, the Comparative Psychologist has yet to suddenly and softly vanish away. Beach was correct in his perception of the advantages of the comparative method. It is indeed a method possessing enormous power in the approach to a

---

Donald A. Dewsbury • Department of Psychology, University of Florida, Gainesville, Florida. Supported by Grant No. BMS75-08658 from the National Science Foundation.

broad range of behavioral questions. That the Comparative Psychologist is still on the scene as in part attributable to "The Snark was a Boojum," a paper that has had a major influence on the decisions of a considerable number of behavioral scientists to pursue comparative study.

It is one thing to assert that the comparative method is a powerful one; but quite something else to demonstrate its power in such a way as to convince the skeptical scientist to forsake his beloved albino companion. It is only when it can be convincingly demonstrated that significant new behavioral facts, and new principles based on these facts, will emerge from comparative study that the average behavioral scientist will be convinced (Schrier, 1969). The bulk of this paper will constitute a review of some of our efforts to use the comparative method in establishing such new behavioral facts and principles. Following this review, an attempt will be made to analyze the status and prospectus for such comparative studies.

1. The Comparative Method in Studies of Copulatory Behavior

Over the course of the last decade, my students and I have been engaged in an extensive program of quantitative descriptive analyses of the copulatory patterns of a variety of species of muroid rodents. The muroids represent a taxonomically restricted, yet diverse, group of species that adapt well to the laboratory and are thus ideal subjects for such a program. Because the results of this program have recently been reviewed elsewhere (Dewsbury, 1975a), just a brief sketch of some of the major findings will be presented here. The intent is to provide a foundation for the discussion of research on pregnancy initiation.

Description of the copulatory pattern of laboratory rats provides a convenient starting point. It is customary to distinguish three major copulation-related male behavioral patterns. *Mounts* are characterized by the male mounting the female, clasping her, and displaying pelvic thrusting; but failing to gain vaginal penetration. *Intromissions* also involve a mounting pattern by the male, but vaginal penetration is achieved, though no sperm are transferred to the female. Each intromission contains a single intravaginal thrust and lasts about one-quarter of a second (Bermant, 1965). *Ejaculations* include mounting, vaginal penetration, and the ejaculation of semen. A trained observer can easily distinguish ejaculations from mounts and intromissions using behavioral criteria (e.g., the thrust is longer and deeper and the dismount slower and quite stereotyped). Mounts, intromissions, and ejaculations occur in series, each of which contains just one ejaculation, and that, by definition, always terminates the series. The male alternates between periods of ejaculatory series (including mounts, intromissions, and ejaculations) and postejaculatory intervals, in which such activity is absent. Copulatory behavior often is studied in satiety tests—tests continued until there is a one-half hour interval with no intromissions or ejaculations. In such tests an average of about seven ejaculatory series typically occur.

The copulatory patterns of other muroid species differ from that of laboratory rats in a number of ways. Dewsbury (1972) proposed that copulatory patterns can usefully be classified with respect to the presence or absence of four major features. First, in some species there is a lock, or mechanical tie between the penis and vagina that is resistant to separation during copulation. Other species do not lock. Second, while the copulations of some species, like laboratory rats, entail a single intravaginal thrust, others display repetitive intravaginal thrusting. Third, while multiple intromissions are prerequisite to ejaculation in some species, like laboratory rats, other species are capable of ejaculating on the first occurrence of vaginal penetration. Finally, while most species attain multiple ejaculations in a given episode, some appear to cease copulating after a single ejaculation. In this system, there are 2^4 or 16 possible patterns of copulatory behavior. The copulatory patterns of the first 31 species we have studied are summarized in Table 4.1. The great diversity in muroid patterns should be readily apparent. Eight of the 31 species display locking copulatory patterns, while 23 species do not lock. Ten species display intravaginal thrusting; 14 species require multiple intromissions before ejaculating; and one species ceases copulating after a single ejaculation.

This bare-bones classification system presents a mere introduction to the great diversity of copulatory patterns displayed by muroid rodents. For example, while house mice of many strains rarely complete two ejaculatory series, Syrian golden hamsters typically complete about ten series. Like laboratory rats, many species reach the satiety criterion between ejaculatory series—with a 30-min postejaculatory interval. For others, however, the criterion is met after an intromission, as an incomplete "series." This effect is so extreme in some species that it is not uncommon for the majority of intromissions to occur after the last ejaculation of an episode. In some species there are patterns of "long intromissions" following the completion of organized series. In golden hamsters the specialized pattern of postejaculatory copulatory activity is characterized by intravaginal thrusting—a feature not observed during organized ejaculatory series. Laboratory rats typically display about ten intromissions before the first ejaculation and fewer intromissions in later series. Among other muroid species the number of intromissions in the first series may be reliably much higher or lower than this. Further, the pattern of change across series varies greatly with different species showing increasing numbers of intromissions per series, no change over a series, and even more complex patterns. We believe that we

Table 4.1. Summary of Patterns of Copulatory Behavior

Species	Lock?	Thrust?	Multiple intromissions?	Multiple ejaculations?	Pattern
Tylomys nudicaudus	Yes	Yes	No	Yes	3
Ototylomys phyllotis	Yes	Yes	No	Yes	3
Onychomys torridus	Yes	No	No	Yes	7
Ochrotomys nuttalli	Yes	No	No	Yes	7
Neotoma albigula	Yes	No	No	Yes	7
Neotoma floridana	Yes	No	No	Yes	7
Onychomys leucogaster	Yes	No	No	Yes	7
Baiomys taylori	Yes	No	No	No	8
Mus musculus	No	Yes	Yes[a]	Yes	9
Microtus montanus	No	Yes	Yes	Yes	9
Microtus ochrogaster	No	Yes	Yes	Yes	9
Peromyscus eremicus	No[a]	Yes	Yes	Yes	9
Microtus pennsylvanicus	No	Yes	No	Yes	11
Peromyscus californicus	No	Yes	No	Yes	11
Microtus californicus	No	Yes	No	Yes	11
Microtus pinetorum	No	Yes	No	Yes	11
Rattus norvegicus	No	No	Yes	Yes	13
Meriones unguiculatus	No	No	Yes	Yes	13
Meriones tristrami	No	No	Yes	Yes	13
Mesocricetus auratus	No	No[a]	Yes	Yes	13
Sigmodon hispidus	No	No	Yes	Yes	13
Oryzomys palustris	No	No	Yes	Yes	13
Peromyscus polionotus	No	No	Yes	Yes	13
Peromyscus leucopus	No	No	Yes	Yes	13
Peromyscus crinitus	No	No	Yes	Yes	13
Rattus rattus	No	No	No	Yes	15
Reithrodontomys megalotis	No	No	No	Yes	15
Peromyscus gossypinus	No	No	No	Yes	15
Peromyscus floridanus	No	No	No	Yes	15
Neotoma lepida	No	No	No	Yes	15
Peromyscus melanophyrys	No	No	No	Yes	15

[a] Qualification due to complicated pattern.

have answered Schrier's charge that "few, if any, unequivocal, consistent, qualitative differences of any kind between species have been uncovered yet by comparative psychologists (Schrier, 1969, p. 682)."

However, as Schrier points out, it is not the mere accumulation of new "behavioral facts" that will lead to increased use of the comparative method, but the derivation of new principles based on these facts. This is a much more difficult task. The primary interest in the present research program has been in the adaptive significance of copulatory patterns. The goal is to determine the ways in which copulatory patterns function in promoting the adaptation of animals in particular environments through increased survival and successful reproduction. Some tentative hypotheses have been formulated. It appears as though a protected nest site is essential for the evolution of a locking pattern. Thus, wood rats build elaborate houses, golden mice and climbing rats are highly arboreal, and grasshopper mice dig a variety of burrows. All offer some kind of protection from predators. By contrast, patterns in which ejaculation can occur on a single, brief insertion with no intravaginal thrusting appear adapted to particularly exposed habitats. There are morphological adaptations correlated with copulatory patterns. Locking species have glans penes that are relatively thicker than those of other simple-baculum muroids. Locking species also tend to show a reduction in the complement of accessory reproductive glands. Research in progress suggests that the latter correlation may be due to the absence of a copulatory plug in locking species. While there is reason to be encouraged by progress in using the "method of adaptive correlation" in an attempt to understand the adaptive singificance of copulatory patterns, that method alone does not seem sufficiently powerful to provide comprehensive solutions to the questions of adaptive significance that must be raised. Thus, we have turned increasingly to the experimental method, particularly in comparative studies of pregnancy initiation.

2. *Patterns of Female Reproductive Physiology*

Before discussing research on pregnancy initiation, a brief review of patterns of female reproductive physiology is essential. Conaway (1971) proposed a system for classifying patterns of female reproductive physiology, the essence of which is represented in Table 4.2. Two pivotal events in the female cycle are considered—ovulation and the initiation of a functional luteal phase. These are regarded as being "spontaneous" if they occur in every cycle, whether or not the female mates, or "induced," if they occur only following copulation or other comparable stimulation. Thus, Conaway regarded the initiation of a luteal phase as spontaneous "if the formation of

Table 4.2. Patterns of Female Reproductive Cycles[a]

Type	Ovulation	Luteal phase	Example
I	Spontaneous	Spontaneous	*Mesembriomys gouldii*
II	Induced	Spontaneous	*Microtus ochrogaster*
III	Spontaneous	Induced	*Rattus norvegicus*
IV	Induced	Induced	*Microtus agrestis*

[a] Based on the results of Conaway (1971), Crichton (1969), Gray *et al.* (1974), and Milligan (1975a).

a functional corpus luteum always follows ovulation" and induced if "activation of the corpus luteum does not obligatorily follow ovulation, but requires a separate stimulation" (Conaway, 1971, pp. 239–240)." These dichotomies are not absolute and patterns can be changed with extreme experimental and environmental conditions. Conaway distinguished three major types of female pattern, Types I, II, and III. Type I, represented among the muroids by *Mesembriomys gouldii* (Crichton, 1969), is characterized by the occurrence of ovulation and a functional luteal phase in every cycle. Type II is characterized by the induction of ovulation by copulatory activity, with the obligatory activation of corpora lutea. This pattern generally is considered to be characteristic of various species of *Microtus*. Laboratory rats and house mice display pattern III, in which ovulation is spontaneous, but copulation is essential for the initiation of a functional luteal phase.

Milligan (1975a) has reported data that necessitate the addition of a fourth pattern to Conaway's scheme, a pattern in which both ovulation and luteal function are induced. Milligan's data will be discussed in more detail below. At this point, however, it should be recognized that many of the species that have traditionally been classified as displaying Type II cycles could indeed turn out to be more appropriately classed within Type IV, with the initiation of luteal function not obligatory on the occurrence of ovulation.

It is copulatory behavior that triggers one pivotal event or another in female reproductive physiology in the majority of muroid species. One way to approach the study of the adaptive significance of copulatory behavior is to consider this relationship between copulatory behavior and the initiation of pivotal events in the female cycle.

3. Pregnancy Initiation in Laboratory Rats and House Mice

The modern era in studies of the behavioral initiation of pregnancy in rodents was ushered in by Wilson, Adler, and Le Boeuf (1965) with the publication of a study conducted in Frank Beach's laboratory on preejaculatory intromissions in pregnancy initiation in rats. Wilson *et al.* found that nine of ten females became pregnant after receiving a relatively normal complement of preejaculatory intromissions (three or more) plus one ejaculation. By contrast, just two of nine other females became pregnant when they received one or two preejaculatory intromissions plus an ejaculation. It thus became apparent that at least one function of the multiple intromissions of laboratory rats was to somehow initiate responses in the female that could lead to successful pregnancy.

The mechanisms underlying the behavioral initiation of pregnancy in laboratory rats have been partially unravelled in a series of rather elegant studies (Adler, 1969, 1974; Adler, Resko, and Goy, 1970; Chester and Zucker, 1970; Edmonds, Zoloth, and Adler, 1972). Two major sequelae have been demonstrated to be initiated in the female upon stimulation from a complete series of intromissions and deposition of an ejaculatory plug. The first is a humoral, neural, or mechanical effect that results in a facilitation of sperm transport and ultimately in the fertilization of ova. The second is the initiation of a neuroendocrine reflex that leads to the secretion of functional amounts of progesterone and the termination of the female's regular estrous cycle pattern. Further, a period of quiescence must follow ejaculation. Copulation too soon after an ejaculation can interfere with pregnancy (Adler and Zoloth, 1970). From these studies, it would appear that a complete ejaculatory series is both necessary and sufficient for the initiation of neuroendocrine reflexes and sperm transport that result in pregnancy at nearly optimal levels in laboratory rats.

The copulatory pattern in house mice resembles that of laboratory rats in that there is no lock and multiple intromissions typically precede ejaculation. However, house mice differ from rats with respect to the presence of intravaginal thrusting and a relatively low ejaculation frequency. In BALB/c mice, for example, the first ejaculation is preceded by a mean of 36 intromissions and 621 thrusts (McGill, 1962). The ability to produce a second ejaculatory series varies with genotype, with some strains showing a median recovery time of 96 hr under the testing conditions used (McGill and Blight, 1963). Like laboratory rats, house mice display a Type III pattern of female reproductive physiology.

When studies of pregnancy initiation were conducted on house mice, it soon became apparent that the mechanisms underlying pregnancy initiation in house mice differ from those in rats. The multiple preejaculatory intromissions of house mice appear neither necessary nor sufficient for the initiation of pregnancy (Land and McGill, 1967; McGill, 1972, Wilson, 1968). Rather, it is the occurrence of the ejaculatory reflex and the associated swelling of the distal end of the male's penis that seem to be the critical events in pregnancy initiation, whether or not a plug is present (McGill, 1970; McGill, Corwin, and Harrison, 1968; McGill and Coughlin, 1970). While artificial mechanical stimulation designed to mimic the pattern of a normal male can be effective in stimulating pseudopregnancy in house mice (Diamond, 1970), the preejaculatory insertions of the male mouse appear ineffective.

4. *Recent Comparative Studies of Pregnancy Initiation*

The promise of research linking male copulatory behavior to female reproductive physiology for understanding the adaptive significance of copulatory behavior led us to conduct a series of pregnancy initiation studies in a variety of species of muroid rodents. Particular attention has been focused on the function of multiple ejaculatory series. An evolutionary hypothesis had been proposed by Kuehn and Zucker (1968), "during evolution, the number of intromissions the male of a species requires for ejaculation may become linked to the amount of stimulation needed by the female for successful pregnancy" (p. 751). However, in commenting on the occurrence of multiple ejaculatory series, Zucker and Wade (1968) noted that "much of this activity seems wasted, since it has been observed that all fertile females become pregnant and deliver and rear litters of normal size after only two ejaculations" (p. 818).

Studies on our first two species led us to a somewhat different evolutionary hypothesis: *There has been, in the course of evolution, a coadaptation between the stimulus requirements of the female of a species for maximal successful pregnancy and the amount of copulatory activity in an episode by the male of the species. In species in which males display much copulatory activity, females have high stimulus requirements, and vice versa.* The variable amount of copulatory behavior associated with variations in female requirements should be particularly apparent with respect to the number of ejaculations and/or the presence and amount of postejaculatory copulation in a species. A series of studies designed to test and extend this hypothesis will now be reviewed.

4.1. Pregnancy Initiation in Four Species of Induced Ovulators

It appears that all voles of the genus *Microtus* may be induced ovulators (Gray, Davis, Zerylnick, and Dewsbury, 1974). Data on pregnancy initiation are available for four species. In research in our laboratory, fertile males and females are placed on opposite sides of a barrier in a divided cage. Vaginal smears are taken daily throughout the experiment. Mating tests are conducted on the second day of consecutive estrus with the females killed and examined, usually on the eighth day following mating.

4.1.1. *Microtus montanus*

Montane voles, *Microtus montanus,* are a species commonly found in the dry, grassland areas of northwestern United States. In a study of copulatory behavior in montane voles it was found that males display a pattern with no lock, intravaginal thrusting, multiple intromissions, and multiple ejaculations (pattern 9 of Dewsbury, 1972). A mean of 15 intromissions and 32 thrusts preceded the first ejaculation, with a mean of 5.0 ejaculations preceding the fulfillment of the standard 30 min satiety criterion (Dewsbury, 1973a).

Pregnancy initiation in *M. montanus* was studied by Davis, Gray, Zerylnick, and Dewsbury (1974). Four groups of females were studied, with one group unmated. Other groups were mated for one ejaculatory series, two series, or for a period of time beyond two series that would typically result in sexual satiety. The results are presented in Table 4.3. In contrast to the relatively high proportion of laboratory rats and house mice that were pregnant after one complete series, just 25% of montane vole females became pregnant after one complete series. By contrast, all females mated to satiety became pregnant. These data provided a clear demonstration of a role for multiple ejaculatory series in pregnancy initiation in montane voles. The hypothesis that multiple ejaculatory series may be important in induced ovulators was first formulated when these results became known.

A mediating step in pregnancy initiation in this species is an increase in LH contingent upon copulatory behavior. Mating female montane voles for one hour after the first intromission produces significant increases in plasma LH and progesterone (Gray, Davis, Kenney, and Dewsbury, 1976).

4.1.2. Microtus ochrogaster

Prairie voles, *Microtus ochrogaster*, inhabit the prairies of central United States. Like montane voles, prairie voles display pattern 9 in their

Table 4.3. Percent of Animals Pregnant, Mean Number of Corpora Lutea, and Implanted Embryos in Two Species of Microtus as a Function of Condition[a]

Condition	% pregnant	Number of corpora lutea mean (S.E.)	Number of embryos mean (S.E.)
M. montanus			
0 ejaculations	0(0/10)	—	—
1 ejaculations	25(3/12)	6,7(1.2)	5.0(1.0)
2 ejaculations	67(8/12)	6,5(0.6)	5.8(0.7)
>2 ejaculations	100(12/12)	5,6(0.4)	4.7(0.5)
M. ochrogaster			
	Experiment I		
0 ejaculations	0(0/10)	—	—
1 ejaculations	90(9/10)	5.3(0.5)	3.7(0.6)
2 ejaculations	90(9/10)	5.0(0.4)	3.1(0.5)
	Experiment II		
Mounts only	0(0/10)	0.0	—
1 intromission	14(2/14)	5.5(0.5)	—
3 intromissions	36(5/14)	5.2(0.4)	—
5 intromissions	64(9/14)	5.6(0.4)	—
8 intromissions	64(9/14)	5.7(0.4)	—
1 ejaculation	100(4/4)	6.3(1.0)	—

[a] Based on the results of Davis *et al.* (1974) and Gray, Zerylnick, Davis, and Dewsbury (1974).

copulatory behavior, but with a mean of 10 intromissions and 43 thrusts in their first series. The mean ejaculation frequency in *M. ochrogaster* is just 2.0 (Gray and Dewsbury, 1973).

Gray, Zerylnick, Davis, and Dewsbury (1974) compared the ovulation and pregnancy rates of female prairie voles mated for zero, one, and two ejaculatory series. Ninety percent of the females mated for one or two ejaculatory series became pregnant (Table 4.3). Thus, although they are induced ovulators, the stimulus requirements of female *M. ochrogaster* more closely resemble those of laboratory rats than *M. montanus*, at least with respect to the number of series required. In a second experiment females received differing numbers of intromissions without ejaculation. As can be seen in Table 4.3, increased numbers of intromissions were associated with increased probabilities of ovulation. The data from this second experiment are remarkably parallel to the "dose–response curve" reported for laboratory rats in Experiment 5 of Adler (1969). These data led us to formulate our major hypothesis, which was tested with data from the next nine species.

4.1.3. *Microtus pennsylvanicus*

Meadow voles, *M. pennsylvanicus*, are distributed over much of northeastern and central United States, as well as in much of Canada. The first ejaculatory series in meadow vole copulatory behavior is typical of many vole species, with a mean of 14 intromissions and 58 thrusts preceding ejaculation (Gray and Dewsbury, 1975). Whereas meadow voles always require multiple intromissions before their first ejaculation, they ejaculate on their first insertion in the majority of subsequent series. Because they can ejaculate on a single mount with insertion, meadow voles are considered to display pattern 11 in the system of Dewsbury (1972). Gray and Dewsbury (1975) found a mean of 5.9 ejaculations preceding satiety.

Data from a pregnancy initiation study of *M. pennsylvanicus* conducted in our laboratory by Gray (1974) are presented in Table 4.4. No females became pregnant when they were either unmated or received just one ejaculatory series that was the second series for a male on a given day. Under these conditions, males ejaculated on their first mount with insertion half of the time. After receiving one complete first series from a male, 70% of the females were pregnant; all females mated to satiety were pregnant. While there was a slight increase in the percentage of females pregnant as a function of multiple ejaculatory series, the difference was not statistically significant. This latter fact could have been fatal for our fledgling hypothesis were it not for an additional fact in the data. Females that mated to satiety and that ovulated had significantly more corpora lutea (6.1 versus 4.6) and implanted embryos (4.9 versus 3.7) than females that received just

Table 4.4. Percent of Animals Ovulating, Pregnant, Mean Number of Corpora Lutea, and Implanted Embryos in Two Species of Microtus as a Function of Condition[a]

Condition	% ovulating	% pregnant or pseudopregnant	Number of corpora lutea mean (S.E.)	Number of embryos mean (S.E.)
M. pennsylvanicus				
0 ejaculations	—	0(0/10)	—	—
1 ejaculation (2nd series)	—	0(0/10)	—	—
1 ejaculation (1st series)	—	70(7/10)	4.6(0.5)	3.7(0.3)
Satiety	—	100(10/10)	6.1(0.4)	4.9(0.2)
M. agrestis				
Mounts only	0(0/5)	0(0/5)	—	—
1 intromission	92(11/12)	25(3/12)	—	—
1 ejaculation	100(6/6)	50(3/6)	—	—
2 ejaculation	100(6/6)	67(4/6)	—	—
4 ejaculation	100(10/10)	100(10/10)	—	—

[a] Based on the results of Milligan (1975a) and Gray (1974).

one ejaculatory series and ovulated. Thus, while the increase in probability of pregnancy produced by mating to satiety in *M. pennsylvanicus* was only marginal, such prolonged copulation increased litter size. These data are thus consistent with our hypothesis.

4.1.4. *Microtus agrestis*

The short-tailed field vole, *Microtus agrestis*, is a British vole, whose copulatory behavior and pregnancy mechanisms have been studied in England by Milligan (1975a; 1975b). In his study of copulatory behavior, Milligan (1975b) observed ejaculation on a single insertion to be fairly common, with a mean of 5.6 intromissions preceding the first ejaculation. While satiety data were not provided, a median of four series were initiated in the 30-min interval following the first intromission. It is thus clear that *M. agrestis* has a relatively prolonged pattern of copulatory activity and that we would expect females to have a relatively high stimulus requirement.

Milligan's (1975a) data on pregnancy initation in *M. agrestis* are presented in Table 4.4. Just 50% of Milligan's females were pregnant after one ejaculatory series. Increasing numbers of series resulted in an increased probability of pregnancy, until all females mated to four ejaculations were found to be pregnant.

Milligan examined many of his females three or four days after mating and found that ovulation and pregnancy were not inevitably linked. For example, whereas 25% of the females receiving one intromission were pregnant, 92% had ovulated. Two-thirds of these animals contained short-lived degenerating corpora lutea. It is this finding that has led us to add the fourth category of female cycle pattern to the three originally proposed by Conaway (see Table 4.2). It now appears that in females given minimal stimulation ovulation occurs, but corpora lutea are not maintained.

In our studies of pregnancy initiation in three American species of voles, females were examined on the eighth day following mating. Rather than report our results in terms of percent ovulating, as in the original papers, it is now clear that in some cases we should report percent pregnant. It is possible that females ovulated short-lived corpora lutea in those studies. We have found this dissociation of ovulation and luteal function in montane voles (Kenney and Dewsbury, 1977), but not prairie voles (Kenney, Lanier, and Dewsbury, 1977). It is clear that the whole area of the relationship between the induction of ovulation and pregnancy in induced ovulators is in need of more work. However, it should also be remembered that regardless of the particular mechanism, the end result is unaffected: The females of species in which males display prolonged copulatory behavior generally have relatively high stimulus requirements for maximal pregnancy.

4.2. Pregnancy Initiation in Seven Species with Spontaneous Ovulation

The question of the generalizability of the proposed relationship between male behavior and female stimulus requirements to rodents with spontaneous ovulation now will be considered. The remaining species to be considered all appear to have Type III estrous cycles (Dewsbury, Estep, and Lanier, 1977). We shall first consider data from five species, all of which display the common muroid pattern of no lock, no thrusts, brief (usually multiple) intromissions, and multiple ejaculations.

The procedural fundamentals are identical for all studies. Males and females of known fertility are used. Vaginal smears are taken daily throughout the experiment. Females are mated for a particular amount of copulatory behavior and are tested only when they have shown two consecutive estrous cycles of seven days or less. Following copulation, females are classified according to whether they are pregnant, pseudopregnant (diestrus of at least eight days), or continuing to cycle. Within-subject designs are generally used.

4.2.1. Mesocricetus auratus

The copulatory behavior of Syrian golden hamsters, *Mesocricetus auratus*, has been studied by Beach and Rabedeau (1959) and Bunnell,

Boland, and Dewsbury (1977). Hamsters typically display about ten ejaculatory series followed by a number of "long intromissions." Clearly, we would expect females to have a high stimulus requirement.

Diamond and Yanagimachi (1968) found an increased probability of pseudopregnancy in female hamsters as a function of increased time of mating with vasectomized males. They also studied the effects of artificial stimulation. These data were then extended by Diamond (1972). As the original data of Diamond and Yanagimachi did not deal with pregnancy and used time, rather than number of ejaculatory series, as the independent variable, the results cannot readily be compared with those of other species. We have recently provided more comparable data (Lanier, Estep, and Dewsbury, 1975). As can be seen in Table 4.5, we found in one experiment that all females mated to satiety were pregnant, whereas just 21% females mated to one ejaculation were pregnant. The functional relationship between the amount of copulation and probability of pregnancy was further delineated in a second experiment. All females mated for seven or more ejaculations became pregnant. Thus, multiple ejaculatory series increase the probability of pregnancy in hamsters.

4.2.2. *Meriones tristrami*

Israeli gerbils, *Meriones tristrami*, are native to the deserts of Israel, and are slightly larger than, but otherwise quite similar to the more familiar congeneric Mongolian gerbil, *M. unguiculatus*. *M. tristrami* display the characteristic pattern 13 with brief vaginal insertions. Dewsbury, Estep, and Oglesby (1977) found a mean of 2.8 ejaculations in satiety tests with *M.*

Table 4.5. Percent of Female Golden Hamsters Pregnant, Pseudopregnant, and Continuing to Cycle as a Function of Condition[a]

Condition	Pregnant	Pseudopregnant	Cycle
Experiment I			
1 ejaculation	21(5/24)	8(2/24)	71(17/24)
Satiety	100(24/24)	0(0/24)	0(0/24)
Experiment II			
Mounts	0(0/10)	0(0/10)	100(10/10)
1 ejaculation	40(4/10)	10(1/10)	50(5/10)
4 ejaculations	64(7/11)	0(0/11)	36(4/11)
7 ejaculations	100(10/10)	0(0/10)	0(0/10)
1 long intromission[b]	100(11/11)	0(0/11)	0(0/11)
Satiety	100(11/11)	0(0/11)	0(0/11)

[a] Based on the results of Lanier, Estep, and Dewsbury (1975)

[b] After a given number of ejaculatory series, male hamsters display a pattern of "long intromissions" with thrusting. Females in this group were mated until the male displayed his first long intromission.

tristrami. However, a mean of 48 intromissions occurred after the last ejaculation. Once more, one would expect a high stimulus requirement in the female. As can be seen in Table 4.6, mating females to satiety rather than one series produced an increase in the percent pregnant from 8% to 75%.

4.2.3. *Peromyscus gossypinus*

Cotton mice, *Peromyscus gossypinus*, are a semiarboreal species, common in southeastern United States. Working in our laboratory, Deirdre Lovecky has found cotton mice to occasionally ejaculate on a single insertion, but generally to show the typical multiple-intromission–multiple-ejaculation pattern. Although the mean ejaculation frequency was just 2.4, much copulatory activity, including long intromissions (without thrusting), followed the last ejaculation.

Data on pregnancy initiation in *P. gossypinus* are presented in Table 4.6 (Dewsbury and Lanier, 1976). Females mated to one ejaculatory series never became pregnant in 19 tests; 63% of the females mated to satiety became pregnant. The difference was statistically significant.

4.2.4. *Peromyscus leucopus*

White-footed mice, *P. leucopus*, are the most closely related species to cotton mice, but are generally more northern in distribution. Dewsbury (1975b) found them to display pattern 13 (no lock, no thrusts, multiple intromissions, and multiple ejaculations) with a mean of 2.7 ejaculations and a considerable amount of copulation following the last complete ejaculatory series.

As can be seen in Table 4.6, no females mated to a single ejaculation became pregnant. When females were mated to satiety, and even when left in the male's home cage overnight, the incidence of pregnancy in this

Table 4.6. Percent of Females of Three Species Pregnant, Pseudopregnant, and Continuing to Cycle as a Function of Condition

Species	Condition	Pregnant	Pseudopregnant	Cycle
Meriones	1 ejaculation	8(1/12)	0(0/12)	92(11/12)
tristrami	Satiety	75(9/12)	17(2/12)	8(1/12)
Peromyscus	1 ejaculation	0(0/19)	5(1/19)	95(18/19)
gossypinus[a]	Satiety	63(12/19)	10(2/19)	26(5/19)
Peramyscus	1 ejaculation	0(0/28)	4(1/28)	96(27/28)
leucopus[a]	Satiety	25(7/28)	4(1/28)	71(20/28)
	Satiety + overnight	14(4/28)	7(2/28)	78(22/28)

[a] Based on the results of Dewsbury and Lanier (1976) and Dewsbury, Estep, and Oglesby (1977).

Table 4.7. Percent of Rattus rattus Females Pregnant, Pseudopregnant, and Continuing to Cycle as a Function of Amount of Copulation in Two Experiments

Condition	Pregnant	Pseudopregnant	Cycle
Experiment I[a]			
1 ejaculation	56(9/16)	13(2/16)	31(5/16)
Satiety	87(14/16)	13(2/16)	0(0/16)
Experiment II			
Mask[b]	0(0/5)	0(0/5)	100(5/5)
1ejaculation (low)	20(2/10)	0(0/10)	80(3/10)
1 ejaculation (normal)	30(3/10)	20(2/10)	50(5/10)
1 ejaculation (high)	70(7/10)	20(2/10)	0(0/10)
3 ejaculations	80(8/10)	20(2/10)	0(0/10)
Satiety	40(2/5)	60(3/5)	0(0/5)

[a] Based on the results of Estep (1975).
[b] Females wore a piece of tape over the perineal region, thus preventing intromission.

species was relatively low (Dewsbury and Lanier, 1976). Although there is no significance difference in this study because of the low incidence of pregnancy in satiety conditions, the facilitative effect of prolonged copulation is apparent against a baseline of zero.

4.2.5. *Rattus rattus*

Roof rats, *Rattus rattus,* are somewhat less familiar to the laboratory scientist that the more common *R. norvegicus,* which have displaced them in many locations around the world. Estep (1975) found roof rat copulatory behavior to be quite similar to that of laboratory rats, with an ejaculation frequency of 4.3 and a mean of 7.6 intromissions in the first series.

Estep's (1975) data on pregnancy initiation are presented in Table 4.7. Although satiety tests increased the pregnancy rate from 56% to 87% relative to a single ejaculation, this difference too was not statistically significant. Considerable variability was apparent in the first series intromission frequency scores. Within the one-ejaculation group, females becoming progestational received significantly more intromissions than those continuing to cycle (11.7 versus 5.4). Thus, there is indeed a relationship between amount of stimulation and incidence of pregnancy. This relationship has been explored more systematically in a second experiment. Within the first series, females received either low (0–5), normal, or high (9–19) intromission frequencies. This was done by using either priming females or priming males to alter intromission frequency. The data relating increased amounts of copulatory behavior to increased probabilities of becoming progestational and quite orderly (Table 4.7). It appears that multiple ejaculatory series are important primarily in tests with low numbers of intromissions in the first series.

4.2.6. *Peromyscus eremicus*

Cactus mice, *P. eremicus,* are found around rocks and cactus in the hot deserts of southwestern United States. Their copulatory pattern is somewhat complex (Dewsbury, 1974). Locks are occasionally observed, but appear to be more a matter of a pair becoming stuck together than the truly functional locks observed in other species. Cactus mice display intravaginal thrusting and always require multiple intromissions before ejaculating. Dewsbury (1974) observed just one ejaculation in 93% of tests with mice from his Rillito population. However, although they rarely attained a second ejaculation, the cactus mice usually resumed copulation after the first ejaculation. They displayed a mean of 15 intromissions and 34 thrusts after the last ejaculation. When one compares this to the first-series intromission frequency of 4.8, it is apparent that females typically receive stimulation equivalent to about four series. Cactus mice provide an interesting experimental preparation because the stimulation beyond the first series in this species, in contrast to that of all others considered, is not associated with increased discharge of additional sperm or plugs.

Data on pregnancy initiation from Dewsbury and Estep (1975) are presented in Table 4.8. It is apparent that the satiety conditions resulted in a considerable enhancement of pregnancy responses. Interestingly, females becoming pregnant in the satiety conditions received significantly more postejaculatory intromissions than those becoming pseudopregnant. This suggests the possibility of an effect of copulatory behavior on sperm transport.

4.2.7. *Onychomys leucogaster*

Northern grasshopper mice, *O. leucogaster* are an aggressive, predatory, carnivorous rodent species, native to the deserts of western United States. They display copulatory pattern 7 of Dewsbury (1972), with a true lock, no thrusting, ejaculation on a single insertion, and multiple ejaculations. Lanier and Dewsbury (1977) observed a mean of 9.3 locks lasting a mean of 33 sec in satiety tests with females in natural estrus.

As can be seen in Table 4.8, no females receiving just one lock became pregnant, even though sperm were ejaculated. As incidence of pregnancy was relatively low in this species, the difference between the one lock and satiety conditions is not statistically significant. However, as one again can compare the incidence of pregnancy relative to a baseline of zero, the effect of multiple locks is apparent. Further, there was a statistically significant

Table 4.8. Percent of Females of Two Muroid Species Pregnant, Pseudopregnant, and Continuing to Cycle as a Function of Condition[a]

Species	Condition	Pregnant	Pseudopregnant	Cycle
Peromyscus	1 ejaculation	5(1/20)	5(1/20)	90(18/20)
eremicus	Satiety	47(9/19)	32(6/19)	21(4/19)
	Satiety + overnight	60(12/20)	20(4/20)	20(4/20)
Onychomys	1 lock	0(0/14)	7(1/14)	93(13/14)
leucogaster	Satiety	21(3/14)	14(2/14)	64(9/14)
	Satiety + overnight	28(4/14)	7(1/14)	64(9/14)

[a] Based on the results of Dewsbury and Estep (1975) and Lanier and Dewsbury (1976).

relationship between number of locks received and probability of pregnancy in the satiety conditions. Pregnant females received significantly more locks than nonpregnant females (19.7 versus 7.5).

4.3. Overview

A summary of data on pregnancy initiation in eleven species of muroid rodents is presented in Figure 4.1. The general trend for probability of pregnancy to be elevated when these rodents are permitted to mate to satiety rather than for just one series is apparent. The sole exception is *Microtus ochrogaster*, that species with induced ovulation and a relatively low ejaculation frequency.

It would appear from these data that in many species multiple ejaculations are not an artifact of animal breeding practices enforced by man, nor are they wasted activity (Zucker and Wade, 1968). Rather, multiple ejaculatory series and other forms of prolonged copulation have an important function in the reproductive biology of the species, often producing substantial increments in the probability of pregnancy or in litter size. The stimulus requirements of females appear generally coadapted with the degree of prolongation of copulatory behavior typical of males of the species.

5. Pregnancy Initiation in Laboratory Rats and House Mice: A Second Look

With the background provided from the eleven species just discussed, we can return to the question of pregnancy initiation in laboratory rats and house mice. These comparative data force us to view the data on these two species in a somewhat different manner.

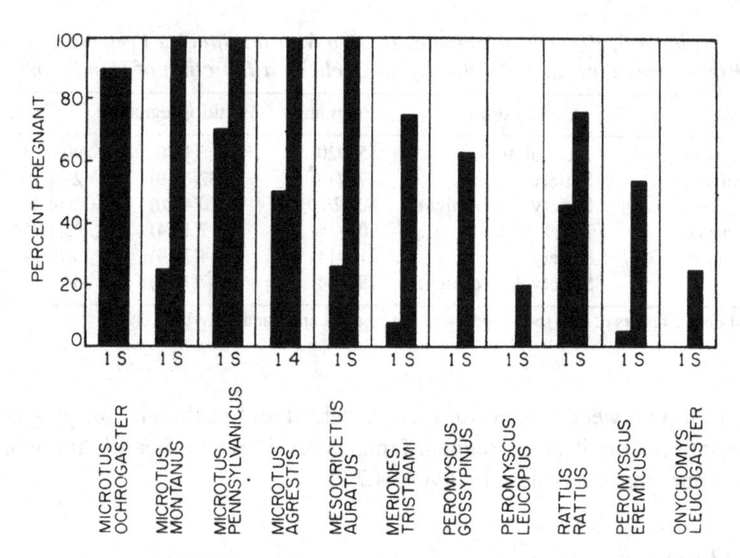

Figure 4.1. Summary of the effects of one-ejaculation and satiety tests of copulatory behavior on per cent of females pregnant in eleven species of muroid rodents. *Microtus agrestis* females were tested to four ejaculatory series rather than to satiety.

5.1. House Mice

It will be recalled that male *Mus musculus* have quite a low ejaculation frequency and that female *M. musculus* have a high probability of becoming pregnant when mated for one complete series. To the best of our knowledge, females in natural estrus have never been tested with males attaining multiple ejaculations. The data regarding ejaculation frequency in house mice appear in agreement with our hypothesis relating prolongation of male copulatory activity to female stimulus requirements. Indeed, house mice are the only spontaneously ovulating species we have found with low ejaculation frequencies and no postejaculatory copulation on most tests. The relatively low stimulus requirement of females is consistent with this aspect of male behavior.

The puzzling feature of the *Mus* literature is the lack of effect of preejaculatory intromissions. It is true that not all behavior, especially all behavior observed in the laboratory, need be adaptive. However, it is difficult to believe that copulatory activity involving 36 intromissions, 621 thrusts, and one hour before sperm are ejaculated is completely without biological function (BALB/c strain; McGill, 1962). The agreement in the comparative data between male behavior and female stimulus requirements led us to look further for a function of the male's behavior in pregnancy initiation.

There are at least three possible reasons for the apparent lack of effect of preejaculatory intromissions in *Mus*.

5.1.1. Domestication

The apparent lack of function for preejaculatory intromissions might be attributable to selective breeding processes that have occurred during the tenure of *Mus* in the laboratory. By selecting "good" breeders over many generations, humans may have so changed the gene pool as to have bred females whose reproductive systems are more active than is typical of wild animals. An hypothesized stimulus requirement for preejaculatory insertions might have been lost in the process. We are presently making our second attempt at conducting a pregnancy initiation study in wild *Mus*, which appear difficult to work with in such studies. Females of wild genotypes breed well in laboratory cages, but not when under observation.

5.1.2. Litter Size

It is possible that preejaculatory intromissions (or even multiple series), while not affecting the probability of pregnancy in *Mus*, may increase litter size. Multiple ejaculations have been shown to affect litter size in meadow voles (Gray, 1974) and laboratory rats (see below). To the best of our knowledge, there are no available data on the effects of variation in intromission frequency and litter size in *Mus*.

5.1.3. Variability in Male Behavior

Males do not always display preejaculatory intromissions. The copulatory behavior of males from a wild genotype maintained at the University of Nebraska is under study at present. In 66 tests completed thus far, three have been observed in which ejaculation was attained without prior mounts with insertion. There was a range of 28–49 thrusts. The relative frequency with which such behavior occurs and the degree to which this frequency is altered by environmental conditions has yet to be determined. McGill (1977) has suggested that ejaculation on one or a few insertions may be characteristic of dominant males, those males likely to be siring most litters in the natural habitat. Mosig and Dewsbury (1976) found ejaculation on a single insertion to be common in some genotypes when males attained their second ejaculation in a day with a female other than the female with which they had attained their first.

From an extension of the "principle of stringency" (Wilson, 1975), adaptation of the female stimulus requirement to match a pattern of low stimulation from the male is not inconceivable.

5.2. Laboratory Rats

Male laboratory rats typically attain a mean of about seven ejaculations in satiety tests (Beach and Jordan, 1956). About 90% of female rats become pregnant when mated to a single complete ejaculatory series (Wilson, Adler, and Le Boeuf, 1965). Laboratory rats, which have been a mainstay of research on pregnancy initiation now appear to be a major exception to our hypothesis relating pregnancy initiation to the persistence of male copulatory behavior. We can propose at least three possible reasons for this apparent exception.

5.2.1. Domestication

As was proposed for *M. musculus*, it is quite possible that the stimulus requirements of females have been changed as a result of domestication. It is to be hoped that Adler and his associates, who have been studying the copulatory behavior of wild *R. norvegicus* (Adler, 1974; McClintock, 1974), will extend their work to include studies of pregnancy initiation in wild rats.

5.2.2. Other Functions

It is possible that multiple ejaculatory series in laboratory rats function not in altering the probability that a female will become pregnant, but that they may have some other important function. There are a number of possible functions, one of which will be discussed at this time.

Under at least some conditions copulatory behavior can affect litter size. The number of ova shed can be altered by the presence or absence of mating at certain times of day and under certain conditions (e.g., Rodgers and Schwartz, 1974). Chester and Zucker (1970) found litter size to vary as a function of (a) amount of stimulation received and time within the estrous phase at which females are mated, and (b) the reversal of the normal ordering of intromissions and ejaculation. Sachs, Warden, and Pollak (1971) found an increase in number of implantation sites as a function of the number of ejaculatory series received. Similar effects were found by Davis (1974) and are discussed below. The general point to be made from these data is that the function of multiple series may lie not with an effect on probability of pregnancy, but with litter size or some other effect on the female.

5.3. Are Female Stimulus Requirements a Constant?

Most of the research that has been conducted on the behavioral initiation of pregnancy has been conducted with relatively young, healthy, well-nourished animals which have been showing regular estrous cycles. Implicit in much of the interpretation of this research has been an assumption that the nature of the female's stimulus requirements is highly species-characteristic and invariant. It is far more likely that stimulus requirements may vary as a function of any number of variables, and that by testing females under optimal conditions, important functions of multiple ejaculations may be obscured.

One indication of the situational dependency of female stimulus requirements appeared in the report of Chester and Zucker (1970), who found the stimulus requirements of their females to vary as a function of the time within the estrous phase at which the females were tested.

In a classical paper on pregnancy initiation in rats, Ball (1934) found that just 65% of females became pregnant when mated for one complete series with a median of nine "intromissions" (the criterion for classification of intromissions in this study is unclear). Although she did not run the appropriate control, it is likely that nearly all of her females would have become pregnant if mated for several series. While the experiment was reported in such a way that prevented full understanding of the reasons for the relative ineffectiveness of a single ejaculatory series, Ball's data again suggest that multiple series may have some function.

The strongest indication of a role for multiple ejaculatory series in laboratory rats comes from data collected in our laboratory by Davis (1974). Eleven of twelve young virgin female rats became pregnant after being mated for one complete series. These data are consistent with those from earlier studies (e.g., Adler, 1969). However, Davis also studied a group of females, 9–10-months old, which had given birth to and reared seven litters. None of the twelve females mated to one ejaculatory series became pregnant. Given three series, 67% of the females became pregnant, while 92% were pregnant after five series. The pregnant females in the five-ejaculation group also had significantly more implanted embryos than the pregnant females of the three-ejaculation group (12.4 versus 9.3). In a separate experiment, Davis found that significantly more ova was shed by old, multiparous females given five series than by females mated for just one series. The mechanism responsible for increasing the probability of pregnancy with multiple series in old females seems quite similar to that seen within the first series in younger females. Davis found that only females receiving more than one series had plasma progestin levels above those of controls. Further, administration of exogenous progesterone significantly increased the incidence of pregnancy in old females mated for just one series.

The change in female stimulation requirements as a function of age and/or parity is not limited to laboratory rats; we have found a similar effect in *M. pennsylvanicus*, an induced ovulator. Milligan (1975a) found effects of parity in *M. agrestis*.

6. Status and Prospectus for Comparative Studies of Pregnancy Initiation

6.1. Contributions of Comparative Analysis

Hopefully, the value of comparative analyses in the study of pregnancy initiation has been established by the preceding review. There are at least three major contributions of such analyses.

First, a new body of information has been developed. Data on pregnancy initiation have been provided for induced ovulators, species with diverse multiple-intromission, multiple-ejaculation patterns, and species with locking and thrusting patterns. These studies have produced a wide range of information on the effects of various behavioral patterns on pregnancy incidence, litter size, and endocrine function. The research has also led to the recognition of a fourth pattern of female reproductive cyclicity.

Second, new principles have emerged from comparative study. Prior to these experiments there was little indication of a biological function for copulatory activity beyond the first ejaculatory series. By contrast, species in which more than one series seems critical for maximal pregnancy now constitute the clear majority. The proposed correlation between persistence in male copulatory activity and stimulus requirements of the female, if confirmed by further data, may help to synthesize some of the diverse facts that have emerged.

Third, the comparative data have led us to reexamine data on the standard laboratory species. Data from laboratory rats and house mice appear consistent with those of other species in some regards and inconsistent in others. These considerations led us to consider possible reasons for such discrepencies and hopefully will lead to further research in which important new questions are asked.

6.2. Pregnancy Initiation as a Communication System

The male–female copulatory interaction can be viewed in part as a tactile communication system. In the study of the evolution of displays in the auditory and visual modalities, evolutionary changes in the characteristics of the signal must be correlated with evolutionary changes in the characteristics of the receiver system (Blest, 1963). The problem of understanding the mechanisms, especially the genetic mechanisms, involved in simultaneously changing two such closely coadapted systems are indeed considerable. The correlated changes in male behavior and female stimulus requirements may be regarded as a further instance of this type of change in a communication system—occurring in this instance via direct contact between male and female.

6.3. Prospectus

There are at least four main directions in which I believe research on pregnancy initiation will go.

6.3.1. Extensions of the Proposed Correlation

The copulatory behavior of males and stimulus requirements of females will be studied in new species, thus providing tests of the proposed hypothesis and development of new hypotheses. Given information on either characteristic, we now are in a position to make predictions about the other. It will be interesting to study species with extreme behavioral patterns. What of *Baiomys taylori*, which, in our laboratory, are the only species never to achieve more than one ejaculation? What of species like *Onychomys torridus*, which in our laboratory displayed a mean of 18.7 locks before satiating (Dewsbury and Jansen, 1972)? Viewed from the other side of the correlation, data on male copulatory behavior in the one muroid species with a Type I female cycle, *Mesembiomys gouldii*, would be of great interest.

6.3.2. Mechanisms

The task of understanding the physiological mechanisms underlying pregnancy initiation in muroid rodents is a difficult, but extremely important one. Studies such as those of Chester and Zucker (1970) and research from the laboratories of Adler (e.g., Adler, 1974; Matthews, 1975), McGill (e.g., McGill, 1970; McGill and Coughlin, 1970), and Diamond (e.g., Diamond, 1972) all provide excellent models.

6.3.3. Female Stimulus Requirements as a Dynamic System

The study of pregnancy initiation may be about to undergo a familiar transition in the development of scientific research areas. Areas that start out appearing simple, change with the addition of more data. Complexities arise and research becomes so esoteric as to be a mystery to all but the specialist. While this is unfortunate, it may be inevitable. The findings that such factors as time within the estrous phase and age and parity of the female can affect female stimulus requirements may represent just the beginning. Nutritional states, season, population density, and a host of other variables are likely to alter female stimulus requirements. It is possible that females in postpartum estrus may have stimulus requirements that differ from those of cycling females. If so, this difference may help to explain the smaller litter sizes that typically follow postpartum matings in rats as compared to matings following a sterile cycle (Bennett and Vickery, 1970). It may be adaptive for this to be the case. Research with females in postpartum estrus is of particular importance because it is likely that most matings in the natural habitat are in postpartum estrus. "In natural populations the nonpregnant cycle is a rarity, and is essentially a pathological luxury which cannot be tolerated." (Conaway, 1974, p. 239.)

It is increasingly being recognized that the female plays a critical role in controlling the timing and amount of copulatory activity, at least in some species under some conditions (e.g., McClintock, 1974). It is likely that the persistence of male copulatory activity is a joint function of changes in the male and the female. It is possible that the female plays an active role in prolonging copulation when she is beneath her threshold for pregnancy initiation and in terminating it when she is above threshold. Although changes for an individual female appear to play little role in determining the timing and/or persistence of male copulatory behavior in species such as cactus mice (Dewsbury, 1976), meadow voles (Gray, 1974), and laboratory rats (Dewsbury, 1968a; Tiefer, 1969), a role in other species appears quite likely (e.g., Clemens, 1969; Mosig and Dewsbury, 1976).

It is to be hoped that the complexities inherent in viewing pregnancy initiation systems in a dynamic rather than static fashion eventually will permit a new synthesis at a higher level. Whether or not this occurs, a dynamic view of pregnancy initiation systems is demanded by the data.

6.3.4. Ultimate Causation

All of these data take us but one step closer to the question that I personally find to be both the most interesting and most difficult in the study of copulatory behavior. What has been the ultimate causation of the variability we find in copulatory behavior and reproductive systems? What have been the selective pressures that have led to such diversity in male behavior and female reproductive physiology? For me, the question of ultimate causation is the ultimate question. It is clear that there will be no simple, univariate answer. Untangling the vast multivariate web of determinants of these systems will be a long and arduous task.

7. *Status and Prospectus for Comparative Psychology in General*

This chapter began with a discussion of Beach's classic paper, "The Snark Was a Boojum" and its impact on comparative psychology in general. A return to that theme is appropriate. In "The Snark Was a Boojum," Beach was unimpressed with the development of comparative psychology in the United States: "There has been no concerted effort to establish a genuine comparative psychology in this country . . ." (Beach, 1950). More recent writers have been anything but optimistic:

> Actually, we have developed no discipline of comparative psychology, either in investigation or in theory, of any substantial proportions." (Schneirla, 1952, p. 561.)
>
> What we once knew as comparative psychology has been overrun by a scientific revolution. In the wake of the revolution lies the debris of what was once a traditional branch of psychology, now a confused scatter of views of nature, problems, and methods. (Lockard, 1971, p. 168.)
>
> . . . Comparative psychology has no real basis in contemporary American psychology and dim prospects for the future. (Boice, 1971, p. 858.)

I take a different view (Dewsbury, 1968; 1973b). I am encouraged by the development of comparative psychology in this country. I see a new interest in the study of a variety of species and behavioral patterns, together with increasing pressures to leave the comfort of the laboratory for field research. Much of this new development is originating exactly where Beach foresaw that it must:

> I do not anticipate that the advanced research worker whose main experimental program is already mapped out will be tempted by my argument to shift to an entirely new field . . . the alert beginner who is looking for unexplored areas in which he can develop new methods of attacking unsettled issues would be wise to give serious consideration to comparative psychology as a field of professional specialization. (Beach, 1950.)

The explosion of journals, expansion of graduate and undergraduate courses and textbooks, and the development of the Animal Behavior Society all provide indications of substantial growth.

In "The Snark was a Boojum" Beach documented the decline of comparative psychology with an analysis of the content of the *Journal of Comparative and Physiological Psychology*. He found (a) no increase in the number of species studied as the number of articles per year increased, (b) a dramatic turn toward the laboratory rat as a subject, and (c) a concentration on studies of conditioning and learning. The persistence of these trends in more recent years has been repeatedly confirmed (e.g., Dukes, 1960; Whalen, 1961; Bitterman, 1965; Yeager, 1973; Lown, 1975).

Along with others (e.g., Gottlieb, 1976), I question the relevance of such analyses to the contemporary situation. Research by comparative psychologists has become scattered in a bewildering array of journals. Further, mere utilization of a species other than laboratory rats (e.g., White Carneaux pigeons) does not make a study comparative in the strictest sense of the term. Nevertheless, the *Journal of Comparative and Physiological Psychology* is regarded as the most prestigious journal in the area by most psychologists (Mace and Warner, 1973; Koulack and Keselman, 1975). Where does it stand now? A new set of editors assumed control of the Journal in 1975 with a new editorial statement, excerpts of which follow:

> Biological psychology is undergoing a renaissance. . . . Beginning this year, JCPP will publish papers that are explicitly comparative in central thrust, i.e., papers in which the species themselves were chosen for theoretically significant and valid comparative reasons . . . a paper should be an empirical report; and the truly comparative species of the contribution, not just the implications for general behavior theory, should be explicitly emphasized. (Thomas, 1975, pp. 1–3.)

It is, of course, too early to judge the results of this policy change as the new editors have produced but one volume and there has been insufficient time for comparative workers to alter their habits of sending comparative studies elsewhere. Nevertheless, a study of Volume 89 is instructive. By my count, Volume 89 contains 123 papers, 50% of which deal exclusively with domesticated forms of the Norway rat. While this percentage is quite high, it may be the lowest in this journal in my lifetime. I count 37 species represented in Volume 89. That may be an all-time record. By my classification, just 34% of the papers dealt with conditioning and learning. While it is too early to properly assess the significance of these trends, they are encouraging.

In 1950, Frank Beach turned to Lewis Carroll for a parable of developments in comparative psychology. Should he write a similar parable 30 years from now he may again turn to Lewis Carroll. Perhaps like Hebb (1958) he will turn to *Alice's Adventures in Wonderland.* I would suggest the section in which Alice (the personification of comparative psychology) is seated in a court room next to the dormouse (the personification of the rest of psychology) as she begins to grow:

> "I wish you wouldn't squeeze so," said the Dormouse, who was sitting next to her. "I can hardly breathe."
>
> "I can't help it," said Alice very meekly: "I'm growing."
>
> "You've no right to grow here," said the Dormouse.
>
> "Don't talk nonsense," said Alice more boldly: "you know you're growing too."
>
> "Yes, but *I* grow at a reasonable pace," said the Dormouse: "not in that ridiculous fashion." And he got up very sulkily and crossed over to the other side of the court. (Carroll, 1960, pp. 147–148.)

References

Adler, N. T. Effects of the male's copulatory behavior on successful pregnancy of the female rat. *Journal of Comparative and Physiological Psychology*, 1969, *69*, 613–622.

Adler, N. T. The behavioral control of reproductive physiology. In W. Montagna and W. A. Sadler (Ed.), *Reproductive behavior*. New York: Plenum Press, 1974, pp. 259–286.

Adler, N. T., and Zoloth, S. R. Copulatory behavior can inhibit pregnancy in female rats. *Science*, 1970, *168*, 1480–1482.

Adler, N. T., Resko, J. A., and Goy, R. W. The effect of copulatory behavior on hormonal change in the female rat prior to implantation. *Physiology and Behavior*, 1970, *5*, 1003–1007.

Ball, J. Demonstration of a quantitative relation between stimulus and response in pseudopregnancy in the rat. *American Journal of Physiology*, 1934, *107*, 698–703.

Beach, F. A. The snark was a Boojum. *American Psychologist*, 1950, *5*, 115–124.

Beach, F. A., and Jordan, L. Sexual exhaustion and recovery in the male rat. *Quarterly Journal of Experimental Psychology*, 1956, *7*, 121–133.

Beach, F. A., and Rabedeau, R. G. Sexual exhaustion and recovery in the male hamster. *Journal of Comparative and Physiological Psychology*, 1959, *52*, 56–61.

Bennett, J. P., and Vickery, B. H. Rats and mice. In E. S. E. Hafez (Ed.), *Reproduction and breeding techniques for laboratory animals*. Philadelphia: Lea and Febiger, 1970.

Bermant, G. Rat sexual behavior: Photographic analysis of the intromission response. *Psychonomic Science*, 1965, *2*, 65–66.

Bitterman, M. E. Phyletic differences in learning. *American Psychologist*, 1956, *20*, 396–410.

Blest, A. D. The concept of "ritualisation." In W. H. Thorpe and O. L. Zangwill (Eds.), *Current problems in animal behavior*. Cambridge: Cambridge University Press, 1963, pp. 102–124.

Boice, R. On the fall of comparative psychology. *American Psychologist*, 1971, *26*, 858–859.

Bunnell, B. N., Boland, B. D., and Dewsbury, D. A. Copulatory behavior of golden hamsters (*Mesocricetus auratus*). *Behaviour*, 1977, *61*, 180–206.

Carroll, L. Alice's adventures in wonderland. In *The annotated alice*. New York: Clarkson N. Potter Inc., 1950.

Chester, R. V. and Zucker, I. Influence on male copulatory behavior on sperm transport, pregnancy and pseudopregnancy in female rats. *Physiology and Behavior*, 1970, *5*, 35–43.

Clemens, L. G. Experimental analysis of sexual behavior of the deermouse, *Peromyscus maniculatus gambeli*. *Behaviour*, 1969, *34*, 267–285.

Conaway, C. H. Ecological adaptation and mammalian reproduction. *Biology of Reproduction*, 1971, *4*, 239–247.

Crichton, E. G. Reproduction in the pseudomyine rodent *Mesembriomys gouldii* (Gray) (Muridae). *Australian Journal of Zoology*, 1969, *17*, 785–797.

Davis, H. N. Maternal age and male behavior in relation to successful reproduction by female rats. Doctoral dissertation, University of Florida, 1974.

Davis, H. N., Gray, G. D., Zerylnick, M., and Dewsbury, D. A. Ovulation and implantation in montane voles (*Microtus montanus*) as a function of varying amounts of copulatory stimulation. *Hormones and Behavior*, 1974, *5*, 383–388.

Dewsbury, D. A. Copulatory behavior in rats: Changes as satiety is approached. *Psychological Reports*, 1968, *22*, 937–943.

Dewsbury, D. A. Comparative psychology and comparative psychologists: An assessment. *Journal of Biological Psychology*, 1968, *10*, 35–38.

Dewsbury, D. A. Patterns of copulatory behavior in male mammals. *Quarterly Review of Biology*, 1972, *47*, 1–33.

Dewsbury, D. A. Copulatory behavior of montane voles (*Microtus montanus*). *Behaviour*, 1973a, *44*, 186–202.

Dewsbury, D. A. Comparative psychologists and their quest for uniformity. *Annals of the New York Academy of Sciences*, 1973b, *223*, 147–167.

Dewsbury, D. A. Copulatory behavior of wild-trapped and laboratory-reared cactus mice (*Peromyscus eremicus*) from two natural populations. *Behavioral Biology*, 1974, *11*, 315–326.

Dewsbury, D. A. Diversity and adaptation in rodent copulatory behavior. *Science*, 1975a, *190*, 947–954.

Dewsbury, D. A. Copulatory behavior of white-footed mice (*Peromyscus leucopus*). *Journal of Mammalogy*, 1975b, *56*, 420–428.

Dewsbury, D. A. Determinants of ejaculatory failure in the copulatory behavior of cactus mice (*Peromyscus eremicus*). *Animal Learning and Behavior*, 1976, *4*, 177–180.

Dewsbury, D. A., and Estep, D. Q. Pregnancy in cactus mice: Effects of prolonged copulation. *Science*, 1975, *187*, 552–553.

Dewsbury, D. A., and Jansen, P. E. Copulatory behavior of southern grasshopper mice (*Onychomys torridus*). *Journal of Mammalogy*, 1972, *53*, 267–278.

Dewsbury, D. A., and Lanier, D. L. Effects of variations in copulatory behavior on pregnancy in two species of *Peromyscus*. *Physiology and Behavior*, 1976, *17*, 921–924.

Dewsbury, D. A., Estep, D. Q., and Oglesby, J. Copulatory behavior and pregnancy initiation in Israeli gerbils (*Meriones tristrami*). *Biology of Behavior*, 1977, in press.

Dewsbury, D. A., Estep, D. Q., and Lanier, D. L. Estrous cycles of nine species of muroid rodents. *Journal of Mammalogy*, 1977, *58*, 89–92.

Diamond, M. Intromission pattern and species vaginal code in relation to induction of pseudopregnancy. *Science*, 1970, *169*, 995–997.

Diamond, M. Vaginal stimulation and progesterone in relation to pregnancy and parturition. *Biology of Reproduction*, 1972, *6*, 281–287.

Diamond, M., and Yanagimachi, R. Induction of pseudopregnancy in the golden hamster. *Journal of Reproduction and Fertility*, 1968, *17*, 165–168.

Dukes, W. F. The snark revisited. *American Psychologist*, 1960, *15*, 157.

Edmonds, S., Zoloth, S. R., and Adler, N. T. Storage of copulatory stimulation in the female rat. *Physiology and Behavior*, 1972, *8*, 461–464.

Estep, D. Q. Copulatory behavior of *Rattus rattus*. Doctoral dissertation, University of Florida, 1975.

Gottlieb, G. Comparative psychology. *American Psychologist*, 1976, *31*, 295–297.

Gray, G. D. Ovulation and implantation in *Microtus* as a function of copulatory behavior. Doctoral dissertation, University of Florida, 1974.

Gray, G. D. and Dewsbury, D. A. A quantitative description of copulatory behavior in prairie voles (*Microtus ochrogaster*). *Brain, Behavior, and Evolution*, 1973, *8*, 437–452.

Gray, G. D. and Dewsbury, D. A. A quantitative description of the copulatory behavior of meadow voles (*Microtus pennsylvanicus*). *Animal Behaviour*, 1975, *23*, 261–267.

Gray, G. D., Davis, H. N., Kenney, A. McM., and Dewsbury, D. A. Effect of mating on plasma levels of LH and progesterone in montane voles (*Microtus montanus*). *Journal of Reproduction and Fertility*, 1976, *47*, 89–91.

Gray, G. D., Davis, H. N., Zerylnick, M., and Dewsbury, D. A. Oestrus and induced ovulation in montane voles. *Journal of Reproduction and Fertility*, 1974, *38*, 193–196.

Gray, G. D., Zerylnick, M., Davis, H. N., and Dewsbury, D. A. Effects of variations in male copulatory behavior on ovulation and implantation in prairie voles, *Microtus ochrogaster*. *Hormones and Behavior*, 1974, *5*, 389–396.

Hebb, D. A. Alice in wonderland, or psychology among the biological sciences. In H. F. Harlow and C. N. Woolsey (Eds.), *Biological and biochemical bases of behavior*. Madison, Wisconsin; University of Wisconsin Press, 1958, pp. 451–467.

Kenney, A. McM. and Dewsbury, D. A. Effects of limited mating on the corpora lutea in montane voles, *Microtus montanus*. *Journal of Reproduction and Fertility*, 1977, *49*, 363–364.

Kenney, A. McM., Lanier, D. L., and Dewsbury, D. A. Effects of vaginal-cervical stimulation in seven species of muroid rodents. *Journal of Reproduction and Fertility*, 1977, *49*, 305–309.

Koulack, D., and Keselman, H. J. Ratings of psychology journals by members of the American Psychological Association. *American Psychologist*, 1975, *30*, 1049–1053.

Kuehn, R. E., and Zucker, I. Reproductive behavior of the Mongolian gerbil (*Meriones unguiculatus*). *Journal of Comparative and Physiological Psychology*, 1968, *66*, 747–752.

Land, R. B., and McGill, T. E. The effects of the mating pattern of the mouse on the formation of corpora lutea. *Journal of Reproduction and Fertility*, 1967, *13*, 121–125.

Lanier, D. L., and Dewsbury, D. A. Studies of copulatory behavior in northern grasshopper mice (*Onychomys leucogaster*). *Animal Behaviour*, 1977, *25*, 185–192.

Lanier, D. L., Estep, D. Q., and Dewsbury, D. A. Copulatory behavior of golden hamsters: Effects on pregnancy. *Physiology and Behavior*, 1975, *15*, 209–212.

Lockard, R. B. Reflections on the fall of comparative psychology: Is there a message for us all? *American Psychologist*, 1971, *26*, 168–179.

Lown, B. A. Comparative psychology 25 years after. *American Psychologist*, 1975, *30*, 858–859.

Mace, K. C., and Warner, H. D. Ratings of psychology journals. *American Psychologist*, 1973, *28*, 184–186.

Matthews, M. Immobility at ejaculation determines fit between copulatory plug and cervix: Facilitation of sperm transport in rats. Paper presented at the Eastern Regional Conference on Reproductive Behavior, Nag's Head, North Carolina, May 1975.

McClintock, M. Sociobiology of reproduction in the Norway rat (*Rattus norvegicus*), estrous synchrony and the role of the female rat in copulatory behavior. Doctoral dissertation, University of Pennsylvania, 1974.

McGill, T. E. Sexual behavior in three inbred strains of mice. *Behaviour*, 1962, *19*, 341–350.

McGill, T. E. Induction of luteal activity in female house mice. *Hormones and Behavior*, 1970, *1*, 211–222.

McGill, T. E. Preejaculatory stimulation does not induce luteal activity in the mouse *Mus musculus*. *Hormones and Behavior*, 1972, *3*, 83–85.

McGill, T. E. Reproductive isolation, behavioral genetics, and functions of sexual behavior in rodents. In J. S. Rosenblatt and B. R. Komisaruk (Eds.), *Reproductive behavior and evolution*. New York: Plenum Press, 1977.

McGill, T. E., and Blight, W. C. Effects of genotype on the recovery of sex drive in the male mouse. *Journal of Comparative and Physiological Psychology*, 1963, *56*, 887–888.

McGill, T. E., and Coughlin, R. C. Ejaculatory reflex and luteal activity induction in *Mus musculus*. *Journal of Reproduction and Fertility*, 1970, *21*, 215–220.

McGill, T. E., Corwin, D. M., and Harrison, D. T. Copulatory plug does not induce luteal activity in the mouse, *Mus musculus*. *Journal of Reproduction and Fertility*, 1968, *15*, 149–151.

Milligan, S. R. Mating, ovulation and corpus luteum function in the vole, *Microtus agrestis*. *Journal of Reproduction and Fertility*, 1975a, *42*, 35–44.

Milligan, S. R. The copulatory behavior of *Microtus agrestis*. *Journal of Mammalogy*, 1975b, *56*, 220–224.

Mosig, D. W., and Dewsbury, D. A. Studies of the copulatory behavior of house mice (*Mus musculus*). *Behavioral Biology*, 1976, *16*, 463–473.

Rodgers, C. H., and Schwartz, N. B. The effect of mating on the afternoon of proestrus on serum LH (and FSH) and ovulation in the rat. *Proceedings of the Society for Experimental Biology and Medicine*, 1974, *147*, 148–150.

Sachs, B. D., Warden, A. F., and Pollak, E. I. Studies of partpartum estrus in rats. Paper presented at the Eastern Regional Conference on Reproductive Behavior, Haverford, Pennsylvania, June 1971.

Schneirla, T. C. A consideration of some conceptual trends in comparative psychology. *Psychological Bulletin,* 1952, *49,* 559–597.

Schrier, A. M. *Rattus* revisited. *American Psychologist,* 1969, *24,* 681–682.

Thomas, G. J. Editorial. *Journal of Comparative and Physiological Psychology.* 1975, *89,* 1–4.

Tiefer, L. Copulatory behaviour of male *Rattus norvegicus* in a multiple-female exhaustion test. *Animal Behaviour,* 1969, *17,* 718–721.

Whalen, R. E. Comparative psychology. *American Psychologist,* 1961, *16,* 84.

Wilson, E. O. *Sociobiology, the new synthesis.* Cambridge, Massachusetts: Harvard University Press, 1975.

Wilson, J. R. Effects of components of the sexual behavior pattern of male laboratory mice upon maintenance of pregnancy in female mice. Doctoral dissertation, University of California, Berkeley, 1968.

Wilson, J. R., Adler, N. T., and Le Boeuf, B. The effects of intromission frequency on successful pregnancy in the female rat. *Proceedings of the National Academy of Sciences, USA,* 1965, *53,* 1392–1395.

Yeager, D. L. Comparative psychology? *American Psychologist,* 1973, *28,* 181–184.

Zucker, I., and Wade, G. N. Sexual preferences of male rats. *Journal of Comparative and Physiological Psychology,* 1968, *66,* 816–819.

PART II

MECHANISMS CONTROLLING REPRODUCTIVE BEHAVIOR

5

Social and Environmental Control of Reproductive Processes in Animals

Norman T. Adler

1. Introduction

Sexual reproduction requires the coordination of two separate organisms with respect to each other and to the physical environment. If the coordination is successful, viable progeny are produced. Within this integrative framework, physiological events subserving reproduction are viewed in two ways: (a) as causes that lead to the performance of appropriate patterns of copulatory behavior (b) as events that are themselves caused by external factors, i.e., stimulation derived from environmental and behavioral events.

This article is organized around the reproductive life cycle of a "typical" mammal. This abstracted organism is followed through its development until it reaches reproductive maturity (puberty). Subsequent sections deal with the control of ovarian cyclicity, the induction of pregnancy, and some aspects of maternity. In each reproductive epoch, the focus is on the way in which exogenous stimulation (from the physical environment as well as from the social environment) affects the physiological status of sexually reproducing organisms to produce the orderly sequence of changes involved in reproduction. In each section of the discussion, after determining what aspects of reproductive behavior are functional in inducing the appropriate physiological changes in the organism, an attempt will be made to analyze the nature of the mechanism, which in turn

Norman T. Adler • Department of Psychology, University of Pennsylvania, Philadelphia, Pennsylvania. Preparation of this chapter was supported by NIH Grant No. HD 04522 and NSF Grant No. BNS 76-10198.

produces the adaptive behavior. This essay thus represents an attempt to analyze the reciprocal relationship by which physiology affects behavior and behavior influences physiology.

2. Maturation of Reproductive Function

2.1. Behavioral Effects on the Qualitative Nature of Sexual Development

The first epoch in the reproductive life of the organism is the ontogenetic period, during which the differentiation of the sexual type (male or female) occurs. Several investigators have shown how neonatal hormones effect this type of maturation. For example, in mammals the presence of androgen early in development predisposes or "organizes" the adult organism to respond in a male-like fashion. In many animals, the pattern of early hormone secretions is determined by the genetic material (Mittwoch, 1973). Thus in mammalian species, males, with an XY chromosome complement, perinatally secrete enough androgen to masculinize the behavior and morphology of the adult. In some animals, however, the sex typing process also depends on the kind of social stimulation to which the animal is exposed (Robertson, 1972). For example, in the labrid fish, *Labroides dimidiatus,* the basic social unit is defined by the male territory, which includes one mature male and a harem of some three to six females. In this group the largest, oldest individual is the male which dominates all of the females in the group; the larger, older females, in turn, dominate the smaller ones. If the male dies or is removed, sex reversal of the dominant female frequently occurs. Within a few hours after the dominant male's departure, this female begins to show male-like behavior (aggression). After two to four days, "she" begins to perform male courtship and spawning behavior. Within 14–18 days, the behavioral sex reversal extends to physiological functioning: "she" apparently can now release sperm. In this species, therefore, sex is not invariably a reflection of chromosomal complement; the presence or absence of a male determines the sex of the other animals.

In this example dealing with coral reef fish, social stimulation has been shown to modify physiological, morphological, and behavioral characteristics of individual organisms. In other species, social and behavioral stimulation has been shown to be a critical factor in the developmental course of sexual behavior and sexual roles even without changes in basic endocrine organization. In humans, there is an impressive amount of evidence suggesting that gender identity is dictated by the social milieu within which the child develops (Money and Erhardt, 1972). Even in nonhuman primates, there are sex-typed behavior patterns that are influenced by the social setting (Goldfoot and Wallen, 1976). Young male and female rhesus monkeys mount conspecifics. In the typical experimental setting in which both males and females are present, the young male displays high levels of one class of mounting behavior—the "double foot clasp mount," in which he wraps both of his hind feet around the mounted animal. Female rhesus monkeys in these groups typically display relatively low levels of this behavior. If young females, however, are raised in isosexual groups, the more "dominant" females within the group display significantly higher frequencies of the "double foot clasp mount" than do the females in heterosexual groups.

2.2. *Behavioral Effects on the Quantitative Nature of Sexual Development*

In the examples discussed in the previous section (i.e., the coral reef fish and rhesus monkeys), social stimulation was shown to affect the qualitative nature of sexual development, i.e., whether male-typical or female-typical patterns of sexuality emerge in the individual. Social stimulation can also affect the *quantitative nature of an organism's development,* the speed with which various reproductive events occur in the life of the animal.

Social stimulation, interacting with other environmental events, has been shown to affect (1) ovarian cyclicity, (2) the onset of annual periods of reproductive function, and (3) the speed with which puberty is attained. The first two phenomena will be documented in the next sections of the chapter; while the remainder of this portion of the discussion will be devoted to the third effect, the advancement of puberty by socially derived stimulation.

In some species, e.g., some mice, the onset of puberty can be advanced significantly by exposure to socially derived stimulation. If an immature female mouse is caged with an adult male, puberty can be advanced by 20 days (Vandenbergh, 1967). The underlying mechanism involves chemical cues emitted by the male and received by the female, for puberty can also be accelerated by housing females on bedding soiled by males (Vandenbergh, 1967). Since the induction of estrus may be accomplished by airborne pheromones emitted by the male (Whitten *et al*., 1968), great interest has developed in the possibility that puberty can be advanced by pheromonal stimulation. Three days of exposure to males is sufficient to induce a first estrus in young females (Bronson and Stetson, 1973).

3. *Behavioral and Environmental Control of Ovarian Cyclicity*

3.1. *Introduction*

In the previous discussion, social stimulation was shown to influence both the quality of adult sexuality (maleness or femaleness) as well as the speed with which adult sexuality (puberty) was achieved.

In considering our "typical" female mammal once again, the change in reproductive status over the lifetime of the animal (the reproductive life cycle) represents only one kind of physiological cyclicity. There are at least two other major reproductive rhythms embedded in the overall life cycle. The single ovarian cycle is perhaps the fundamental form of reproductive cyclicity. During this cycle, the mammalian follicle grows, ovulates, and becomes transformed into a corpus luteum. This series of events comprising the single ovarian cycle can be repeated, and the ovarian cycle (or series of cycles) is in turn embedded in another periodic fluctuation: annual or seasonal rhythmicity, in which there are periods of reproductive quiescence alternating with periods of ovarian activity.

In the next section, the social control of events *within* an ovarian cycle will be documented. Here, data will be presented to show that the onset of reproductive activity at the start of the breeding season can be influenced by social stimulation.

3.2. The Advancement–Synchrony of Breeding

3.2.1. The Forms and Functions of Reproductive Synchrony

A stable seasonal rhythmicity in sexual activity is quite important in tying the reproductive output of populations to the most advantageous conditions in the environment. This is especially true for those groups living in temperate regions where seasonal fluctuations in the physical environment are great (Sadleir, 1969). (In Philadelphia, for example, there are approximately 9 hr 22 min of sunlight per day at the end of December while, toward the end of June, there are 15 hr of light per day). The variation in this particular geophysical variable leads to correlated changes in other ones: the average monthly temperature can vary over the year by 30°C in some regions, with rainfall showing similar, wide annual fluctuations. Because of such annual cycles in geophysical variables, availability of food and other biotic resources in turn fluctuates annually (cf. Ricklefs, 1973, for review).

Although breeding activities for a given group of animals occur at the same *general* season every year, behaviorally derived stimulation can in some species advance or "fine tune" the time at which an animal first comes into reproductive condition during its annual cycle. In general, such modifications result in conspecific organisms within a population becoming reproductively more synchronous than they would be without such stimulation.

Male rhesus monkeys show some degree of testicular regression during the nonbreeding season (Sade, 1964); however, both behavioral and testicular gonadal function increase following exposure to an endocrinologically patent female (Vandenbergh, 1969). Breeding synchrony may be operating in a study dealing with *Lemur catta* (Jolly, 1967), in that all nine females in the breeding group came into estrus within 10 days of each other. If a ram is introduced to a group of ewes early in the breeding season, the females tend to come into estrus together (Schinkel, 1954; Sinclair, 1950; Thibault *et al.*, 1966). In birds, the observation that reproductive success is higher in large colonies than in small ones might be the result of the greater "social stimulation" augmenting and synchronizing breeding.

There are probably several biological functions served by the advancement–synchrony of annual breeding cycles: (1) lowering intragroup aggression, (2) saturation of predator populations with birth peaks, and (3) maximizing the number of breeding animals in response to local increases in food supply (at the start of the breeding season) (Brown, 1975; McClintock, 1974).

3.2.2. *An Experimental Model of Breeding Synchrony and Its Physiological Mechanism*

Despite specific variations in form and function of breeding synchrony, there is often a common aspect of the mechanism bringing about such synchrony: one organism, already in a breeding condition, "broadcasts" its condition; the "signal" then facilitates the appearance of the same reproductive state in a second, "recipient," organism.

A recently completed series of studies in our laboratory on female rats may provide an analog for studying reproductive synchronization (McClintock and Adler, in preparation). The purpose of these studies was to determine if one animal could "broadcast" its reproductive condition to a second organism and thereby affect the reproductive status of the recipient.

In this study, groups of six to eight female rats were placed in one of three compartments. Tubing was placed between the three boxes so that the rats could communicate via olfactory cues. An exhaust fan system was set up so that air was drawn from the first box (*upwind*), through the second (*middle*) box, into the third box (*downwind*), and then exhausted out of the room. Normal, cycling female rats were placed in the "upwind" and "downwind" boxes. In these two boxes, the animals received 14 hr of light per day (LD 14:10).

In Part I of the experiment, female rats that had been brought into constant estrus by prior exposure to constant light were placed in the middle box. During the 20 days in the middle box, the constant light condition (LL) was continued. Thus, in the first phase of the study, animals in the upwind and downwind boxes lived under an LD cycle while animals in the middle box were placed in LL.

In Part II of the experiment, the light-induced constantly estrous female rats in the middle box were replaced by normal cycling females, and an LD (14:10) cycle was then imposed in this environment. In Part II of this study, therefore, all three groups of female rats were housed in LD cycles. Through both phases of I and II of the experiment, the estrous cycles of all subjects were monitored. The results of this study are presented in Figure 5.1.

Eighty percent of the animals living under constant light in the middle box in Part I were in estrus (i.e., exhibited both cornified vaginal smears and a lordosis reflex to manual stimulation). Over the 20 days in this part of the study, the percentage of animals in estrus in the downwind box grew steadily to a maximum, in the last few days, of 83% (over the last nine days, the average was 56%). The percentage of females in the downwind box displaying estrus was significantly higher than the percentage of estrous females living in the upwind box, where the average during the last 9 days was only 25% ($p < .02$; chi square test).

In Part II of the study, when the constantly estrous animals were replaced with animals displaying normal cycles (16% in heat on the average), the number of downstream animals in estrus dropped to approximately 24%, which was not significantly different from the percentage of upwind animals in heat. Since the only difference between Parts I and II of this study was the presence of either constantly estrous or nor-

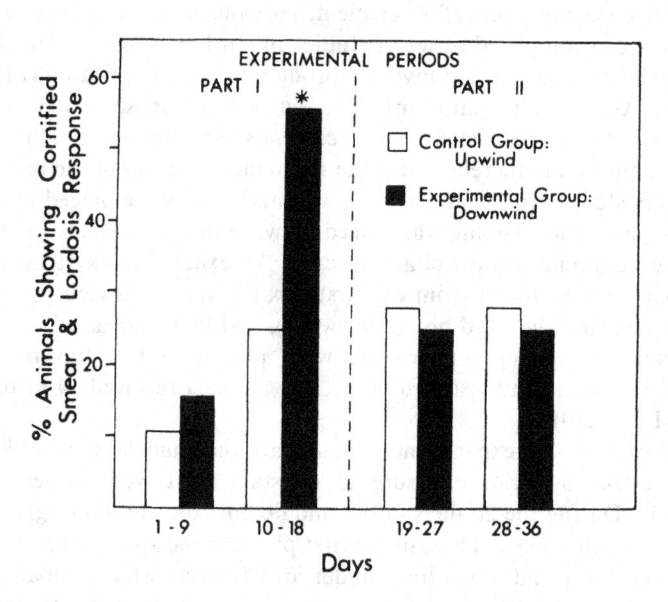

Figure 5.1. The effects of odor on estrus. (* $p \leq .02$; χ^2 test. All other comparisons are not significant.)

mally cycling females in the middle box, the changes in the vaginal smear picture in females living in the "downwind" boxes was ascribable to stimuli emanating from constantly estrous female rats. These data support the notion that *olfactory* cues from animals in constant estrus are sufficient to induce this state in other female rats, although other cues—like sound—might have played a role. This demonstrates, at least, that one organism can "broadcast" its reproductive condition to another organism and can thereby influence the reproductive state of the "recipient" organism. It is this general paradigm that may underlie breeding synchrony, although for any particular species, the modality of the stimulus "message" may vary.

3.2.3. Some Restrictions on the Concept of Pheromonal Regulation

The demonstration that one female rat may influence the reproductive status of another female is, at this point, pharmacological, i.e., it is not known whether this phenomenon plays a role in breeding synchrony of wild rats or indeed whether it plays any role at all in the natural history of this species. By the early definitions of the term "pheromone," olfactory stimulation must be shown to be species-specific and should be ascribed to phenomena that play a role in the normal ecology of a species (cf. Adler, 1977; and Doty, 1976, for reviews). Currently, there is great interest in the question of whether or not mammals have pheromonal communication to the specialized extent that insects have been shown to display. It may be that the alteration of reproductive states as a function of social stimulation is a property of most mammalian species and that only in some cases is this basic property refined into a truly adaptive signaling system. According to this view, pheromones would be somewhat akin to hormones, which often represent biologically "common" compounds that in some cases become refined, through evolutionary time, into biologically adaptive systems of integration (Gorbman and Bern, 1962).

A recent experiment by Weizenbaum, Adler, and McClintock (1977) reinforces the notion that olfactory influences on reproductive physiology is a common occurrence that may, in fact, be a general aspect of mammalian reproduction. The purpose of the experiment was to determine whether the female rat's vaginal cycle could be influenced by stimuli emanating from members of another rodent species with which it normally does not have contact. The method involved housing female rats and hamsters together and assessing the effects of hamsters on the reproductive cycle of the rat. Seventy mature female rats were placed into one of two rooms. In each of the rooms, there was a light-dark cycle (LD 14:10). There were no obvious qualitative or quantitative differences between the cycles in the two rooms.

After four weeks in the laboratory, 16 hamsters were introduced into one room ("rat–hamster" room); in the other room, there were only rats ("rat room"). The spatial arrangement of cages was standardized in the two rooms so that effects due to housing densities were minimized.

The rats' vaginal smears were classified according to a standardized scoring scheme. The most common sequences were four- and five-day cycles. There were, in addition, cycles of six days in which the sequence of cell types in the vaginal smears were generally NCCCLL (where N = nucleated, C = cornified, and L = leucocytic). In addition to the regular estrous cycle, there were periods of acyclicity. The criterion for scoring a bout of acyclicity was five or more consecutive days of either C or L smears. With the exception of one acyclic period, the acyclicity always consisted of seven or more days of a single smear type. The data were collected for 28 consecutive days.

An overall comparison of the vaginal cycles exhibited by rats in the room with hamsters versus the vaginal cycles exhibited by rats in the room with only members of their species revealed significant differences between the two groups (Figure 5.2).

Rats in the rat–hamster room generated significantly fewer days of acyclicity than did rats in the rat room. The greater number of acyclic days in the rat room could have been due either to an increase in the number of animals exhibiting acyclicity, or to an increase in the duration of acyclicity for a given animal. The data indicated that the increased

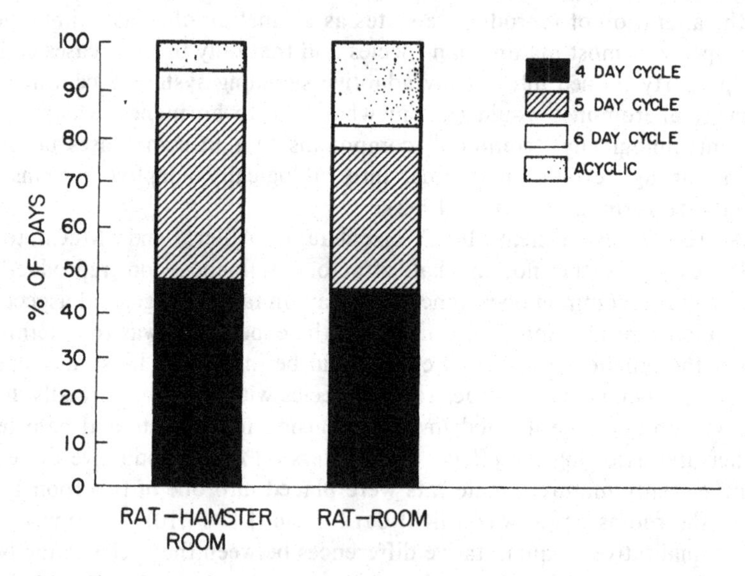

Figure. 5.2. The effect of the presence or absence of female hamsters on the vaginal cyclicity of female rats.

acyclicity in the rat room was due to the significantly greater number of animals exhibiting periods of acyclicity: 13 rats in the rat room versus 4 rats in the rat–hamster room. Given that an individual rat showed acyclicity, the average number of days of acyclicity generated by the rat in the rat room did not differ statistically from the number of acyclic days generated by the rat in the rat–hamster room.

The increase in acyclicity in the rat room might have been produced by a decrease in the number of 6-day cycles. Although there were no differences in the number of 4- and 5-day cycles, or in the frequency of the various subtypes of the 4- and 5-day cycles across rooms, there were significantly fewer 6-day cycles in the rat room than in the rat–hamster room.

According to Long and Evans, (1922), 4-, 5-, and 6-day estrous cycles are normal events. In most colonies of female rats, there are also runs of acyclic days. In this study, such periods of acyclicity were significantly reduced as a consequence of female hamsters being present in the same room. Since it has been shown, in other cases, that interspecific chemical stimulation can affect the reproductive status of female mammals (Michael, 1972; Vandenbergh *et al.*, 1975), the use of the term "pheromone," for any particular case, should be used with caution until the normal ecological function for such chemical stimuli can be determined. It may be that sensitivity to olfactory-gustatory stimulation is a basic property of the mammalian neuroendocrine system which, only in some cases, becomes refined into a biologically meaningful integrating mechanism.

3.3. *Control of the Ovarian Cycle*

In the previous section, studies were reviewed which showed that (a) the attainment of reproductive maturity could be influenced by social stimulation and (b) once reproductive maturity had been reached, the synchronization–advancement of seasonal reproductive cycles could be achieved by social interaction. One possible mechanism—olfactory communication between female rats—was introduced as a possible model for breeding synchrony in mammals.

Although the advancement of breeding seasons (and the consequent synchronization of breeding cycles) resembles the acceleration of puberty in that both effects can bring organisms into reproductive condition faster, the former effect does not merely involve "turning on the reproductive switch" earlier in the life of the organism. Ovarian cyclicity is, in fact, a biological rhythm in which there are factors, endogenous to the animal, which are capable of maintaining a *self-sustaining oscillation.* This self-sustaining biological oscillation, the ovarian cycle, can, however, have its precise periodicity affected by external events. Indeed, the jamming of the estrous cycle by constant light with resulting runs of constant estrus (Section 3.2.) is just such an environmental event.

In this section, studies will be presented showing how each stage of the estrous cycle can be influenced by environmental events (including social stimulation). In addition, some work will be presented which examines the way in which the animal's endogenous clock and external cues interact to produce ovarian cyclicity.

The generalized ovarian cycle of female mammals can be divided into three phases: (a) an initial *follicular phase* during which the ovarian follicles grow; (b) *ovulation*, when the mature follicle releases the egg; and (c) the *luteal phase*, when the ruptured follicle is transformed into the corpus luteum, which in turn secretes progesterone. In different species, stimulation derived from behavior plays an important role in many, if not all, of these stages.

3.3.1. Induced Ovulation

One of the most striking examples of behavioral control in the estrous cycle is seen in some species (e.g., cat, rabbit, ferret, mink, some species of mice, and perhaps tree shrews) in which ovulation is not spontaneous but depends upon stimuli delivered by the male during courtship and copulation (Rowlands, 1966).

In one series of studies, Dewsbury and his colleagues examined the relationship between copulatory activity and ovulation in montane voles, *Microtus montanus*, a species in which the females are classified as induced ovulators (Cross, 1972). They demonstrated that copulatory behavior was necessary and sufficient to cause ovulation (Gray *et al.*, 1974; Davis *et al.*, 1974). Following radioimmunoassay of plasma samples, it was found that within one hour after mating, plasma levels of both LH and progesterone were elevated in female voles (Gray *et al.*, 1976).

Diamond (1970) proposed that there is a species-specific vaginal code by which the pattern of copulatory stimulation provided by the male of a particular species corresponds to the parameters required to stimulate endocrine events in females of the same species. In a broad and careful phylogenetic survey of muroid rodents, Dewsbury and his co-workers have related the patterns of copulatory behavior of male rodents in closely related species to the induction of ovulation and pregnancy in the females (Dewsbury, 1975; also see Chapter 4, this volume).

Behaviorally derived stimuli can even influence ovulation in species that normally ovulate spontaneously (Aron *et al.*, 1966). For example, in female rats displaying normal estrous cycles, copulatory stimulation can advance the time when ovulation will occur and can sometimes increase the number of eggs released (Rodgers, 1971; Rodgers and Schwartz, 1972, 1973). Behavioral stimuli exert an even stronger effect under some experimental conditions. When ovulation is inhibited in female rats by placing these animals under conditions of constant light (Brown-Grant *et al.*, 1973; Hoffmann, 1973) or by injecting a number of pharmacological agents during a critical period of proestrus (Everett and Sawyer, 1950), the animals become "induced ovulators": they do not ovulate spontaneously but will do so if they receive enough copulatory stimulation (cf. Adler, 1974, for review). Although the female rat is normally a "spontaneous ovulator," she nonetheless possesses the neuroanatomical and neuroendocrine machinery capable of producing induced ovulation.

These data imply that the differences between various categories of reproductive function in related species (e.g., the difference between female rodents that are "induced" ovulators versus those that are "spontaneous" ovulators) can perhaps be attributed to slight quantitative differences in the neuroendocrine thresholds for triggering pituitary–ovarian events (Everett, 1964). The sufficiency of a female's endogenous endocrine cycles for inducing ovulation, independent of copulatory stimulation, will determine the categorization of a species as a "spontaneous" or "induced" ovulator.

3.3.2. Induced Luteal Phase

In the mammalian ovarian cycle, once successful ovulation has occurred, either of two sequences can follow. A female can go through the luteal phase of the cycle and return to the start of another estrous cycle, or she can become fertilized and enter the prolonged luteal phase characteristic of pregnancy. In the females of some species (e.g., in some primates), the spontaneous luteal phase of each ovarian cycle is long; and there is enough spontaneously secreted progesterone to permit uterine implantation of a fertilized egg. If no egg implants, the uterus, which has been developing under the influence of progesterone, sheds its inner lining (the endometrium), and another cycle begins. In the rat, the hamster, and several species of mice, however, there is no spontaneous luteal phase of any functional consequence; in these organisms, some aspect of the male's copulation triggers the progestational state underlying pregnancy (Adler, 1974; Dewsbury, 1975; Diamond, 1970; Diamond and Yanagimachi, 1968; McGill, 1970).

In Section 4 of this chapter, the induction of pregnancy in the female rat will be considered in detail and will be related to the total pattern of copulatory behavior in this species. In the remainder of Section 3, some general properties of the estrous cycle will be considered from the point of view of how environmental events can influence this biological rhythm.

3.3.3. The Estrous Cycle as an Endogenously Generated Biological Rhythm

Much of the preceding discussion has focused on the ways in which exogenous stimulation can influence the ovarian cycle of mammals. The advancement of puberty in mice illustrates how the physiological competence necessary for cyclicity can be accelerated, the "jamming" of estrous cycles into constant estrus represents a disruption of the endogenous ovarian cycle, and the induction of ovulation and luteal activity represent *triggers* by which copulatory stimulation facilitates progression from one component of the ovarian cycle to the next.

To describe a periodic biological phenomenon as representing the output of a true endogenous biological rhythm, the periodic behavior must be generated by a "self-sustaining biological oscillation" (Pittendrigh, 1974; Rusak and Zucker, 1975). One operational way of demonstrating such endogeneity is to show that the biological rhythm continues in constant environmental conditions. The estrous cycle of the female rat displays such characteristics. In continuous darkness (DD) the female rat often continues to show estrous cyclicity for several months (Browman, 1937; Hoffmann, 1967). (Similar phenomena occur in female hamsters; for review, see Rusak and Zucker, 1975.)

Since the estrous cycle of the female rat often takes the form of a 4- or 5-day rhythm, a well-documented hypothesis has been that the estrous cycle bears an intimate relationship to the approximately 24-hr (or "circadian") rhythm characteristic of many periodic functions, e.g., running activity. In the estrous cycle of the female rat, the timing of estrogen and progesterone secretions, the surge in pituitary LH release, the time of ovulation, and the onset of sexual receptivity are all normally synchronized to a particular phase of the light–dark cycle and hence are correlated with a particular phase of the entrained circadian running rhythm (cf. Adler, 1977, for review; Schwartz, 1969).

There are several other sources of evidence that the 4-day estrous cycle of rats is related to the circadian clock. First, an increase in general activity accompanies sexual receptivity and the presence of cornified epithelium in the vaginal smear (Wang, 1923; Gerall *et al.*, 1973). Second, the 24-hr timing component of the estrous cycle is revealed by an approximate 24-hr delay in ovulation following pentobarbital administration administered during a "critical period" during the afternoon of proestrus (Everett, 1964) and by circadian surges of LH in ovariectomized rats given exogenous estrogen (Burnet and MacKinnon, 1975; Caligaris, *et al.*, 1971; Legan, *et al.*, 1975; Mennin and Gorski, 1975). Finally, if the LD cycle is "phase shifted," then both running activity and estrous cyclicity shift until the normal phase angle with respect to the environmental light cycle is restored (Hemmingsen and Krarup, 1937; Alleva *et al.*, 1970).

Because of these similarities in the response of circadian activity rhythms and the 4-day estrous cycle of female rats, it seemed quite possible that the estrous rhythm is controlled by a biological oscillator that is either identical to, or strongly coupled to, the circadian oscillator. This hypothesis has been strongly supported in hamsters (Rusak and Zucker, 1975). Accordingly, all the examples of social control of ovarian events, as reviewed in this chapter, could have been produced by modifying a basic circadian oscillatory system.

A second way of conceptualizing the relationship between circadian and estrous rhythms is that, although the two are normally correlated, they are in fact generated by two separate oscillators (or sets of oscillators) which could be experimentally dissociated. Under this model, the two oscillators are *normally* coupled, resulting in the natural coherence of circadian and ovarian events. One way of dissociating the operations of these two hypothetical rhythmic processes would be to place the animals under environmental conditions that are known to produce disruptions of both the circadian running rhythm and the ovarian cycle and then noting whether the disruptions could occur independently in a single animal. For example, placing female rats in constant light (LL) leads to both a *free-running* activity cycle and to *constant estrus.* If it can be shown that under environmental conditions approaching constant light (LL)—e.g., 23 hr of light and 1 hr dark—that some female rats showed free-running activity rhythms while continuing to display regular estrous cycles, then this would be evidence for the separability of estrous and circadian oscillators. Similarly, if female rats continued to show an entrained circadian activity rhythm while displaying constant estrus, the separability of the endogenous oscillators for activity and estrous cycles would be supported.

In a recent study (Weber and Adler, unpublished) we placed female rats in environmental conditions that disrupt cyclicity. There were two basic procedures. In one study, the light–dark ratio within a 24-hr cycle was increased beyond the "normal" LD 12:12. (The light ratios in this study consisted of the following: LD 12:12, LD 20:4, LD 22:2, and LD 23:1). The second method for disrupting entrained cyclicity was to impose light cycle lengths that deviated from 24 hr. In this study, animals were placed in 22 hour "days" (11L:11D), 23-hr "days" (LD 11.5:11.5), 24-hr "days" (LD 12:12), 25-hr "days" (LD 12.5:12.5), and 26.5-hr "days" (LD 13.25:13.25). It has been documented that the characteristics of the estrous cycle and circadian running activity change under both (a) non-24-hr photoperiods and (b) in 24-hr photoperiods in which there is more than 20 hr of light per day (Besch, 1969; Chen and Besch, 1975; DeCoursey, 1972; Hoffmann, 1970; Pittendrigh and Daan, 1976).

The basic results of our study were that, over several weeks, disruptions in entrained circadian running activity (as defined by the occurrence of a free-running rhythm) and disruptions in ovarian cyclicity (as defined by the occurrence of constant vaginal estrus) could occur separately within a single animal. Over the entire group, the Spearman rank correlation coefficient between persistent estrus and free-running activity cycle was only 0.06 (n = 64, n.s.). Figure 5.3 illustrates the development of a free-running activity cycle with the simultaneous maintenance of estrous cyclicity for one female rat. This animal was placed under a LD 23:1 light cycle. In this record,

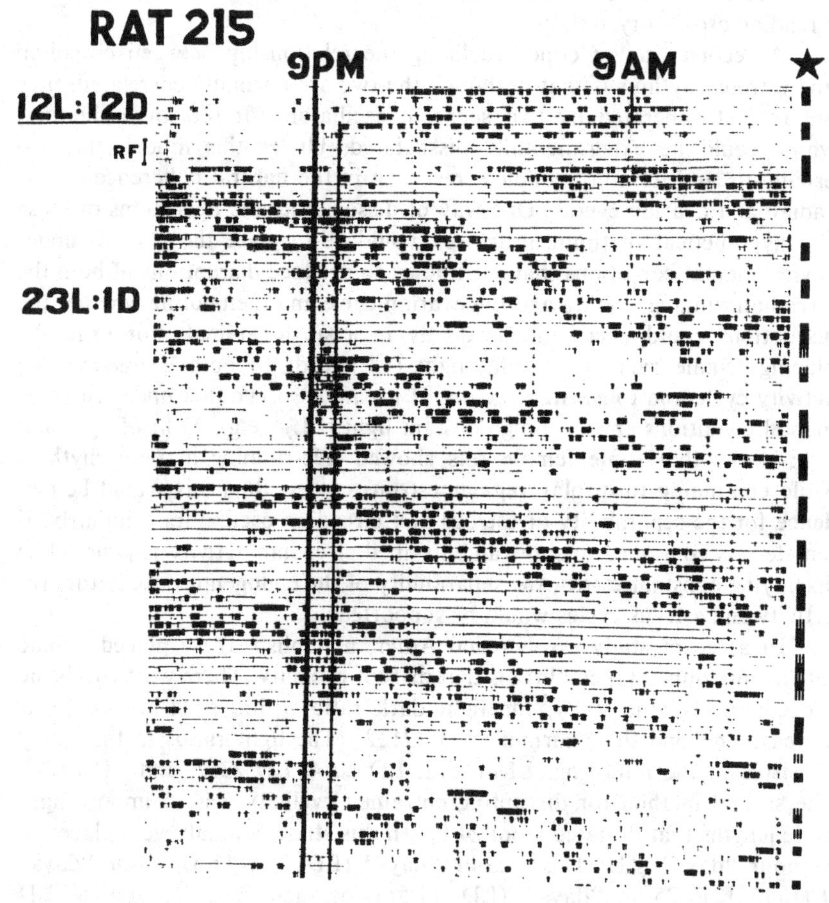

Figure 5.3. Activity and vaginal smear records of a female rat housed under a light cycle of LD23:1. The animal exhibited a free-running activity cycle and regular estrous cycles. Successive 24-hr periods are plotted one below the other. The time of day of the activity record starts at 3 PM at the left edge to 3 PM of the following day at the right edge. Thick black line (at 9 PM) indicates the onset of lights-off; thin black line the onset of lights-on. The light-dark ratio during a particular time span is indicated to the left of the activity record. RF is recording failure. The vaginal smear record is shown below the star in the top right-hand corner. The black square indicates predominantly cornified and/or nucleated vaginal smear taken within the 24-hr period shown to the left; blank indicates all other smear types; striped square indicates that no smear was taken during that 24-hr period due to the quasirandom smearing schedule.

successive 24-hr periods are plotted one below the other. The activity cycle was initially entrained to a LD 12:12 ratio; the onset of activity occurred about 3 hr after lights-off. After the increase in the light–dark ratio to LD 23:1, the activity cycle began free-running with a period longer than 24 hr and continued to do so for the remaining 13 weeks of the study.

The vaginal smear record for this animal is shown below the star in the right-hand column of the figure. Black square indicate that the vaginal smear was predominantly cornified and/or nucleated on a given day. For the last 13 weeks in the LD 23:1 cycle, the rat displayed 4- to 6-day estrous cycles (with one exception of a 7-day bout of leucocyctic smears during the second week) in spite of the free-running activity cycle.

The results of this study indicate that although the estrous cycle and circadian running activity *normally* show a strong coherence, it was experimentally possible to separate the two events in a single female rat. The implication is that there are at least *partially separable* biological oscillators subserving these two events. This is in accordance with the view that the rhythmic behavior of a single biological function may in fact be controlled by multiple endogenous oscillators (Pittendrigh, 1974). Working within this framework of "multiple clocks," we have been able to show that a single rat can show two bouts of running activity, each influenced by a different periodicity of food presentation (Edmonds and Adler, 1977a,b).

Since it was possible, at least in the studies reported here, to dissociate experimentally estrous and circadian running cyclicity, there is now a question of the mechanism by which social stimuli (as discussed in previous sections) affect the occurrence of reproductive events. It remains to be determined whether olfactory cues from conspecifics, for example, modify ovarian cyclicity without modifying the circadian cycle or whether such social stimulation modulates both kinds of cyclicity.

4. *Behavioral Induction of Pregnancy in Female Rats: An Integrative Case Study*

4.1. *Progestational Hormone Secretion*

4.1.1. *A Neuroendocrine Reflex*

In this section of the chapter I will present data on the way in which the male rat's copulatory behavior facilitates the induction of pregnancy in the female rat and will describe some of the mechanisms that produce the adaptive behavioral patterns. By concentrating on one species, an attempt will be made to integrate behavioral and physiological aspects of reproduction into one coherent picture.

The copulatory behavior of the male rat consists of a series of mounts and dismounts from the female. On some of these mounts, penile intromission occurs; and on the final intromission, the male ejaculates sperm and a vaginal plug. Following a period of behavioral inactivity (the postejaculatory interval), the male resumes copulating. The pattern of mount, intromission, ejaculation (comprising an ejaculatory series), and the subsequent refractory period can be repeated several times in one mating session.

Since one of the most consistent features of this species' copulatory pattern is the series of multiple intromissions preceding each ejaculation (Dewsbury, 1975), the question arises: what function do these multiple intromissions possess for successful reproduction?

In several experiments the probability of pregnancy was determined for females that received a *normal complement of intromissions* preceding the male's ejaculation (high-intromission group); these data were compared to the probability of inducing pregnancy in an experimental group of females that were permitted only *a reduced number of preejaculatory intromissions* (low-intromission group), (Adler, 1969; Wilson *et al.*, 1965). Twenty days after copulation, the females in both groups were sacrificed and their uteri examined for the presence of viable fetuses. In one study (Wilson *et al.*, 1965), approximately 90% of the females in the high-intromission group were pregnant while only 20% of the females in the low-intromission group were pregnant. Thus, multiple intromissions appear to be necessary for the induction of pregnancy.

How do multiple copulatory intromissions stimulate pregnancy? Part of the answer is to be found in the organization of the female rat's 4- or 5-day estrous cycle. Unlike most primates, the ovarian cycle of the female rat does not automatically go through a functional luteal phase; that is, the mature follicle ovulates but does not secrete enough progesterone to permit uterine implantation of a fertilized egg. If the female is to become pregnant, some event must trigger the progestational state. We hypothesized that the stimulation derived from the male rat's multiple copulatory intromissions triggers a neuroendocrine reflex that results in the secretion of progesterone. This hypothesis was supported by progesterone assays (Adler *et al.*, 1970). Within 24 hr after mating, females in a high-intromission group had significantly more progesterone in their peripheral blood than did females in the low-intromission group. It is now known that cervical stimulation induces twice-daily surges of pituitary prolactin release and that this daily prolactin is necessary for secretion of the ovarian progesterone necessary for pregnancy (Smith *et al.*, 1976).

We wanted to study the dynamics of this behaviorally initiated neuroendocrine reflex. The first step was to determine what aspects of male intromissions are responsible for initiating pregnancy. Since females in both high- and low-intromission groups received ejaculations with sperm and vaginal plugs, ejaculation was not *sufficient* for the induction of the progestational state; on the other hand, stimulation from ejaculation might have been a *necessary* part of the triggering stimulus. The results of several experiments, however, suggest otherwise. In one study, males were treated with guanethidine sulfate before being placed with females. These males could copulate but could not deposite any semen (sperm or plug material) during the ejaculatory response. Nonetheless, females mated to these males became progestational (Adler, 1969).

Normally, the stimulation derived from one ejaculatory series is sufficient to trigger the progestational state in the female rats studied in our laboratory. In one experiment, 13–16 intromissions—even without an ejaculation—induced the secretion of gestational hormones (Adler, 1969). These female rats are generally young (90–120-days old) virgins. Recently, Davis *et al.* (1977) have shown that there is an important effect of the female's age. While older females that are near the end of their reproductive lives still become progestational following copulation, they require much higher levels of this stimulation. Instead of one ejaculatory series' providing sufficient stimulation, these females require multiple series (Davis *et al.*, 1977). It seems that as the female rat's neuroendocrine system ages, it becomes less "efficient" in responding to copulatory stimulation.

We are now attempting to follow the behavioral stimulus through successive levels of the female's neuroendocrine system to determine the mechanism by which copulatory intromissions trigger the secretion of the gestational hormones.

From electrophysiological recordings of genital sensory nerves, we determined that it was the pelvic nerve that innervated the vagino-cervical area (Komisaruk *et al.*, 1972). These data support the hypothesis that the pelvic nerve is the afferent channel for the induction of progesterone secretion. Further evidence for this conclusion comes from experiments in which the pelvic nerves were cut. In these studies, the female rats did not become progestational after mechanical stimulation of the cervix (Kollar, 1953; Carlsson and De Feo, 1965; Spies and Niswender, 1971).

4.1.2. Behavioral Mechanisms in the Promotion of Cervical Stimulation

Since the entire neuroendocrine reflex resulting in progesterone secretion is initiated by a nerve that innervates the vagino-cervical area (i.e., the pelvic nerve), there is consequently a behavioral "problem" that must be solved if the female is to become pregnant: the copulatory sequence of male and female rat must result in cervical stimulation. The pattern of multiple pre-ejaculatory penile intromissions does provide the cervical stimulation that triggers activity in the pelvic nerve, but the pattern of multiple intromissions in turn requires a precise integration of male and female response: (1) male and female must be attracted to each other, (2) this attraction leads to an initial orientation and body contact (male mount), which finally produces (3) the properly oriented penis inserted into the vagina.

Each of these three stages depends on precise synchronization between male and female. Although the male obviously plays an active role in initiating contact in that he mounts the female, the female is not "passive" but rather displays an active role in the copulatory sequence, a role that Beach (1976) has termed "proceptive" behavior. (McClintock, 1974; McClintock and Adler, unpublished). The active solicitation by the female becomes obvious if the testing cage is large enough (7 × 7 ft in McClintock's experiment). In this situation, the female orients to the male, approaches him (to within two body lengths), and then quickly darts away. This pattern of female orientation–approach–rapid withdrawal is usually followed by the male's pursuit of and mounting the female (McClintock, 1974).

In this sequence, the combination of male and female "proceptivity" ensures that male and female are (1) attracted to each other and (2) make an initial orientation and contact of their bodies (during the mount). The third requirement for pelvic nerve stimulation, penile contact with the vagino-cervical area, requires coordination of the two organisms' genital apparatus, on the *reflexological level.* Careful behavioral analyses (sometimes involving high-speed films of the copulatory sequence) have begun to unravel the complexity of the relation between mount, lordosis, and penile insertion (Diakow, 1975; Pfaff *et al.*, 1973; Pfaff and Lewis, 1974). When the male rat initially makes contact with the female, she displays the lordosis reflex in which she raises her rump and deflects her tail laterally. Concurrently, the male develops an erection and contracts his somatic musculature so that the penis can enter the vaginal orifice.

During the male's mount, his penis makes several thrusts around the vaginal orifice before finally achieving insertion (Adler and Bermant, 1966; Bermant, 1965). This area of the female's perineum that the male initially contacts with his ventrum is the same general area innervated by the pudendal nerve (Komisaruk *et al.*, 1972); this nerve receives afferent input ipsilaterally from the perineal region surrounding the vagina. Its sensory field extends from the base of the clitoral sheath to the base of the tail in the midline, and laterally along the inner surface of the thigh.

There are two characteristics of the pudendal nerve's sensory field that may be relevant to the ultimate stimulation of the vagino-cervical receptors, which is necessary for the induction of progesterone secretion. First, the total size of the sensory field of the pudendal nerve was significantly greater (approximately 30%) in castrated females given exogenous estrogen injections than in uninjected castrate controls (Komisaruk *et al.*, 1972; Kow and Pfaff, 1973/74). Second, the most sensitive portion within the sensory field was not invariably the perivaginal area but a spot that was approximately 1 cm caudal and 1 cm lateral to the tip of the clitoral sheath (Adler *et al.*, in press; Komisaruk *et al.*, 1972).

The enlargement of the pudendal nerve's sensory field by exogenously administered estrogen may be an experimental analog of events that normally occur during the estrous cycle of female rats (Adler *et al.*, in press). The sensory field of the pudendal nerve was significantly larger in estrous female rats than in diestrous rats, the increase in total area approximately 30%. In addition to measuring the change in field size (by strongly stimulating the skin with a probe), we determined the threshold for neural responses of six points within the sensory field (Figure 5.4). To determine the threshold for each point, von Frey hairs of increasing

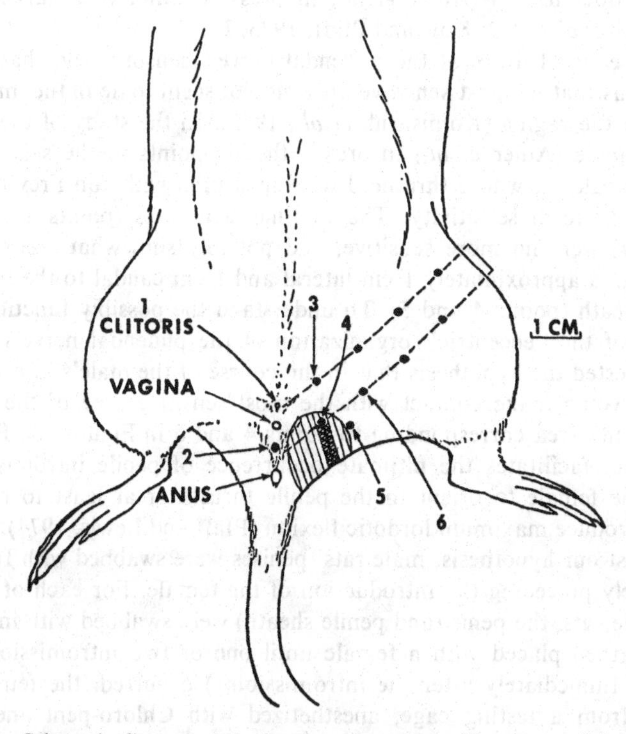

Figure 5.4. Schematic diagram of the female rat's ventrum. The two parallel rows of dots were orientation points used in mapping the pudendal nerve's sensory field. The numbered spots were the test points where threshold of the pudendal nerve to von Frey hair stimulation was measured. The hatched area of the genital region corresponds to the points, averaged over the females in this study, that the male rats contacted during copulation.

diameters were applied to each point. The average sensitivity (1/threshold) for each of the six points was determined in each rat, and it was found that sensitivity increased 84% in estrus over its diestrous value.

The statistically significant increase in both size and sensitivity of the estrous field compared to the diestrous field may have a function in the mating sequence. From preliminary high-speed motion picture analysis, the female rat seemed to move her rump toward the male and open the vaginal orifice before penile insertion occurred. Since the pudendal nerve's sensory field is larger in estrus (when the female normally mates) than diestrus, one of the effects of the cyclic changes in hormone levels during estrus may be to enlarge, and make more sensitive, the sensory field of the pudendal nerve. A larger, more sensitive field would facilitate the female's perineal adjustments to the male's mount. Although this functional hypothesis remains to be documented precisely, the increase in a nerve's sensory field size over the estrous cycle or as a consequence of exogenous hormone administration is now well documented (Adler *et al.*, in press; Bereiter and Barker, 1974; Komisaruk *et al.*, 1972; Kow and Pfaff, 1973/74).

The second feature of the pudendal nerve's sensory field that seemed unusual was that its most sensitive area did not seem to lie in the immediate vicinity of the vagina (Komisaruk *et al.*, 1972). In the study of the normal cycling female (Adler *et al.*, in press), the six points in the sensory field of each female rat whose threshold were measured with von Frey hairs did not show uniform sensitivity. The midline structures (points 1 and 2 in Figure 5.4) were no more sensitive, and possibly somewhat less sensitive, than the area approximately 1 cm lateral and 1 cm caudal to the tip of the clitoral sheath (points 4 and 5). To understand the possibly functional significance of the "eccentric" organization of the pudendal nerve's sensory field, we tested the hypothesis that in the course of the male's intromission, his penis would make contact with the most sensitive area of the sensory field, i.e., the area corresponding to points 4 and 5 in Figure 5.4. Possibly, this contact facilitates the ultimate occurrence of penile intromission by causing the female to orient to the penile thrusts or at least to raise her rump to produce maximum lordotic flexion (Pfaff and Lewis, 1974).

To test our hypothesis, male rats' penises were swabbed with India ink immediately preceding the introduction of the female. For each of ten different male rats, the penis (and penile sheath) were swabbed with ink. Each male was then placed with a female until one or two intromissions were delivered. Immediately after the intromission(s) occurred, the female was removed from a testing cage, anesthetized with Chloro-pent anesthesia, placed on her dorsum, and her legs positioned and taped down in the same position as in our electrophysiological studies of the pudendal nerve (Komisaruk *et al.*, 1972). The pattern of ink was noted and reproduced on a scaled drawing of the rat (Figure 5.4). This drawing represents a composite of the pattern of ink deposition on the perineal region of the ten female rats in this study. The drawing was created by marking the shape and location of the ink marks on each female. The numerical dimensions of the mark were described in Cartesian coordinates, with the tip of the clitoral sheath serving as the origin. The top, center, and bottom of each female's perineal ink stain was calculated; and an average for each of these three values was calculated over the ten animals. This "average" stain was then drawn in heavy strippling on the schematic in Figure 5.4. The lighter hatched area represents the range of the stains on the ten females. (Following the intromission, there was also ink deposited in the immediate vicinity of the vaginal orifice; for the sake of graphic clarity, however, this perivaginal stain was not represented in the figure). The extravaginal stain coincided, on the average, to points 4 and 5 in Figure 5.4. These points were the most highly sensitive loci within the sensory field of the pudendal nerve, as measured by electrophysiological responses in the nerve, as described above. Thus, penile stimulation of these highly sensitive points within the sensory field of the female's pudendal nerve may facilitate the elevation of the female's rump to produce a maximum lordosis (Pfaff, *et al.*, 1973; Pfaff and Lewis, 1974). It may also help orient the female's vaginal orifice with respect to the male.

4.1.3. Central Mechanisms in the Induction of Pregnancy

These studies on the mechanisms by which behavioral interactions result in penile intromissions are propadeutic to understanding how the vagino-cervical area (and its innervation, the pelvic nerve) comes to be stimulated, thereby triggering the neuroendocrine reflex resulting in progesterone. In trying to follow this initial cervical stimulation through the various stages of the neuroendocrine reflex, we investigated the "storage properties" of the central nervous system–pituitary complex. Since a number of intromissions were required to trigger the progestational response (Adler, 1969, 1974), it seemed that there must be some sort of mnemonic device by which the stimulation from each brief (about 250 msec) penile insertion was retained and combined with stimulation from succeeding intromissions.

To determine the limits of the storage capacity of this system, an experiment was performed in which the rate of stimulation was varied (Edmonds *et al.*, 1972). For a given group of females, both the number of intromissions permitted (two, five, or ten) and the interval between intromissions varied: the interintromission-interval values ranged from the control rate of ad libitum copulation (approximately 40 sec) up to one hour. The results are summarized in Figure 5.5.

With 10 intromissions, 100% of the females became progestational.

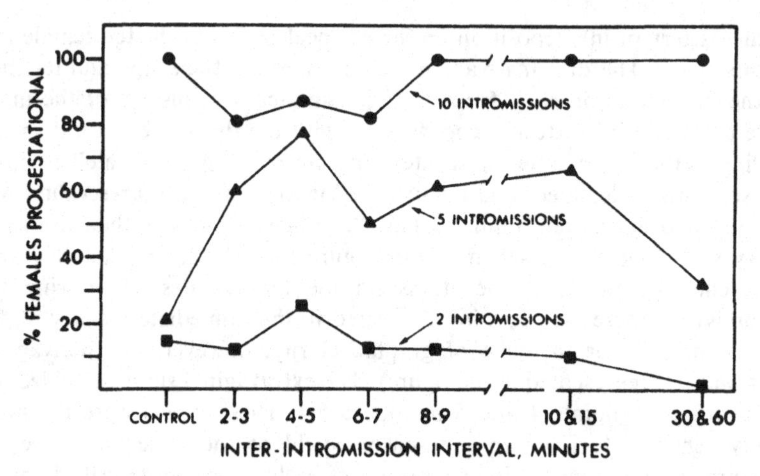

Figure 5.5. Effects of different interintromission intervals on the induction of progestational hormone secretion in female rats ($n = 213$).

This result is especially striking because the intromissions could be spaced one every half hour without a diminution of their effectiveness. The results for females receiving five intromissions also indicate that spaced intromissions are as effective as intromissions delivered at the control rate and, in addition, that intromissions at the rate of one every four or five minutes may be *more* effective in stimulating the progestational response than the control rate of one approximately every 40 sec. One intromission delivered every half hour provided a density of stimulation of only one part per 7,200 (one 250-msec intromission per half hour). This potency points to an exquisitely adapted form of neuroendocrine integration by which species-specific stimuli (the multiple intromissions) are stored by an adaptively specialized neural mechanism.

The operation of this entire pregnancy system can be conceptualized as a biological amplifier in which each stage lengthens the temporal characteristics of the previous one. Stimulation from the brief penile intromission is stored and accumulated. The stimulation leads to a central nervous system event, the duration of which is undetermined but which may run for several hours (Barraclough and Sawyer, 1959). Pituitary involvement may last 12 days (Pencharz and Long, 1933); the ovary secretes large amounts of progesterone for 19 days (Morishige *et al.*, 1973). The final stage, uterine pregnancy, continues for more than 20 days. In this sequence, the intromissions trigger the tonic secretion of progesterone.

The storage of the initial copulatory stimulation in the female rat resembles the other examples of reproductive mnemonic devices. Adaptive specialized storage systems are, in fact, frequently occurring phenomena in reproductive physiology. The phenomenon of "delayed pseudopregnancy" represents a classic example (Everett, 1968); if proper parameters are used, electrical stimulation of the brain results in the induction of a progestational state in female rats—but only after a delay of hours or days. Another example of reproductive storage occurs in the hamster, in which copulatory stimulation facilitates the subsequent parturition more than 20 days after the stimulus (Diamond, 1972). Although exogenous progesterone injections can trigger the progestational response and permit pregnancy to follow artifical insemination, parturition is not normal unless copulatory stimulation from a male hamster is provided at the time of insemination. The mechanism by which the stimulus is stored in the female's neuroendocrine system is not yet known precisely in any of these examples, but phenomenologically they all involve the storage and amplification of a triggering stimulus.

4.2. Behavioral Control of Sperm Transport in Female Rats

When first studying the copulatory influences on pregnancy in rats, we concentrated on the function of multiple intromissions in stimulating the neuroendocrine reflex that resulted in progesterone secretion and subsequent gestation. Since all of the female rats in our experiment, in both the *high* and *low intromission groups,* had sperm and plug deposited in their vaginas, it was an easy assumption that sperm transport and fertilization were normal (Adler, 1969). In one experiment, however, we checked the condition of ova in the fallopian tubes of the female rats three days after copulation and found that females that had received many preejaculatory intromissions all had developing ova in the blastocyst stage. In contrast, females that had received an ejaculation preceded by only one intromission had unfertilized and degenerating ova (Adler, 1968; 1969); even though these females received an ejaculation that resulted in the depositions of sperm and plugs in their vaginas, transcervical sperm transport did not occur in the absence of preejaculatory intromission. Multiple intromissions before ejaculation, therefore, have two functions: they trigger the neuroendocrine reflex which results in progesterone secretion, and they are necessary for the transport of sperm from the vagina through the tightly closed cervix into the uterine lumen. Without a few preejaculatory intromissions, there can be no sperm transport and therefore no fertilization.

There are two other ways in which copulatory behavior facilitates sperm transport in the female rat. Of obvious importance is the ejaculation. When the male ejaculates, there is a seminal coagulate, the vaginal plug, which forms in the female's vagina. The deposition of a vaginal plug is necessary for subsequent sperm transport; male rats surgically deprived of their seminal vesicles (and thus rendered incapable of depositing vaginal plugs) leave no sperm in the uterus after ejaculation (Blandau, 1945). In a series of experiments (Matthews and Adler, 1977), we have discovered that the mere *deposition* of a vaginal plug is not sufficient to ensure full sperm transport. In one study, 29 female rats were permitted *ad libitum* copulatory stimulation through one ejaculation. After receiving the ejaculate, the female rats were anesthetized and laporotomized. The reproductive tract, including vagina, cervix, and uteri, was carefully dissected out. Estimations were made (a) of the total number of sperm having reached the uterus and (b) the number of sperm remaining in the vagina and cervix (Matthews and Adler, unpublished). Comparing the number of sperm in the *uterus* to the number of sperm remaining in the *vagina and cervix* yielded a measure of the relative success of sperm transport. For each female, the degree of sperm transport was related to the tightness of the vaginal plug. (To quantify the tightness of the plug, the extent to which the walls of the vaginal plug adhered to the vagino-cervical juntion was noted). Females were classified as having 0, ¼, ½, ¾, or 4/4 of the circumference of the vagino-cervical canal adhering to the vaginal plug.

The results of this study are presented in Figure 5.6. The greater the degree to which the vaginal plug adhered to the vagino-cervical junction, the higher the *uterine* sperm counts (and, consequently, the lower the number of the sperm remaining in the vagino-cervical area). Although the transport of sperm through the reproductive tract of female rats was proportional to the tightness of the vaginal plug, there was no significant correlation between

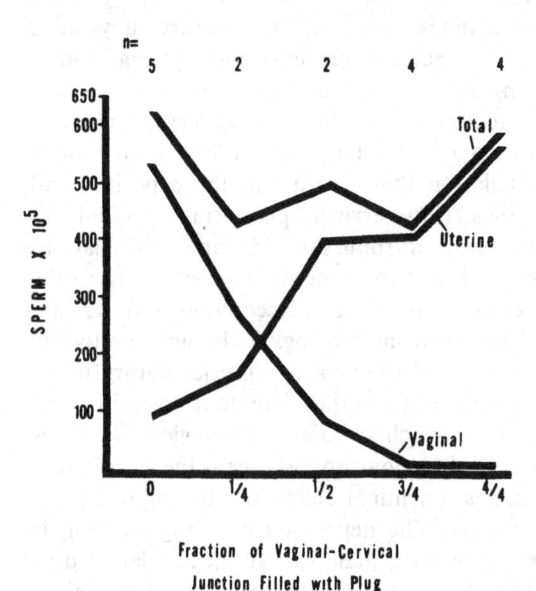

Figure 5.6. Transport of sperm from the female rat's vagina into the uterus as a function of the amount of the vagino-cervical junction filled with "plug."

"plug fit" and the *total* number of sperm in the female's reproductive tract (total combined uterine, cervical, and vaginal sperm count).

The results imply that the *deposition* of a vaginal plug is not sufficient to induce maximal sperm transport through the reproductive tract of female rats. Variations in the tightness with which the plug adheres to the vagino-cervical junction are correlated with variations in the number of sperm reaching the uterus. The question then becomes, "What are the factors responsible for variations in the tightness of the vaginal plug?" It is here that behavior plays a role. Even when the number of *pre-ejaculatory intromissions* is sufficiently large to stimulate sperm transport in the female, the *behavior at the moment of ejaculation* can vary, and it is this variability of the ejaculatory response that leads to differences in the placement of the vaginal plugs (Matthews and Adler, 1977).

While male rats *typically* assume an immobile posture during the ejaculation, there is considerable individual variation in the extent and duration of this immobility. On some ejaculations, a male may fail to maintain a forward movement of his genital region after the "convulsive" thrust; he may not demonstrate a convulsive thrust at all; or he may step backward at various times after the thrust. After some ejaculations, a rapid "stepping backward" appeared to cause the penis to slip out of the vagina more quickly than during the ejaculation during which the male assumed a prolonged "immobile" posture. To test the hypothesis that variations in the latency to "step backward" might be related to the tightness with which the vaginal plug adhered to the vagino-cervical area (and consequently to variations in sperm transport in the female), a second experiment was performed (Mathhews and Adler, unpublished). The ejaculations of 13 males were observed. If the male stepped back within two seconds after the "convulsive thrust" of ejaculation, he was categorized in the "ejaculation mobile group." If he remained on the female for at least two seconds after the convulsive thrust, he was categorized in the "ejacultion immobile group." When grouped in this fashion, 63% (7 of 11) males in the ejaculation immobile group produced plugs that adhered tightly to the female's vagino-cervical junction (adhering on $\frac{3}{4}$ or $\frac{4}{4}$ of the circumference). In contrast, none of the four males in the ejaculation mobile group produced plugs that lodged tightly in the vagina. This difference was statistically significant ($p < .05$, Mann–Whitney U test). In addition, the percentage of sperm in the reproductive tract, which had been transported into the uterus, was significantly greater in the ejaculation immobile group than in the ejaculation mobile group (median percentages were 99.2 versus 21.4 for the two groups).

The results of these experiments support the notion that a prolonged ejaculatory response facilitates the deposition of a tightly fitting vaginal plug and that a tightly fitting plug in turn expedites the migration of sperm into the female's uterus. Dissections of the vaginal plug have revealed that the plug mass is in fact not solid throughout its length. At the cervical end is a hollow area, the "sperm cup," which after ejaculation contains a pool of sperm-rich semen. When this plug was dissected free from the cervix shortly after an ejaculation, semen was often expelled as if under considerable pressure. This suggests that the "sperm cup" acts as a reservoir from which sperm is supplied for intrauterine transport. Figure 5.7 is a photograph of a vaginal plug, pulled away from the cervix, but still lying in the vaginal canal. The opaque white mass lying to the right of the plug is a pool of semen which has leaked out of the "plug cup" following experimental dislocation of the plug from the cervix.

A general picture is emerging of the ways in which the copulatory

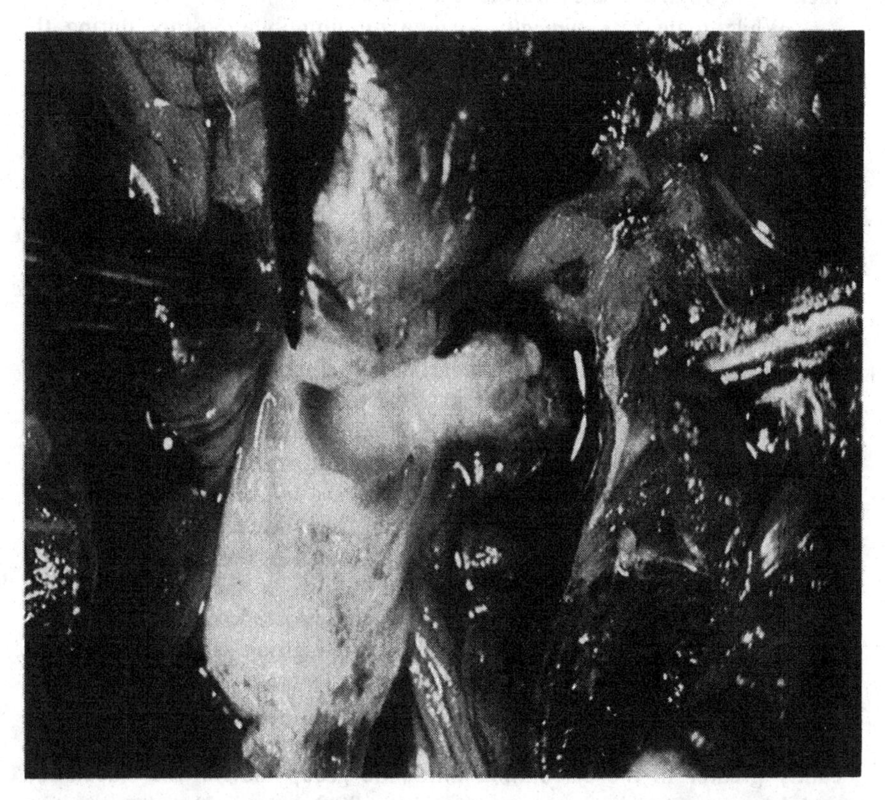

Figure 5.7. Photograph of a ventral view of the female rat's reproductive tract, taken through a dissecting microscope. The vagina has been completely cut from opening to cervix. The vaginal plug has been loosened. The probe points to the left wall of the plug cup protruding into the vagino-cervical junction. The white mass emanating from the right side of the cup is sperm and uncoagulated seminal material that flowed from the cup when the anterior wall was cut.

behavior of the rat facilitates sperm transport. The multiple preejaculatory intromissions delivered by the male prepare the female for the transport of sperm subsequent to the ejaculation. The mechanism of this priming may involve neural, mechanical, or neuroendocrine factors (e.g., by causing release of oxytocin). On ejaculation, the male does not leap back, as he does after an intromission, but remains immobile, maintaining penile penetration. The immobility during ejaculation permits the formation of a vaginal plug, tightly adhering to the vagino-cervical area. At the rostral end of the solidified plug is the "plug cup," which may act as a reservoir for sperm. Because of the tight seal between plug and cervix, the sperm is under great pressure, this pressure possibly facilitating transport through the cervix, into the uterus.

In addition to requiring the preejaculatory intromission and a prolonged ejaculatory response for successful sperm transport, there is a third behavioral requirement for this transport through the female rat's reproductive tract; a period of copulatory quiescence following the ejaculation.

Maximal numbers of sperm do not reach the uterus until 6–8 min following ejaculation (Adler and Zoloth, 1970; Matthews and Adler, 1977). If, during this time, a female rat receives copulatory intromissions, the number of sperm in the uterus is reduced, and the size of her litter diminished (Adler, 1974).

The potential disruption of sperm transport by postejaculatory copulatory stimulation may have functional significance for the organization of reproduction in this species. Theoretically, one male could "cancel" the effects of another male's copulation. If the second male begins copulating too soon after the previous male has completed an ejaculatory series, his multiple intromissions may prevent the first male's sperm from reaching the uterus, presumably by dislodging the plug. The second male's sperm would then be deposited upon *his* ejaculation and could fertilize the ova. To test this possibility, we arranged a series of mating tests involving albino female rats (Adler and Zoloth, 1970). An albino female was allowed to mate first with an albino male and was then mated with a pigmented male for a second ejaculatory series. Paternity could be established soon after the pups are born since pups sired by albino fathers are light colored and have clear eyes, whereas pups of pigmented fathers have dark skins and pigmented eyes. If the substitution of a pigmented for an albino male is made within the first few minutes after the initial ejaculation, approximately 60% of the offsprings are pigmented. If, however, 45–60 min elapse between copulation, only 23% of the offspring are pigmented. Therefore, a male rat can, if he begins copulating soon enough, prevent the insemination of a given female by another male that had copulated previously (Adler and Zoloth, 1970).

One of the themes of this chapter is that considerations of physiological organization often lead to behavioral investigation. The transport of the male rat's gametes through the female's reproductive tract provides an example of this approach. For effective sperm transport to occur, the female rat must have no cervical stimulation for several minutes after receiving an ejaculation. What then prevents a male that has just ejaculated from resuming his intromission during this critical time? We suggest that it is the male's *postejaculatory refractory period* that prevents him from resuming copulation too soon (Adler and Zoloth, 1970). The male rats in our laboratory require an average of 4.5 min after ejaculating before they deliver the first intromission of their next ejaculatory series. Since three intromissions on the average are sufficient to completely dislodge the vaginal plug (Lisk, 1969), and since sperm transport is relatively complete within six to eight minutes, the postejaculatory refractory period is of the correct magnitude to permit effective sperm transport into the uterine lumen.

The hypothesized function of the postejaculatory refractory period may have ecological significance for rodent population dynamics. Under conditions of crowding, laboratory rats show persistent social pathology (Calhoun, 1962a). One type of behavioral abnormality is a kind of pansexual behavior in which males mount at a much higher rate than usual. In these crowded colonies reproduction decreases, partly because of the pathological increase in mounting. Even where crowding is not a problem colonies of rats with a higher percentage of males produce fewer pups than colonies with fewer males (Calhoun, 1962b). One of the reasons for reduced reproductive performance in such colonies may be copulatory interference with sperm transport.

Perhaps the main conclusion from all of these experiments on rats is that the induction of successful pregnancy in this species requires the precise temporal integration of behavioral and physiological events into a coherent pattern of reproduction. Given that a pair of male and female rats is in a state of copulatory readiness, stimulation arising from the sexual behavior of the pair produces physiological effects, which are necessary for the occurrence of pregnancy. The penile stimulation received by the male during his copulatory intromissions yields excitation, which eventually triggers his ejaculation (Adler and Bermant, 1966; Carlson and Larsson, 1964). In the female, the cervical stimulation from the preejaculatory intromissions stimulates the neuroendocrine reflex resulting in progesterone secretion and also primes the female so that sperm will be transported into her uterus following ejaculation. The ejaculatory response, in turn, ensures a tightly fitting vaginal plug; and the male's postejaculatory interval insures the tranquility necessary for the passage of asymptotic numbers of sperm into the uterus. These relationships are summarized in part of Figure 5.8.

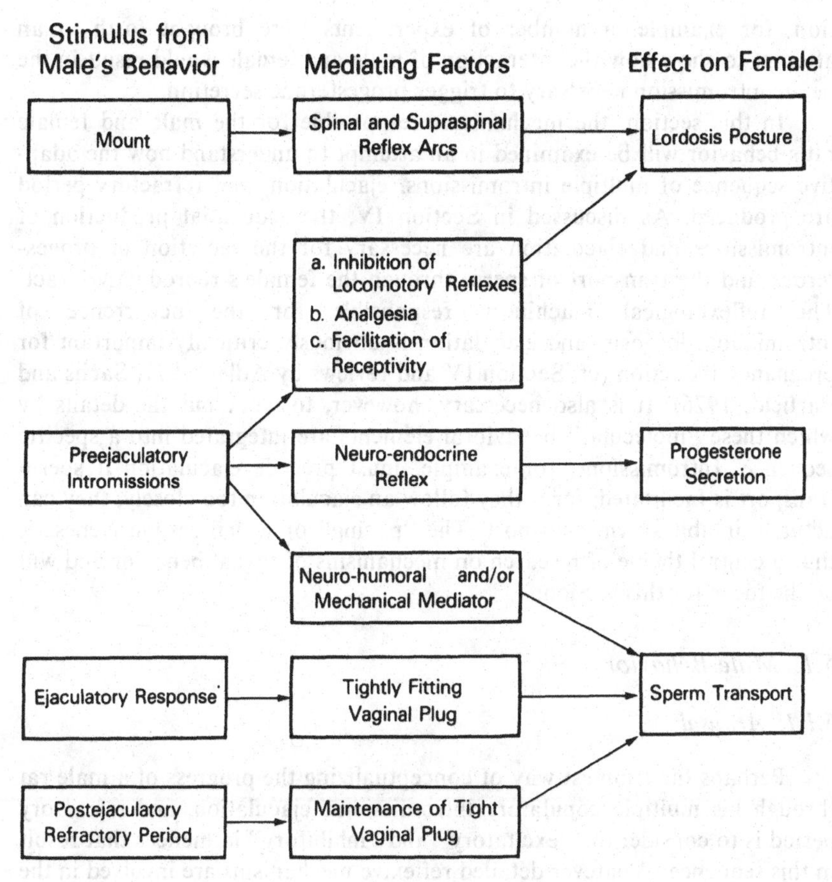

Figure 5.8. Summary of the effects of behavioral stimulation on reproductive physiology in the female rat.

5. *Some Mechanisms Responsible for the Behavioral Adaptations Controlling Pregnancy in Rats*

The studies reviewed in the previous section dealt with the behavioral control of reproductive physiology in rats. The orientation of this discussion was that the orderly induction of pregnancy required specific behavioral adaptations on the part of both males and females. Once the adaptive constraints of a behavior have been outlined, it becomes possible to analyze the physiological machinery responsible for producing the behavior. In the discussion of the triggering of progesterone secretion by cervical stimulation, for example, a number of experiments were brought forth in an attempt to show how the interaction of male and female would result in the penile intromission necessary to trigger progesterone secretion.

In this section, the mechanisms responsible for the male and female rat's behavior will be examined in an attempt to understand how the adaptive sequence of multiple intromissions, ejaculation, and refractory period are produced. As discussed in Section IV, the sequential production of intromissions and ejaculation are necessary for the secretion of progesterone and the transport of sperm through the female's reproductive tract. The reflexological machinery responsible for the occurrence of intromission, lordosis, and ejaculation is obviously critically important for pregnancy induction (cf. Section IV and reviews by Adler, 1977; Sachs and Barfield, 1976). It is also necessary, however, to work out the details by which these "molecular" behavioral elements are integrated into a specific sequence. Intromissions, for example, must precede ejaculation if sperm transport is facilitated; for if they follow an ejaculation too closely, they can actually inhibit sperm transport. The "pacing" of sexual responsiveness is thus a central theme of research on mechanisms of sexual behavior and will be the focus for this section.

5.1. Male Behavior

5.1.1. Arousal

Perhaps the simplest way of conceptualizing the progress of a male rat through his multiple copulatory intromissions, ejaculation, and refractory period is to consider the "excitatory" and "inhibitory" influences that result in this sequence. Whatever detailed reflexive mechanisms are involved in the sequence of the male rat's copulatory behavior, there is now general agreement that two different physiological processes must be interacting in order to account for the quantitative data of the male rat's ejaculatory series. (For a masterful review of the literature, see Sachs and Barfield, 1976). There may be a neurochemical analog of this "two-process" theory in terms of the interactions between norepinephrine and dopamine in male sexual behavior (Antelman and Caggiula, 1977) although it is not yet possible to define a strict isomorphism between behavioral and chemical processes.

An arousal process is necessary because the male and female rat, normally separated over several body lengths (McClintock, 1974; McClintock and Adler, unpublished) must approach each other and make body contact. Data presented in Section IV ascribe this contact to the stimulation of the female rat's solicitations. On the part of the male, the pursuit, mount, and intromission sequence has been thought to arise from an "appetitive" or "arousal" system (Kurtz and Adler, 1973). In this study, the male's copulatory sequence was correlated with concommitant changes in his EEG. We observed the occurrence of hippocampal theta activity when the male oriented, chased, mounted, intromitted, and ejaculated into female rats. The theta activity showed a temporary lowering of frequency during the mount but resumed its higher frequency following this act. In contrast, following an intromission or ejaculation, the theta frequency showed a steady decline in frequency and subsequent desynchronization. The behavioral-electrophysiological data were interpreted according to the following scheme (Kurtz and Adler, 1973). When the male is aroused, he is under the control of an "appetitive" system (Craig, 1918), indicated electrophysiologically by hippocampal theta and behaviorally by pursuit of the female. This arousal leads him to mount the female. If the mount does not result in an intromission, his arousal remains high (reflected by the continuation of the hippocampal theta after the mount). The attainment of an intromission or ejaculation represents a consummatory response (reflected by the desynchronization of hippocampal theta), and the sequence is terminated. (Sachs and Barfield, 1976, postulate the existence of a "mount bout" as a basic timing device in controlling the temporal patterning of these copulatory acts; the relation of mount bouts to theta activity is not yet worked out.)

Almost by definition, an ejaculation is a consummatory response—since it terminates an organized sequence of behavior. According to a conception of ejaculation as consummatory behavior, an ejaculation should be able to serve as reinforcement in a conditioning paradigm. To test this theory, Silberberg and Adler (1974) conditioned a group of male rats by removing the female after the seventh intromission. (In our laboratory male rats typically require ten intromissions to ejaculate). After several weeks, the males learned to ejaculate on or before the seventh intromission. We interpreted these results as showing that ejaculation could act as a reinforcement and could cause the male rat to ejaculate sooner than he normally would without the experimental manipulation.

The desynchronization of hippocampal theta following intromission as well as ejaculation implied that the intromission might also be considered a form of consummatory behavior. This view is supported by the fact that intromissions can act as reinforcement for male rats and support both learning and the continuation of mating activities, while the opportunity to mount without intromission does not do so (Whalen, 1961).

5.1.2. *Inhibition*

The data presented above support the notion that some aspects of the male rat's copulatory behavior are under the control of an arousal (or appetitive) process. There are other behavioral events, however, which seem to be produced by an inhibitory process. The concept of an active inhibition has most often been applied to the occurrence of the postejaculatory refractory period (PEI). Recall that the sperm transport in the female rat is highly susceptible to disruption in the period immediately following ejaculation. Hence a quiescent period on the part of the male is adaptive. It has been assumed that sexual arousal is depressed during the period following ejaculation (Beach, 1956; Beach and Jordan, 1956); however, in recent years, it has become increasingly popular to talk in terms of an active postejaculatory inhibition (Kurtz and Adler, 1973; McGill, 1965; Sachs and Barfield, 1976). According to the opponent-process model proposed by Kurtz (Kurtz and Adler, 1973), when a male and female rat are placed together, stimuli from the female generate sexual arousal in the male until he begins to copulate. Subsequent copulatory activity leads to an increasing rate of growth of this excitation. It has also been proposed that activity within the arousal system induces a competing sexual inhibitory process, which increases in strength, as does the arousal process, throughout the series of events that ultimately results in ejaculation. Upon ejaculation, the arousal process falls sharply to minimal values, leaving a net inhibition. Copulatory behavior is thereby restrained. As the postejaculatory period develops, arousal begins to build again until the male rat is capable of resuming copulation (thus terminating the postejaculatory interval). The increase in inhibition through the ejaculatory series was reflected behaviorally by the increased duration of "rest" (Dewsbury, 1967) as the animal approached ejaculation and thereafter in the initial portions of the postejaculatory interval. Electrophysiologically, there was a parallel increase in the amount of cortical and hippocampal "spindling." Gluck and Rowland (1959) have suggested a relationship between EEG synchronization and spindling, like that reported here, and behavioral inhibition.

The development of an active behavioral inhibition following ejaculation is adaptive since successful sperm transport in the female requires a period of rest, uninterrupted by cervical stimulation. The period required for the transport of sperm from the vagina into the uterus is approximately 6–8 min, the same magnitude as the postejaculatory interval (Matthews and Adler, 1977).

In addition to the data on EEG spindling, which support the notion of an active inhibition, there are behavioral data that also strengthen this interpretation. After ejaculation, male rats emit a characteristic 22-kHz vocalization (Barfield and Geyer, 1975). These authors suggest that the calling reflects a sexual absolute refractory period. When the EEG and vocalizations were recorded simultaneously, a correlation of approximately .95 was found between the occurrence of the "inhibitory" EEG spindles and the emission of the 22-kHz vocalization (Barfield and Geyer, 1975). In an extention of Barfield's and co-workers work on the relationship of ultrasonic calling and sexual refractoriness, we recorded the occurrence of this call in the period immediately *preceding* ejaculation (Anisko *et al.*, unpublished). Although the length of the individual pre-ejaculatory "calling bouts" were not as long as those occurring after ejaculation, all 7 rats in this study exhibited *some* preejaculatory calling; and 45% of all ejaculatory series contained some preejaculatory calling. The occurrence of the preejaculatory calling is consistent with the Kurtz opponent-process model of sexual excitation, which postulated the growth of sexual inhibition to substantial levels before ejaculation occurred.

From all of these data, it seems that the functionally significant sexual refractory period in male rats that follows ejaculation is not due to a passive fatigue or to a simple diminution in arousal, but rather it is the product of an active inhibition. Although the gross motor movements of the male are reduced during this period, electrophysiologically there are signs of active inhibition. Furthermore, he is not really behaviorally quiescent—he is vocalizing at 22 kHz. Recently, we discovered that the male rat is also urinating at this time (Adler, unpublished). Six male rats were each given three ejaculatory series. Four of them released significant amounts of urine in the three postejaculatory intervals (1.4 cc on the average). The mean amount of urine released by these four animals over the 24 hr spanning test was 18.8 cc. These four males released, on the average, 26% of their daily output of urine in the 30 min that comprised three post-ejaculatory intervals. If male rats released urine at a constant rate over the 24-hr day, the thirty minutes of sexual refractoriness should account for only 2.1% of the daily output. These urine measurements were made with the rats housed in small testing cages when we tested the animals. Recording urination in a large seminatural environment (cf. Section 4), we found that the male rats tended to return to the same spot after successive ejaculations and that they would urinate in that spot (Anisko *et al.*, in press). It is not known at this time whether this micturition serves a signaling function; it does reinforce the notion, however, that during his sexual refractory state, the male rat is neither passive nor inactive.

5.2. Female Behavior

In the preceding paragraphs, data on the *mechanisms* controlling sexual responsiveness in the male rat were presented. These data were organized in order to emphasize that there were both arousal and inhibitory processes controlling the male rat's copulatory sequence. Data on the *function* of the male's behavior also reinforced the notion that there are both excitatory and inhibitory effects of stimulation on the induction of pregnancy of the female.

5.2.1. *Arousal*

A pattern of multiple mechanisms is also found in the sexual behavior of the female rat. There is, first, an arousal mechanism which leads her to solicit mounts from the male (Section 4). The aroused male in turn pursues the female, mounts her, and intromits. Stimulation derived from the intromission further augments her responsiveness and thereby makes her more receptive to subsequent mounts by the male. In one study, in which the experimenter delivered approximately 2 sec of mechanical stimulation to the vagina and cervix of female rats, the probability of a female's displaying lordosis to a male rat was increased for several hours after the stimulation (Rodriguez-Sierra *et al.*, 1975).

5.2.2. *Inhibition*

The male's mount elicits the lordosis reflex (Pfaff *et al.*, 1973), which is a hormone-sensitive reflex. As already mentioned, the occurrence of this reflex is necessary if the male is to penetrate the vaginal orifice and provide the cervical stimulation necessary for pregnancy. Komisaruk (1974) suggests that lordosis is an extensor reflex, which is the classical approach reflex. The occurrence of this reflex would stabilize the female in her position on the substrate and maximize the stimulation from penile intromission. Superimposed upon the reflex mechanisms responsible for the "elementary" lordosis response are other mechanisms that would potentiate the cervical stimulation the female receives during her lordosis. First, cervical contact during the intromission potentiates the lordosis reflex (Diakow, 1970). Such stimulation also induces a momentary behavioral inhibition on the part of the female and blocks the transmission of painful stimuli; both of these phenomena would again keep the female in place and increase the duration of the cervical stimulation necessary for the induction of pregnancy (Komisaruk, 1974). It seems adaptive that the female rat should be inhibited from moving and should be less responsive to certain classes of exteroceptive, sexually irrelevant, stimulation while receiving an intromission.

The momentary inhibition of locomotion and the blockade of painful stimulation seemed to imply that the female was in a "relaxed" or parasympathetic state during cervical stimulation. The observation that a female rat "goes limp" in the experimenter's hands when glass rod stimulation is applied to the cervix reinforced this notion of parasympathetic activity (Adler, 1974). To examine more precisely the nature of the autonomic response to cervical stimulation, Henry Szechtman, Barry Komisaruk, and I have begun a series of studies of studies which recorded changes in pupillary diameter of female rats as a function of cervical stimulation during mating (Szechtman *et al.*, unpublished). Although the interaction of parasympathetic and sympathetic systems is complex and varies from one target organ to another, the pupillary muscles of the eye represent a simple but classic bioassay for sympathetic and parasympathetic interaction. The musculature of the iris consists of two muscles, the sphinctor (*parasympathetically* innervated) and the dilator (*sympathetically* innervated). The aperture of the iris, the pupil, is controlled by the relative states of contraction of these two muscles. A relatively contracted pupil is produced by parasympathetic activity (and/or inhibition of the sympathetic input to the iris muscle) while dilation is produced by sympathetic stimulation (and/or parasympathetic inhibition) (Westheimer, 1968). In our study of pupillary changes as a function of copulatory stimulation, male and female rats were placed in a Plexiglas mating cage. During the fifteen minute adaptation period, the female was brought to one side of the cage every minute. At this spot, a Zeiss operating microscope was focused. The image was conveyed through the microscope to a video television camera, and the image was recorded at high speed (30 frames per second) onto a video taperecorder. (These measurements provided a baseline.) Following the adaptation period, a male was introduced into the cage and mating allowed to begin. Following each mount, intromission, and ejaculation, the diameter of the pupil was recorded on the videotape (following ejaculation, measurements were taken once a minute for ten minutes). Pupillary diameter was expressed as a percentage change over the precopulatory baseline. Following mounts without penile intromission, there was a 24% increase over baseline for the twelve females in this study. Following intromission, there was a 40% increase; and following ejaculation, there was a 54% increase. (During the postejaculatory interval, there seemed to be a rebound effect: the pupil showed an 11% decrease in diameter over baseline.) Control stimulation—tail pinch—did not produce a significant increase in pupillary diameter.

Instead of finding a parasympathetic response of the eye's musculature, we found the opposite—extreme sympathetic activity. Although other organ systems could presumably show parasympathetic activity during cervical stimulation, this "ocular assay," at least, indicated a high degree of sympathetic arousal during copulation.

Since cervical stimultaion is critically involved in the induction of pregnancy, we tried to correlate the magnitude of the pupillary response with the probability of successful pregnancy following copulation in these tests. The 12 females in the study were divided into two groups: in the one group of six females, the male was pulled off immediately upon ejaculation. For the other six, the male was allowed to remain for the complete "ejaculatory embrace." Since the probability of inducing progesterone secretion and the probability of sperm transport are both dependent on optimal levels of cervical stimulation (Matthews and Adler 1977), we expected to find differences in the rate of pregnancy. Of the six females whose male partners were removed immediately following the ejaculatory response, only one became pregnant (17%). Of the six females that were permitted a complete ejaculatory response, five became pregnant (83%). The average increase in the size of the pupil for females in the first group was only 33% while those in the latter group showed a 75% increase. Utilizing the iris musculature as an assay in this experiment, the level of sympathetic activation (and/or parasympathetic inhibition) reflected the intensity of genital stimulation and was closely correlated with the probability of pregnancy.

Taken together, the immediate consequences of cervical stimulation seem a bit "paradoxical": the female's gross motor activity is inhibited, sensitivity to pain is blocked, cortical EEG is "aroused," the pupillary response indicates sympathetic activation. The animal seems to be in a state of highly activated inhibition. This constellation of events bears a strong resemblance to the state of "paradoxical sleep" (Cohen, 1977); while the active inhibition observed during mating-elicited lordosis may not be identical to the state of PS periodically observed in a sleeping animal, the initial similarity warrants further investigation.

The research discussed in these two sections has emphasized the interaction between arousal and inhibition in producing both male and female behavioral sequences, which are, in turn, adaptively integrated into the process by which pregnancy is initiated in rats. These interactions are schematized and summarized in Figure 5.8. There is one final aspect of inhibition that is applicable to the females of many species, i.e., the inhibition of receptivity following copulation. Unlike the males of non-monogamously mating species, it is not always adaptive for the female to continue mating after she received the male's ejaculation (Adler, 1977). Once successful mating has occurred, it seems natural that subsequent copulatory activity (which, in rats, might disrupt the pregnancy set up by the preceding mating) be prevented. The kind of cortical spindling that was described in male rats (Kurtz and Adler, 1973) has also been found in female rats (Kurtz, 1975). In the female this develops as she approaches sexual satiety. A diminution of receptivity following copulation is found not only in rats but has been found in females of a wide variety of species including hamsters, guinea pigs, turkeys, lizards, fruit flies, and some cockroaches (Barth, 1968; Carter and Schein, 1971; Crews, 1973; Goldfoot and Goy, 1970; Hardy and DeBold, 1972; Manning, 1967; Schein and Hale, 1965). This inhibition of receptivity is a phylogenetically widespread occurrence, presumably reflecting a biologically important behavioral mechanism.

6. *Maternal Arousal*

The overall organization of this essay has been provided by sequentially treating the major epochs in the reproductive life cycle of a "typical" female mammal: early sexual development, the attainment of puberty, ovarian cyclicity, and the induction of pregnancy. In this discussion, a major theme has been the way in which behaviorally derived stimulation affects physiological processes. To conclude the sequence of reproductive epochs, I would like to present some recently collected data concerning the interaction between behavioral stimulation and the maternal state in female rats (Zoloth and Adler, unpublished). The results of this experiment reinforce the basic concept that behavior can affect physiology, but these results also provide an illustration of another principle, viz. the response to exteroceptive stimulation is a function of the physiological–motivational state of the recipient organism.

In this experiment, three classes of stimuli (pups' calls, a "neutral" tone, and somatic shock) were presented to sleeping female rats. There were two variables that affected the frequency of arousal: (1) the stage of sleep during which stimulation was presented [either slow wave sleep (SWS), or paradoxical sleep (PS)] and (2) the motivational condition of the animal (either maternal female rats or virgin female rats).

Some of the results of this study are presented in Figure 5.9. The pup's call was produced by holding a 5–10-day-old pup approximately 2 ft above the subject, and pinching the pup for two seconds. This caused the pup to emit distress vocalizations (both audible and ultrasonic). For the tone trials, a loudspeaker was placed approximately 2 ft above the sleeping rat. A 2-sec tone of 3 kHz was presented (the intensity of the stimulus measured at the rat's head was approximately 20 dB above ambient noise—80 dB re 20 N/m^2).

In this study, the five maternal female rats (whose litters had been removed shortly before testing) showed a greater frequency of arousal than did the five virgin females. (The greater arousability of maternal females occurred from both SWS and PS.) We interpret these data as demonstrating that the motivational condition of the organism (maternal state) can make her more responsive to an ecologically relevant stimulus (pups' calls).

This heightened arousability in maternal females compared to virgins, however, might not be specific to pup-related stimuli but might be attributable instead to a generalized increase in responsiveness to virtually all environmental input. The results from presenting the "neutral" tones, however, did not support this hypothesis of generalized arousal. There were no statistically significant differences in arousability between the females in the two motivational states in response to the 2-sec tone presentations. The virgin

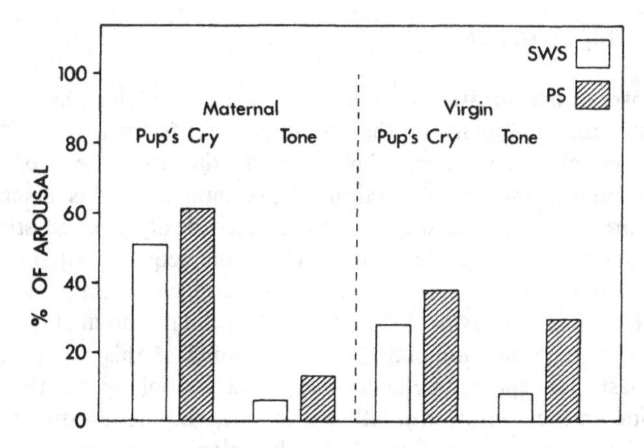

Figure 5.9. The effects of the stimulus (pup's cry versus "neutral" tone") and motivational state (maternal versus virgin) on the frequency of arousal from sleep. The responses from both slow wave sleep (SWS) and paradoxical sleep (PS) were measured.

females were, if anything, more arousable in response to tones than were the animals in the maternal state.

It is important to note that we did not intend the tone to serve as a totally comparable control condition for the pups' cry. Although the intensity of the tone was adjusted so that it would be 50 dB above threshold for this frequency (Gourevitch and Hack, 1966), the tone was not intended to match the frequency, patterning, or even intensity of the pups' vocalization. The question asked was simply, "are maternal females more responsive than virgin females to any auditory stimuli in their environment?" The answer was "no."

In this study, the sleeping organism was used as an assay; its purpose was to evaluate the effectiveness of environmental stimulation as a function of motivational state. An additional result of this work, however, demonstrated an interesting property of the stage of sleep. A classic finding is that paradoxical sleep (PS) is a "deeper" sleep than slow wave sleep (SWS), i.e., it is more difficult to arouse an animal from the former state than the latter (cf. review by Cohen, 1977). In this study, however, we found that for both virgin and maternal female rats and for both pups' calls and tones, paradoxical sleep was not deeper than SWS. On the contrary, the frequency of arousal from PS was generally greater.

We *were* able to replicate the classical finding that PS is "deeper" sleep than SWS when we used stimuli from another modality, i.e., somatic shock delivered to the feet. With this stimulus, the three female rats tested were aroused more frequently from SWS than PS. This finding further reinforces the general conclusion from this study, i.e., that the physiological response of an organism (in this case arousal from sleep) is a function of both the nature of the exogenous stimulation and the physiological-motivational state of the organism at the time of stimulation.

7. Retrospect and Prospect

The main purpose of this volume is to pay tribute to a great scientist, Professor Frank A. Beach. The manner of accomplishing this very pleasant task has been to have those of us fortunate enough to have been Frank's students contribute a sample of our current work and thought, thereby indicating the way in which our friend and teacher has influenced us and our work. For the sake of temporal symmetry, we were asked to engage in some secular prophecy and try to predict the directions in which our individual fields of scientific enquiry are going, or should be going.

Such a prospectus is highly personal and is even more idiosyncratic than the flavor of past research. However, in this final section, I would like to point out three directions in which I hope studies of behavioral endocrinology will proceed. The main philosophical and strategic concern of the work presented in this essay has been to demonstrate that behavior, and the stimulation derived from behavior, is critically important in controlling physiological processes involved in reproduction. Like stimulation from the physical environment, stimulation from the organism's social environment influences reproduction. Sometimes this influence takes the form of "modulating" the physiological variable (e.g., advancement of breeding, acceleration of puberty). Sometimes the effect is stronger in that behaviorally derived stimulation "triggers" an event that would not otherwise occur (the evocation of ovulation, induction of pregnancy). In all of these cases, however, one is led to consider behavior as a *biological adaptation,* which has been shaped by evolutionary-genetic forces into the pattern that it takes in presently living members of the species.

7.1. The Study of Adaptations

If one seriously investigates behavior as an adaptation, there must be an increased emphasis on studying behavior "in the real world." The broad-based species comparison that Dewsbury and his colleagues have begun are critically important in this regard. In addition, I would hope that more investigators would move out of the laboratory and into the field so that the putatively adaptive forms of behavior can be studied under the ecological conditions in which they are normally generated.

One should possibly combine such studies of behavioral taxonomy with an evaluation of their natural "survival value," possibly in the classic form that Tinbergen and students investigated the survival function of certain patterns of gull's behavior.

The study of behavioral adaptations also leads to an examination of the genetic machinery which has been modified through natural selection and now controls the development of the behavior in individual organisms. The work of McGill, discussed in this volume, illustrates this general approach. I would also hope that there will be more work on the developmental genetics of these reproductively adaptive patterns.

7.2. *Extension to the Human Animal*

The behavioral control of reproductive processes is the basic paradigm of psychosomatics (i.e., a behavioral or psychological state influencing physiological organization). It may be time for a rapprochement between studies of behavioral endocrinology in animals and clinical psychosomatics in humans. Psychogenic amenorrhea, false pregnancy (pseudocyesis), and a host of other reproductive psychosomatic syndromes in humans may be either physiologically homologous to similar phenomena in nonhuman mammals or at least formally analogous to them. In either case (homology or analogy) the systematic development of animal models for human psychosomatic syndromes should be profitable.

7.3. *Behavioral Control of Physiological Processes as a Research Tool*

The research presented in this chapter often treats behavior as the independent variable and physiology as the dependent variable. This "psychosomatic" paradigm is, of course, only one way of relating behavior and physiology. The other dominant paradigm in current research is the opposite: examine the neuroendocrine system of animals as the cause, and the resulting behavior as the effect. Even when operating with this paradigm, it is often useful to use behavioral stimulation as a technique for unravelling the complexities of reproductive physiology. Recall that high levels of progestational hormone secretion in the female rat last for 18 days; this major physiological change is produced by only three seconds of copulatory stimulation from the male. If one were interested in progesterone metabolism, or the neurochemical stimulation of pituitary secretion, the proper application of behavioral stimulation could provide a powerful tool for controlling the physiological events being investigated.

In conclusion, I find the investigation of behavioral influences on reproductive function to be an important technique for integrating many of the subdivisions of the life sciences: behavior, ecology, endocrinology, neurophysiology, genetics, and evolution. My own personal philosophy of science—largely derived from the example and teaching of Frank Beach—is that behaviorists have a special obligation to continue emphasizing the fact that the organism, its environment, and its physiological machinery are all integrated into a coherent and biologically adaptive system. By treating behavior as the independent variable, both physiological mechanism and evolutionary-ecological factors are kept in perspective. Although we may use highly specialized research techniques for a specific experiment we should remain, at least some of the time, "general practitioners" in the biological and behavioral sciences.

References

Adler, N. T. Effects of the male's copulatory behavior in the initiation of pregnancy in the female rat. *Anatomical Record*, 1968, *160*, 305.

Adler, N. T. Effects of the male's copulatory behavior on successful pregnancy of the female rat. *Journal of Comparative and Physiological Psychology*, 1969, *69*(4), 613–622.

Adler, N. T. The behavioral control of reproductive physiology. In W. Montagna and W. A. Sadler (Eds.), *Reproductive behavior*. New York: Plenum Press, 1974.

Adler, N. T. On the mechanisms of sexual behavior and their evolutionary constraints. In J. B. Hutchison (Ed.), *Biological determinants of sexual behavior*. New York: Wiley, 1977.

Adler, N., and Barfield, R. The correlation of behavioral inhibition, ultrasonic calling, and urination following ejaculation in the rat, unpublished.

Adler, N. T., and Bermant, G. Sexual behavior of male rats: Effects of reduced sensory feedback. *Journal of Comparative and Physiological Psychology*, 1966, *61*, 240–243.

Adler, N. T., and Zoloth, S. R. Copulatory behavior can inhibit pregnancy in female rats. *Science*, 1970, *168*, 1480–1482.

Adler, N. T., Davis, P. G., and Komisaruk, B. K. Variation in the size and sensitivity of a genital sensory field in relation to the estrous cycle in rats, *Hormones and Behavior*, (in press).

Adler, N. T., Resko, J. A., and Goy, R. W. The effect of copulatory behavior on hormonal change in the female rat prior to implantation. *Physiology and Behavior*, 1970, *5*, 1003–1007.

Alleva, J. J., Waleski, M. V., and Alleva, F. R. The Zeitgeber for ovulation in rats: Non-participation of the pineal gland. *Life Science*, 1970, *9*, 241–246.

Anisko, J. J., Suer, S. F., McClintock, M. K., and Adler, N. T. The relationship between 22 kHz ultrasonic signals and sociosexual behavior in the rat, *Journal of Comparative and Physiological Psychology* (in press).

Antelman, S. M., and Caggiula, A. R. Norepinephrine–dopamine interactions and behavior. *Science*, 1977, *195*, 646–654.

Aron, C., Asch, G., and Roos, J. Triggering of ovulation by coitus in the rat. *International Review of Cytology*, 1966, *20*, 139–172.

Barfield, R. J., and Geyer, L. A. The ultrasonic postejaculatory vocalization and the postejaculatory refractory period of the male rat. *Journal of Comparative and Physiological Psychology*, 1975, *88*, 723–734.

Barraclough, C. A., and Sawyer, C. H. Induction of pseudo-pregnancy in the rat by reserpine and chloropromazine. *Endocrinology*, 1959, *65*, 563–571.

Barth, R. H., Jr. The comparative physiology of reproductive processes in cockroaches. Part I. Mating behavior and its endocrine control. *Advances in Reproductive Physiology*, 1968, 167–207.

Beach, F. A. Characteristics of masculine sex drive. In M. R. Jones (Ed.), *Nebraska symposium on motivation,* Vol. 4. Lincoln: University of Nebraska Press, 1956, pp. 1-32.
Beach, F. A. Sexual attractivity, proceptivity, and receptivity in female mammals. *Hormones and Behavior,* 1976, *7,* 105-138.
Beach, F. A., and Jordan, L. Sexual exhaustion and recovery in the male rat. *Quarterly Journal of Experimental Psychology,* 1956, *8,* 121-133.
Bereiter, D. A., and Barker, D. J. Facial receptive fields of trigeminal neurons: Increased size following estrogen treatment in female rats. *Neuroendocrinology,* 1975, *18,* 115-124.
Bermant, G. Sexual behavior of male rats: Photographic analysis of the intromission response. *Psychonomic Science,* 1965, *2,* 65-66.
Besch, E. L. Activity responses to altered photo-periods. *Aerospace Medicine,* 1969, *40,* 1111-1114.
Blandau, R. J. On the factors involved in sperm transport through the cervix uteri of the albino rat. *American Journal of Anatomy,* 1945, *77,* 253-272.
Bronson, F. H., and Stetson, M. H. Gonadotropin release in prepubertal female mice following male exposure: A comparison with the adult cycle. *Biology of Reproduction,* 1973, *9,* 449-459.
Browman, L. G. Light in its relation to activity and estrous rhythms in the albino rat. *Journal of Experimental Zoology,* 1937, *75,* 375-388.
Brown, J. L. *The evolution of behavior.* New York: Norton, 1975.
Brown-Grant, K., Davidson, J. M., and Grieg, F. Induced ovulation in albino rats exposed to constant light. *Endocrinology,* 1973, *57,* 1-16.
Burnet, F. R., and MacKinnon, P. C. B. Restoration by oestradiol benzoate of a neural and hormonal rhythm in the ovariectomized female. *Journal of Endocrinology,* 1975, *64,* 27-35.
Calhoun, J. B. A "behavioral sink." In E. L. Bliss (Ed.), *Roots of behavior.* Washington: Department of Health, Education and Welfare, 1962a, pp. 295-315.
Calhoun, J. B. The ecology and sociology of the Norway rat. United States Public Health Service Publication No. 1008., 1962b.
Caligaris, L., Astrada, J. J., and Taleisnik, S. Release of luteinizing hormone induced by estrogen injection into ovariectomized rats. *Endocrinology,* 1971, *88,* 810-815.
Carlsson, R. R., and De Feo, V. J. Role of the pelvic nerve vs. the abdominal sympathetic nerves in the reproductive function of the female rat. *Endocrinology,* 1965, *77,* 1014-1022.
Carlsson, S. G., and Larsson, K. Mating in male rats after local anesthetization of the glans penis. *Zeitschrift fur Tierpsychologie,* 1964, *21,* 854-856.
Carter, C. S., and Schein, M. W. Sexual receptivity and exhaustion in the female golden hamster. *Hormones and Behavior,* 1971, *2,* 191-200.
Chen, C. L., and Besch, E. L. Effect of altered photo-period on estrous cycle, pregnancy, and onset of puberty in the rat. *Biology of Reproduction,* 1975, *12,* 396-399.
Cohen, D. B. *Sleep and dreaming: Origin, nature and functions.* Oxford: Pergamon Press, 1977.
Craig, W. Appetites and aversions as constituents of instincts. *Biological Bulletin,* 1918, *34,* 91.
Crews, D. Coition-induced inhibition of sexual receptivity in female lizards (*Anolis carolinensis*). *Physiology and Behavior,* 1973, *11,* 463-468.
Cross, P. Observations on the induction of ovulation in *Microtus montanus. Journal of Mammology,* 1972, *53,* 210-211.
Davis, H. N., Gray, G. D., Zerylnick, M., and Dewsbury, D. A. Ovulation and implantation in montane voles (*Microtus montanus*) as a function of varying amounts of copulatory stimulation. *Hormones and Behavior,* 1974, *5,* 383-388.
Davis, H. N., Gray, G. D., and Dewsbury, D. A. Maternal age and male behavior in relation

to successful reproduction by female rats (*Rattus norvegicus*). *Journal of Comparative and Physiological Psychology*, 1977, *91*, 281–290.

DeCoursey, P. J. LD ratios and the entrainment of circadian activity in a nocturnal and adiurnal-rodent. *Journal of Comparative Physiology*, 1972, *78*, 221–235.

Dewsbury, D. A. A quantitative description of the behavior of rats during copulation. *Behavior XXIX*, 1967, *2-4*, 154–178.

Dewsbury, D. A. Diversity and adaptation in rodent copulatory behavior. *Science*, 1975, *190*, 947–955.

Diakow, C. Effects of genital desensitization on the mating pattern of female rats as determined by motion picture analysis. *American Zoologist*, 1970, *10*, 486.

Diakow, C. Motion picture analysis of rat mating behavior. *Journal of Comparative and Physiological Psychology*, 1975, *88*, 704–712.

Diamond, M. Intromission pattern and species vaginal code in relation to induction of pseudopregnancy. *Science*, 1970, *169*, 995–997.

Diamond, M. Vaginal Stimulation and progesterone in relation to pregnancy and parturition. *Biology of Reproduction*, 1972, *6*, 281–287.

Diamond, M., and Yanagimachi, R. Induction of pseudopregnancy in the golden hamster. *Journal of Reproduction and Fertility*, 1968, *17*, 165–168.

Doty, R. L. *Mammalian olfaction, reproductive processes, and behavior.* New York: Academic Press, 1976.

Edmonds, S. C., and Adler, N. T. Food and light as entrainers of circadian running activity in the rat. *Physiology and Behavior*, 1977, *18*, 915–919.

Edmonds, S. C., and Adler, N. T. The multiplicity of biological oscillators in the control of circadian running activity in the rat. *Physiology and Behavior*, 1977, *18*, 921–930.

Edmonds, S., Zoloth, S. R., and Adler, N. T. Storage of copulatory stimulation in the female rat. *Physiology and Behavior*, 1972, *8*, 161–164.

Everett, J. W. Central neural control of reproductive functions of the adenophypophysis. *Physiological Review*, 1964, *44*, 373–431.

Everett, J. W. Delayed pseudopregnancy in the rat, a tool for the study of central neural mechanisms in reproduction. In M. Diamond (Ed.), *Perspectives in reproduction and sexual behavior.* Bloomington and London: Indiana University Press, 1968.

Everett, J. W., and Sawyer, C. H. A 24-hour periodicity in the LH-release apparatus of female rats, disclosed by barbiturate sedation. *Endocrinology*, 1950, *47*, 198–218.

Gerall, A. A., Napoli, A. M., and Cooper, U. C. Daily and hourly estrous running in intact, spayed, and estrone implanted rats. *Physiological Behavior*, 1973, *10*, 225–229.

Gluck, H., and Rowland, V. Defensive conditioning of electrographic arousal with delayed and differentiation and auditory stimuli. *EEG Clinical Neurophysiology*, 1959, *11*, 485–496.

Goldfoot, D. A., and Goy, R. W. Abbreviation of behavioral estrus in guinea pigs by coital and vagino-cervical stimulation. *Journal of Comparative and Physiological Psychology*, 1970, *72*, 426–434.

Goldfoot, D. A., and Wallen, K. Development of gender role behaviors in heterosexual and isosexual groups of infant rhesus monkeys. In D. J. Chivers (Ed.), *Recent advances in primatology-behaviour, Vol. 1, Proceedings of the International Primatological Society.* New York: Academic Press, 1976.

Gorbman, A., and Bern, H. A. *A textbook of comparative endocrinology.* New York: Wiley, 1962.

Gourevitch, G., and Hack, M. Audibility in the rat. *Journal of Comparative and Physiological Psychology*, 1966, *62*, 280–291.

Gray, G. D., Davis, H. N., Zerylnick, M., and Dewsbury, D. A. Oestrus and induced ovulation in montane voles. *Journal of Reproduction and Fertility*, 1974, *38*, 193–196.

Gray, G. D., Davis, H. N., Kenney, A. McM., and Dewsbury, D. A. Effect of Mating on plasma levels of LH and progesterone in montane voles (*Microtus montanus*). *Journal of Reproduction and Fertility*, 1976, *47*, 89–91.

Hardy, D. F., and DeBold, J. F. Effects of coital stimulation upon behavior of the female rat. *Journal of Comparative and Physiological Psychology*, 1972, *78*, 400–408.

Hemmingsen, A. M., and Krarup, N. B. Rhythmic diurnal variations in the oestrous phenomena of the rat and their susceptibility to light and dark. *Kongelige Danske Videnskabernes Selskab Biologiske Meddleelson*. 1937, *13*, 1–61.

Hoffmann, J. C. Effects of light deprivation on the rat estrous cycle. *Neuroendocrinology*, 1967, *2*, 1–10.

Hoffmann, J. C. The influences of photoperiods on reproductive functions in female mammals. In R. O. Greep (Ed.), *Physiology of the female reproductive tract. Handbook of physiology section on endocrinology*, Vol. II. Washington, D.C.: American Physiological Society, 1973, pp. 57–77.

Jolly, A. Breeding synchrony in wild *Lemur catta*. In S. Altman (Ed.), *Social communication among primates*. Chicago: University of Chicago Press, 1967.

Kollar, E. J. Reproduction in the female rat after pelvic nerve neurectomy. *Anatomical Record*, 1953, *115*, 641–658.

Komisaruk, B. R. Neural and hormonal interactions in the reproductive behavior of female rats. In W. Montagna and W. A. Sadler (Eds.), *Reproductive Behavior*. New York: Plenum Press, 1974.

Komisaruk, B. R., Adler, N. T., and Hutchison, J. Genital sensory field: Enlargement by estrogen treatment in female rats. *Science*, 1972, *178*, 1295–1298.

Kow, L. M., and Pfaff, D. W. Effects of estrogen treatment on the size of receptive field and response threshold of pudendal nerve in the female rat. *Neuroendocrinology*, 1973/74, *13*, 299–313.

Kurtz, R. G. Hippocampal and cortical activity during sexual behavior in the female rat. *Journal of Comparative and Physiological Psychology*, 1975, *89*, 158–169.

Kurtz, R. G., and Adler, N. T. Electrophysiological correlates of copulatory behavior in the male rat: Evidence for a sexual inhibitory process. *Journal of Comparative and Physiological Psychology*, 1973, *84*, 225–239.

Legan, S. J., Coon, G. A., and Karsch, F. J. Role of estrogen as initiator of daily LH surges in the ovariectomized rat. *Endocrinology*, 1975, *96*, 50–56.

Lisk, R. D. Cyclic fluctuations in sexual responsiveness in the male rat. *Journal of Experimental Zoology*, 1969, *171*, 313–326.

Long, J. A., and Evans, H. M. The estrous cycle in the rat and its associated phenomena. *Memoirs of the University of California*, 1922, *6*, 1–128.

Manning, A. The control of sexual receptivity in female *Drosophilia*. *Animal Behavior*, 1967, *15*, 239–250.

Matthews, M., and Adler, N. T. Facilitative and inhibitory influences of reproductive behavior on sperm transport in rats. *Journal of Comparative and Physiological Psychology*, 1977, *92*, 727–742.

Matthews, M., and Adler, N. T. The relationship between mating behavior, the vaginal plug, and sperm transport in the rat, unpublished.

McClintock, M. Sociobiology of reproduction in the Norway rat (*Rattus norvegicus*). Estrous synchrony and the role of the female rat in copulatory behavior. Unpublished doctoral dissertation, University of Pennsylvania, 1974.

McClintock, M. K., and Adler, N. T. The sexual behavior of the wild and domestic Norway rat: II. Effect of cage size and observation techniques, unpublished.

McClintock, M. K., and Adler, N. T. The sexual behavior of the wild and domestic Norway rat: I. The role of the female and the effect of domestication, unpublished.

McGill, T. E. Induction of luteal activity in female house mice. *Hormones and Behavior*, 1970, *1*, 211–222.

McGill, T. E. Studies of the sexual behavior of male laboratory mice: Effects of genotype, recovery of sex drive, and theory. In F. A. Beach (Ed.), *Sex and behavior*. New York: Wiley, 1965.

Mennin, S. P., and Gorski, R. A. Effects of ovarian steroids on plasma LH in normal and persistent estrous adult female rats. *Endocrinology*, 1975, *96*, 486–491.

Michael, R. P. Determinants of primate reproductive behavior. *Acta Endocrinology*, 1972, *166*, 322–363.

Mittwoch, U. *Sex differentiation*. New York: Academic Press, 1973.

Money, J., and Erhardt, A. A. *Man and woman, boy and girl*. Baltimore: Johns–Hopkins University Press, 1972.

Morishige, W. K., Pepe, G., and Rothchild, I. Serum luteinizing hormone, prolactin and progesterone levels during pregnancy in the rat. *Endocrinology*, 1973, *92*, 1527–1530.

Pencharz, R. I., and Long, J. A. Hypophysectomy in the pregnant rat. *American Journal of Anatomy*, 1933, *53*, 117–135.

Pfaff, D. W., and Lewis, C. Film analyses of lordosis in female rats. *Hormones and Behavior*, 1974, *5*, 317–335.

Pfaff, D., Lewis, C., Diakow, C., and Keiner, M. Neurophysiological analysis of mating behavior responces as hormone sensitive reflexes. In E. Stellar and J. Sprauge (Eds.), *Progress in physiological psychology*, Vol. 5. New York: Academic Press, 1973.

Pittendrigh, C. S. Circadian oscillations in cells and the circadian organization of multicellular systems. In R. C. Schmitt and F. G. Worden (Eds.), *The neurosciences*, Third Study Program, Cambridge, MIT Press, 1974, pp. 437–458.

Pittendrigh, C. S., and Daan, S. A functional analysis of circadian pacemakers in nocturnal rodents, IV. Entrainment: Pacemaker as clock. *Journal of Comparative Physiology*, 1975, *106*, 291–331.

Ricklefs, R. E. *Ecology*. Newton, Massachusetts: Chrion Press, 1973.

Robertson, D. R. Social control of sex reversal in a coral-reef fish. *Science*, 1972, *17*, 1007–1009.

Rodgers, C. H. Influence of copulation on ovulation in the cycling rat. *Endocrinology*, 1971, *88*, 433.

Rodgers, C. H., and Schwartz, N. B. Diencephalic regulation of plasma LH, ovulation, and sexual behavior in the rat. *Endocrinology*, 1972, *90*, 461–465.

Rodgers, C. H., and Schwartz, N. B. Serum LH and FSH levels in mated and unmated proestrous female rats. *Endocrinology*, 1973, *92*, 1475.

Rodriguez-Sierra, J. F., Crowley, W. R., and Komisaruk, B. R. Vaginal stimulation in rats induces prolonged lordosis responsiveness and sexual receptivity. *Journal of Comparative and Physiological Psychology*, 1975, *89*, 79–85.

Rowlands, I. W. (Ed.). Comparative biology of reproduction in mammals. Zoological Society of London Symposia, Number 15, 1966.

Rusak, B., and Zucker, I. Biological rhythms and animal behavior. *Annual Review of Psychology*, 1975, *26*, 137–171.

Sachs, B. D., and Barfield, R. J. Functional analysis of masculine copulatory behavior in the rat. In J. S. Rosenblatt, R. A. Hinde, E. Shaw, and C. Beer (Eds.), *Advances in the study of behavior*. New York: Academic Press, 1976.

Sade, D. S. Seasonal cycle in size of testes of free-ranging *Macaca mulatta*. *Folia Primatology*, 1964, *2*, 171–180.

Sadleir, R. M. F. S. (Ed.). *The ecology of reproduction in wild and domestic mammals*. London: Methuen, 1969.

Schein, M. W., and Hale, E. B. Stimuli eliciting sexual behavior. In F. A. Beach (Ed.), *Sex and behavior.* New York: Wiley, 1965.

Schinkel, P. G. The effect of the ram on the incidence and occurrence of oestrus in ewes. *Australian Vetenary Journal,* 1954, *30,* 189–195.

Schwartz, N. B. A model for the regulation of ovulation in the rat. *Record of Progress in Hormone Research,* 1969, *25,* 1–55.

Silberberg, A., and Adler, N. T. Reduction of intromission frequency in male rats by experimental manipulation. *Science,* 1974, *185,* 374–376.

Sinclair, A. N. A note on the effect of the presence of rams on the incidence of oestrus in maiden Merino ewes during spring mating. *Australian Vetenary Journal,* 1950, *26,* 37–39.

Smith, M. S., McLean, B. K., Neill, J. D. Prolactin: The initial luteotropic stimulus of pseudopregnancy in the rat. *Endocrinology,* 1976, *98,* 1370–1378.

Spies, H. G., and Niswender, G. D. Levels of prolactin, LH and FSH in the serum of intact and pelvic-neurectomized rats. *Endocrinology,* 1971, *88,* 937–943.

Szechtman, H., Komisaruk, B., and Adler, N. T. Changes in pupillary diameter of the female rat as a function of cervical stimulation, unpublished.

Thibault, C., Courit, M., Martinet, L., Mauleon, P., DuMesnil, DuBuisson, F., Ortavant, P., Pelletier, J., and Signoret, J. P. Regulation of breeding season and oestrous cycles by light and external stimuli in some mammals. *Journal of Animal Science,* 1966, *25,* 119–142.

Vandenbergh, J. G. Effect of the presence of a male on the sexual maturation of female mice. *Endocrinology,* 1967, *81,* 345.

Vandenbergh, J. G. Endocrine coordination in monkeys: Male sexual responses to females. *Physiology and Behavior,* 1969, *4,* 261–264.

Vandenbergh, J. G., Whisett, J. M., and Lombard, J. Partial isolation of a pheromone accelerating puberty in female mice. *Journal of Reproduction and Fertility,* 1975, *43,* 515–523.

Wang, G. H. The relation between "spontaneous" activity and oestrous cycle in the white rat. *Comparative Psychological Monograph,* 1923, *2,* 1–27.

Weber, A. L., and Adler, N. T. The estrous and activity cycles of rats in light-dark zeitgebers of varying strength, unpublished.

Weizenbaum, F., McClintock, M., and Adler, N. Decreases in vaginal acyclicity of rats when housed with female hamsters. *Hormones and Behavior,* 1977, *8,* 342–347.

Weizenbaum, F. A., Adler, N. T., and Gamjam, V. K. Serum testosterone concentrations in pregnant female rats, unpublished.

Westheimer, G. *The eye.* In V. B. Mountcastel (Ed.), *Medical physiology* 12th ed. Vol. 2, Saint Louis: Mosby, 1968.

Whalen, R. E. Effects of mounting without intromission and intromission without ejaculation on sexual behavior and maze learning. *Journal of Comparative and Physiological Psychology,* 1961, *54*(4), 409–415.

Whitten, W. K., Bronson, F. H., and Greenstein, J. A. Estrus-inducing pheromone of male mice: Transport by movement by air. *Science,* 1968, *161,* 584–585.

Wilson, J. R., Adler, N., and Le Boeuf, B. The effects of intromission frequency on successful pregnancy in the female rat. *Proceedings of the National Academy of Science,* 1965, *53,* 1392–1395.

Zoloth, S. R., and Adler, N. T. Arousal in the female rat: Effects of stimulus relevance and motivational state, unpublished.

6

Genotype–Hormone Interactions

Thomas E. McGill

1. Introduction

I once heard Frank Beach tell Ethel Tobach, one of the editors of *The Biopsychology of Development*, that if he had had more time he could of written a shorter chapter. The chapter in question was Beach's famous "Ramstergig" paper. There Beach noted that while the Ramstergig followed precisely certain current theories concerning the action of gonadal hormones, the rat, hamster, and guinea pig did not. In a recent paper (McGill, 1977), I carried Beach's analysis a step further and discussed instances in which "species differences" could in fact be duplicated when genotype was varied *within* a particular species. My purpose at present is to enlarge upon that theme by describing specific experiments that show that different genotypes within a species differ in such items as hormonal induction of receptivity in females and retention of sexual behavior after castration in males. I shall also discuss the response of certain castrated males to the injection of different amounts of steroid hormones, although in this case within-species differences have not as yet been documented. If the chapter appears to be short, it is not because I had plenty of time to write it, but rather because there is relatively little work that has been done to date in the area of within-species genotype–hormone interactions. Thus the chapter closes with some speculations as to what future research may bring.

Thomas E. McGill • Department of Psychology, Williams College, Williamstown, Massachusetts. Research from the author's laboratory described in this chapter was supported by U.S.P.H.S. Research Grant No. GM-07495.

2. Hormonal Induction of Receptive Behavior in Females

In 1961, W. C. Young reviewed several important species differences in the hormonal treatments required to induce a state of receptivity in females. Previously, Goy and Young (1957) had shown that strains of guinea pigs differed in their response to different amounts of estrogen followed by a constant amount of progesterone. The differences occurred for such items of behavior as latency and duration of maximum lordosis response and male-like mounting behavior. The strain differences persisted over a wide range of estrogen treatments. The results for female guinea pigs supported the findings previously established for male guinea pigs in that suprathreshold quantities of gonadal hormones did not alter the characteristic response of the animals. (In a later section of this chapter we shall present evidence that suprathreshold amounts of androgen can produce large changes in certain temporal measures of sexual response in at least some genotypes of male house mice.) Working with house mice, Thompson and Edwards (1971) demonstrated strain differences in the speed with which females would respond to hormonal treatments. Their results are thus similar to those found by Goy and Young for the guinea pig.

The experiments of Goy and Young and of Thompson and Edwards dealt primarily with response latency and with temporal measurements of behavior after treatment. More recently, Gorzalka and Whalen have demonstrated that the *presence or absence* of response to particular hormones may depend upon the genotype of the animals receiving the hormonal treatments. In their initial experiment (1974) they first injected ovariectomized groups of CD-1 and Swiss–Webster mice with estrogen followed some hours later by progesterone. Weekly tests indicated that both strains gradually increased the proportion of lordotic responses when mounted by male mice. Then, dihydroprogesterone was substituted for progesterone and testing resumed. Once again, CD-1 females progressively increased in responsiveness. However, the Swiss–Webster mice failed to respond to dihydroprogesterone and their ratio of lordotic responses to mounts did not change over five weeks of testing.

In their second study, Gorzalka and Whalen (1976) repeated the experiment with the original two strains and added their reciprocal F_1 hybrids. All four strains showed a progressive increase in lordosis responses as a result of progesterone treatment over six weeks. When dihydroprogesterone was substituted for progesterone, an increase in receptivity occurred for the CD-1 females and for both groups of hybrids. But once again, the Swiss–Webster females failed to respond to this hormone. As a matter of fact, the response level of Swiss–Webster mice receiving dihydroprogesterone was not significantly different from control groups receiving only estrogen and the oil vehicle. Figure 6.1 shows the results for the four groups over the six-week test period.

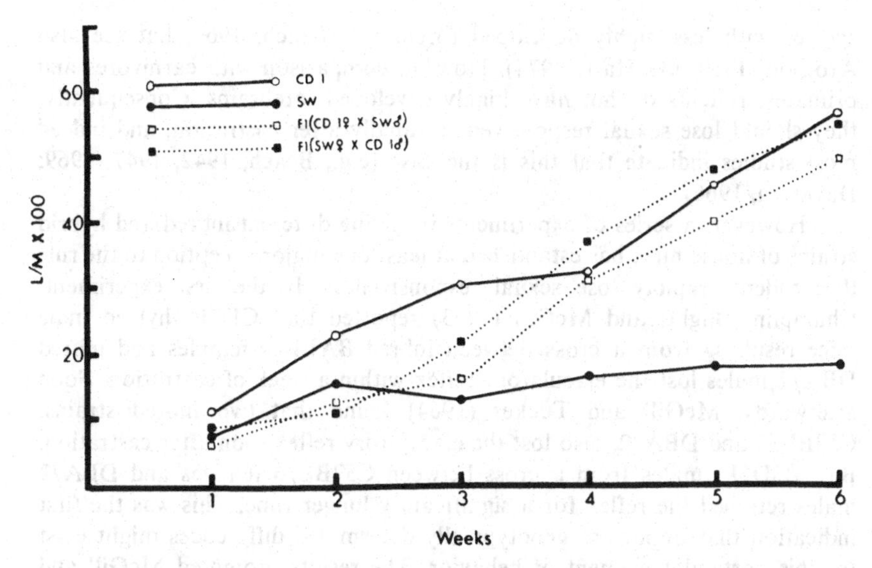

Figure 6.1. Effects of 5α-dihydroprogesterone (DHP) on sexual receptivity (lordosis responses/mounts with pelvic thrusting × 100) in estrogen-primed, ovariectomized Swiss-Webster (SW), CD-1, SWCD1F$_1$, and CD1SWF$_1$ mice. From Gorzalka and Whalen (1976), reprinted by permission of Plenum Press.

Gorzalka and Whalen's experiments show that certain genotypes within a species may differ in the presence or absence of response to particular hormones. The next section reviews experiments demonstrating genotypically determined differences in response to the *absence* of particular hormones.

3. Retention of Postcastration Sexual Responsiveness in Male Mice of Different Genotypes

W. C. Young (1961), considering the response of male mammals to castration, wrote: "There is probably no other phase of the problem of the hormones and mating behavior in which our progress has been so negligible." Young was prompted to make that statement because some males retain sexual responsiveness after castration and because the extent to which they do so varies greatly both between species and within species. The two relatively consistent findings as a result of castrating sexually experienced male mammals are: (1) large individual differences have been found to exist and (2) species with more highly developed forebrains tend to persist in the performance of sexual responses (on the average) for a longer time than do species with less highly developed forebrains (Beach, 1969, but see also Aronson, 1959 and Hart, 1974). Now, in comparison with carnivores and primates, rodents do not have highly developed forebrains. Consequently, they should lose sexual responsiveness rapidly after castration, and indeed most studies indicate that this is the case (e.g., Beach, 1942, 1947, 1969; Davidson, 1966).

However, a series of experiments involving different inbred and hybrid strains of house mice has established at least one major exception to the rule that rodents rapidly lose sexual responsiveness. In the first experiment, Champlin, Blight, and McGill (1963) reported that $CD2F_1$ hybrid male mice resulting from a cross between inbred BALB/c females and inbred DBA/2 males lost the ejaculatory reflex within a week of castration. Soon afterwards, McGill and Tucker (1964) found that two inbred strains, C57BL/6 and DBA/2, also lost the ejaculatory reflex soon after castration. But, $B6D2F_1$ males from a cross between C57BL/6 females and DBA/2 males retained the reflex for a significantly longer time. This was the first indication that important genotypically determined differences might exist for this particular element of behavior. The results prompted McGill and Haynes (1973) to use degree of heterozygosity as an independent variable in studies of retention of the ejaculatory reflex after castration. By examining F_2 and backcross groups from the long-retaining $B6D2F_1$ genotype, they confirmed the results of the McGill and Tucker study and found support for the hypothesis that retention of the ejaculatory reflex after castration is associated with heterozygosity at particular genetic loci.

That was the approximate stage of the development of research in this area when I took a sabbatical leave as a Visiting Scientist at the University of Edinburgh. My colleague and sponsor for that year was Professor Aubrey Manning, a Behavioral Geneticist who had specialized in insect behavior patterns. Manning and I combined the results from the three studies cited above and made specific predictions as to how certain genotypes, heretofore unstudied, should respond in terms of retention of the ejaculatory reflex after castration. First, the fact that hybrid $CD2F_1$, males from a cross between BALB/c females and DBA/2 males quickly lost the ejaculatory reflex suggested that DBA/2 and BALB/c strains did not differ at the relevant genetic loci, with the result that the hybrids remained homozygous. If that were so, then BALB/c males should be similar to DBA/2 males and cease to exhibit the ejaculatory reflex shortly after removal of the gonads. Further, if DBA/2 and BALB/c strains are genetically identical for this particular trait, then males resulting from a cross between BALB/c and C57BL/6 should be similar to males resulting from a cross between DBA/2 and C57BL/6. In other words, these males should retain the ejaculatory reflex for a considerable period of time following castration. To test these hypotheses we used a diallelic design involving the three inbred strains BALB/c, DBA/2, C57BL/6, and hybrids from the six possible crosses among them.

In brief, the method used in our study (McGill and Manning, 1976) was as follows: Males and females of the inbred strains BALB/c (abbreviated C), C57BL/6 (abbreviated B6), and DBA/2 (abbreviated D2) were paired in various combinations to produce the three inbred strains and the six possible hybrid groups. In designating hybrid genotypes, it is customary to place the abbreviation of the strain of the dam before that of the strain of the sire. For example, animals resulting from a cross between C57BL/6 females and BALB/c males are designated $B6CF_1$; those resulting from the reciprocal cross where BALB/c females are dams and C57BL/6 males are sires are designated $CB6F_1$. The other four hybrid groups studied were $B6D2F_1$, $D2B6F_1$, $CD2F_1$, and $D2CF_1$.

In adulthood, ten or eleven animals of each group were given precastration sexual experience that amounted to at least three ejaculations. One day after the last ejaculation the males were castrated. Postcastrational testing was begun two or three days later and continued on a twice-per-week basis until one of two circumstances obtained. First, for certain males, the criterion for loss of the ejaculatory reflex was met. This criterion was five tests without ejaculation with the last four of these tests negative with respect to intromissions and mounts. Other males that were still exhibiting regular sexual responses six months to one year after castration were placed on an irregular testing schedule, generally once a week, until death. Four levels of sexual responsiveness were scored during each test: Zero indicating that no sexual responses occurred in thirty minutes with three different extrous females; M, that mounting but not intromission occurred; I, that intromission responses began but the male ceased intromitting for at least 40 min, or 4 hr passed from the first intromission, at which point a test was terminated even though the male might still be exhibiting sexual responses; and E, that the male exhibited the rather dramatic 20–30-sec ejaculatory reflex that is characteristic of the species.

Table 6.1 presents the results for the three inbred strains studied. The prediction regarding rapid loss of the ejaculatory reflex after castration in the BALB/c strain was confirmed. Note that eight of the eleven males in this group did not even mount after castration. The three other males ejaculated only once before meeting criterion. Results for the DBA/2 inbred strain agree with those from previous studies (McGill and Tucker, 1964; McGill and Haynes, 1973). That is to say, males of this genotype lost the capacity to exhibit the ejaculatory reflex quite rapidly following removal of the gonads. While the performance of C57BL/6 males shown in Table 6.1 was better than previously reported, it does not equal that of the $B6D2F_1$ strain to be described momentarily.

Table 6.2 presents the results for four of the six hybrid groups. The

Table 6.1. Postcastration Sexual Performance of Three Inbred Strains[a,b,c]

Ss	BALB/c	*Ss*	DBA/2	*Ss*	C57BL/6
1	0 0 0 0 0	1	M 0 0 0 0	1	0 0 0 I 0M 0 0M 0 0 0 0
2	0 0 0 0 0	2	M 0 0 0 0	2	M 0 I IM IM 0M 0 0 0 0
3	0 0 0 0 0	3	M 0 0 0 0	3	EM I I I IMM I I IM I (died)
4	0 0 0 0 0	4	MM 0 0 0 0	4	M E 0M 0M 0 0M I 0 0 0 0
5	0 0 0 0 0	5	E 0 0 0 0 0	5	E 0M 0 0 0MMM IMMMMM 0M 0 0 0 0
6	0 0 0 0 0	6	E 0 0 0 0 0	6	IMM I IMMMM 0 0 0MM 0MMMMMM 0MM 0 0 0 0
7	0 0 0 0 0	7	E I 0 0 0 0	7	EMM IM I IMMMMMMMMM 0M I IMMM IM 0 IM 0 0 0 0
8	0 0 0 0 0	8	M E 0 0 0 0 0	8	m 0 I I IE IE I I I IMMM 0MMMMMMMM 0MM E 0 E 0 E 0 E 0M 0 E 0 (died)
9	E 0 0 0 0 0	9	E 0 0MMM 0 0 0 0	9	I I E I E I IM IMMM I I IMM IMMM IMMM 0 0MM 0 0M 0 0 M 0M 0 0 0 0
10	E 0 I 0 0 0 0	10	E E 0M 0 0MMM 0 0 0 0	10	E EMM I I IMM IMMMMMMMM EM IM IM I 0 EMM I 0 E I I 0 0MM 0 IM IM 0M 0 0 (died)
11	0 0 0 0 E 0 0 0 0 0	11	M IM E 0 EM I 0 0M 0 0 0 0	11	E 0 0M IMM IM IMM IMMMM E 0 IM E EM E 0 IM 0 E E 0 0MMM M 0 E 0MMMMM I I 0 (died)

[a] From McGill and Manning (1976). Reprinted by permission of Baillière Tindall.
[b] E is the ejaculatory reflex; I is an intromission-like response; M is mounting behavior; and 0 is no sexual response.
[c] Two letters represent one week of testing.

Table 6.2. Postcastration Sexual Performance of Four Hybrid Groups[a,b,c]

Ss	CD2F₁	*Ss*	D2CF₁	*Ss*	CB6F₁	*Ss*	B6CF₁
1	0 0 0 0 0[b]	1	0 0 0 0 0	1	0 0 0 0 0	1	0 0 0 0 0
2	0 0 0 0 0	2	0 0 0 0 0	2	0 0 0 0 0	2	0 0 0 0 0
3	0 0 0 0 0	3	M 0 0 0 0	3	0 0 0 0 0	3	0 0 0 0 0
4	0 0 0 I 0 (died)	4	M 0 0 0 0	4	M 0 0M 0 0 0 0	4	0 0 0 0 0
5	E 0 0 0 0 0	5	E0 0 0 0 0 0	5	EM 0 I 0 0 0 0	5	0 0 0 0 0
6	E E 0 0 0 0 0	6	E 0 0 0 0 0	6	E 0 0 0M I 0 0 0 0	6	MM I 0 0 0 0
7	0 0 E 0 0 0 0 0	7	E 0 0 0 0 0	7	E 0 0M IM 0 0 0 0	7	EM EM 0 0 0 0
8	E 0 E 0 0 0 0 0	8	E 0 0 0 0 0	8	0M EMMM E EM E E I 0 0 0 0	8	E 0 E 0 0 0 0 0
9	0 0 0 E 0 0 0 0 0	9	M E 0 0 0 0 0	9	0 0 E 0 0 0 E 0 0 I I 0 0 0 0	9	0 0 0 0 I 0 0 0 0
10	E E 0 0 0M 0 0 0 0	10	E 0 0M 0 0 0 0	10	E 0MM I 0 0 0 I 0 0 0M 0 0 E 0 0 0 0 0	10	E 0MM 0 0 I 0 0 0 0
11	E E 0 I E 0 0 0 0 0	11	E E E 0M 0 0 0 0	11	E 0M 0 0 E 0 0 0 E 0 E 0 0M 0M 0 0M 0 0 0M 0 0 0 0	11	E E I I 0M 0 I IMM 0 0 0 0

[a] From McGill and Manning (1976). Reprinted by permission of Baillière Tindall.
[b] E is the ejaculatory reflex; I is an intromission-like response; M is mounting behavior; and 0 is no sexual response.
[c] Two letters represent one week of testing.

findings for $CD2F_1$ agree with those previously reported for this genotype, and the reciprocal $D2CF_1$, also lost the ejaculatory reflex soon after castration. However, the results for the $CB6F_1$ and the reciprocal $B6CF_1$ genotypes did not support the hypothesis of genetic equivalence between the inbred strains BALB/c and DBA/2. Males of both these hybrid groups lost the reflex rapidly after castration, and furthermore, a considerable number of them did not even mount the female in postcastrational tests.

Table 6.3 shows the findings for the two remaining F_1 genotypes: $B6D2F_1$ and $D2B6F_1$. The last three columns of this table summarize the results for tests occurring more than one year after castration. The results for $B6D2F_1$ males agree with previous findings; a large proportion of males of this genotype continued to exhibit the ejaculatory reflex for many months following gonadectomy. As the bottom half of Table 6.3 shows, the reciprocal $D2B6F_1$ males, in general, did not perform as well as the $B6D2F_1$ hybrids. The difference between these two reciprocal crosses in terms of the proportion of each group ejaculating six months or more after castration just failed to reach statistical significance (χ^2, < 0.06). It may be worth noting that three subsequent studies using these two groups yielded similar findings. Sex-linkage and/or maternal effects are strongly suggested by this outcome.

The results of these experiments indicate that, for house mice at least, manipulation of genotype within a species can result in differences in the loss of sexual responsiveness, which are similar to those seen when several species are compared. Certain inbred strains and hybrids lost sexual responsiveness quite rapidly following castration. Other genotypes, perhaps best exemplified by C57BL/6 males, apparently maintained arousal components but seldom, if ever, exhibited the ejaculatory reflex and finally ceased to mount the females. Finally, most $B6D2F_1$ males exhibited ejaculatory reflexes regularly up to a few weeks, or days, or in one case hours, of death. McGill and Haynes (1973) autopsied one $B6D2F_1$ male who was still exhibiting the ejaculatory reflex 14 months after castration and they claimed "on his behalf both the absolute record for rodents and the relative record (postcastrational performance ÷ life span) for all mammalian species." Examination of the last column in Table 6.3 shows that this male's performance was far exceeded by certain males in the McGill and Manning study. It is also worth noting that adrenalectomy does not prevent postcastration retention of the ejaculatory reflex in this genotype (Thompson, McGill, McIntosh, and Manning, 1976).

But it would be misleading to report that the sexual behavior of $B6D2F_1$ males was "unaffected" by castration. The next section describes an intriguing phenomenon that has been observed several times when the postcastration sexual behavior of these males is examined in detail.

4. The "Difficult Period" in Postcastration Sexual Behavior of $B6D2F_1$ Male Mice

Close examination of Table 6.3 indicates that those males that retained the ejaculatory reflex for several weeks after castration went through a "difficult period" in the early postcastration weeks. During this time they ejaculated relatively rarely. When ejaculation did occur during the difficult period, ejaculation latency, the time elapsing from the beginning of the first intromission until the beginning of the ejaculatory reflex, was greatly increased.

Figure 6.2 from McGill and Manning (1976) presents these results graphically. The data in Figure 6.2 are based on the $B6D2F_1$ males shown in Table 6.3 as well as the males from two other treatment groups studied by McGill and Manning but not described in the present chapter. As the figure shows, mean ejaculation latency increased from about 20 min in the first postcastration week to about 80 min for weeks 4–7. During weeks 7–12, ejaculation latency declined rapidly, and thereafter more gradually, until it returned to the level of week 1.

The broken line in Figure 6.2 shows that the proportion of tests in which males ejaculated declined from about 73% in the first week to about 18% in the third week. By week 12, the proportion of tests including the ejaculatory reflex had returned to 62.5%.

What is the mechanism underlying the difficult period? There would seem to be two possibilities which are perhaps not mutually exclusive. First, the effect may be due to experience; that is, animals may be learning to persist in sexual responses, or to alter them, so as to maintain the sensory input necessary for triggering the ejaculatory response. Indeed, Manning and I noted that sexual behavior did change in that we observed occasional intromission-like responses, and even ejaculatory reflexes, when the castrated males were mounted to the side or to the head of the stimulus female. Intromission *per se*, even if it is possible in long-term castrates, is certainly not necessary as a final trigger to the ejaculatory response.

The second possible underlying mechanism is a physiological process, which might be largely independent of experience. This would involve a change, presumably primarily in neural tissue, from a system that is relatively dependent on gonadal hormones to one that might operate in the absence of such hormonal influences (which is not to say that gonadotrophic or other hormones might not be involved). Behaviorally, this "switchover" would be accompanied by a temporary increase in mating time and a decrease in the proportion of tests in which the ejaculatory reflex occurred.

Table 6.3. Postcastration Sexual Performance of Two Hybrid Groups[a,b,c]

Ss	B6D2F$_1$[c]	I.[d]	II.[e]	III.[f]
1	EM I I E I 0MMMMMM 0 0 0 0			
2	E IMM 0 0MM 0 0 0M 0 EM 0 0 0 0			
3	0 I 0M IMM I I 0 IMMM IMMMMM E 0 EMM 0 0 0 0			
4	E I 0 IMME E I 0M 0 EMM E EMM 0M I E E I E 0 0 0 I 0 E E E E E 0MM E EM EM EMMMMMM EM EMM MMMMM 0 I E 0 0 EM I 0M E 0 EM I E I E E E E 0 E E I	33.3	90.5	90.5
5	E IM I 0M 0 E 0 0M I 0 0 0M E 0 0 E 0 E 0 0 E 0 0 0M 0M 0 E 0 E 0M E IMM 0 E 0M E 0 EMM E EMM 0 E E E 0 0 E E E EM E 0 E 0 0 0 0 0 0 E E E E M I E E E E 0	44	94	111
6	E E E EMM EM IM 0M E I 0M E 0M E 0M I 0 E 0 E E E I I E E E E E EM I E E E E EM E E E E E E E E I E E E E 0 E E E E 0 E E E E E E E E E E E I E E E E EM E E E E E I E 0 E	45	99	107
7	E EM IMM I 0 0M E E 0 E E E E E E 0 EM E 0 I E E E E 0 E E I E I E EM E E I E E E E E E E E E E I EM E I E E E E I E E E E E E EMMM E EM EE E E E I M I E M	32.5	102.5	105.5
8	E E EMM 0 EM 0 0 0 E 0 I E E I EM E 0 E E 0 E E E I E 0 E E EM E E E E E 0 E I 0 E E E E E E I E E E E E 0 E E E 0 E 0 0 E 0 0 0 E E I E IMM E E 0 I E E 0 IM I I I M	68.4	104	106
9	E E E I 0MM 0M IM E I E EMM E 0 E E I E E E E E E E E EM I E I E E E E E E E EI I E E E E I E I E E E E E E E E I E I E E E E E E E E E E E E E EM E E E EM E E I E M I 0	21.4	120	137
10	E EM I E E E I E I EMM I IM I I E E IM E I EM E 0 EMM EMM 0M E 0 E E 0 EM EMMM 0 E I 0MM 0 0 M E 0 0M E E E EM MM EM E E E E E E E E E E E E E E	60	124	137

Ss	D2B6F$_1$	I.[d]	II.[e]	III.[f]
1	E 0 I 0 0 0 0			
2	E 0M I 0 0 0 0			
3	E 0 0M 0 0 0 0			
4	I 0 0 I 0M 0 0 0 0			
5	E E 0M 0 IM 0 0 0 0			
6	E E IMM IM I I 0 I 0 0 0 0			
7	E EM E 0 0 0 0M 0 I 0 0M 0 0 0 0			
8	E E 0 I 0M 0M I 0M 0 I 0 0M 0 0 0 0			
9	EM 0 I 0 E 0 IM I E I 0 E 0MM I 0 I 0 0MMM 0M I 0M 0 0 0 0			
10	E EM I 0 I 0M 0 0 0M 0 0M 0MM 0 E 0 0 0M 0E 0 IMM 0 I 0 I E 0 E E 0 E 0M E IMM EM 0 E 0M 0 I 0MMM E 0 0 EM 0 E 0 E 0 0 0 (died)			
11	EM E I IM IM I 0M 0 0 I 0 IM E 0 EM E IM E 0 0 I 0M E E 0 EMM E E E E E E0 E E I 0 0 0 0 I 0 E 0 E 0M I 0 E E E I E E 0 0 E 0MM EM 0 0 EM E M M E E E E E E E	88.5	82	83

[a] From McGill and Manning (1976). Reprinted by permission of Baillière Tindall.
[b] E is the ejaculatory reflex; I is an intromission-like response; M is mounting behavior; and 0 is no sexual response.
[c] Two letters represent one week of testing.
[d] Percent tests with ER after the first year of testing.
[e] Postcastration age, in weeks, at last ER.
[f] Postcastration age, in weeks, at death.

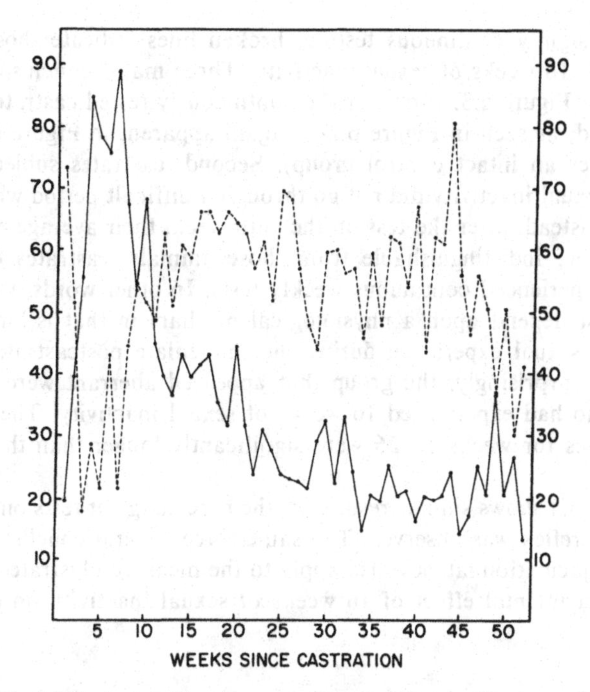

Figure 6.2. The "difficult period" and recovery from it. Mean weekly postcastration ejaculation latency (solid line) and mean weekly percentage of tests ending with the ejaculatory reflex (broken line). From McGill and Manning (1976), reprinted by permission of Baillière Tindall.

A straightforward test of these two possibilities has been conducted by simply denying sexual contact to experienced, castrated $B6D2F_1$ males during the difficult period. Stephen M. McIntosh, Karl A. Mertz, and I permitted a group of 180 $B6D2F_1$ male mice to mate with receptive females once a week for three consecutive weeks. The subjects were then placed into four groups matched for ejaculation latency. One group of 40 males was castrated, and the other left intact; and one group of 50 males was castrated and the other left intact. For the two groups of 40 animals, weekly sex tests began two to three days after the operation and were continued for 26 weeks. The other groups of 50 castrates and the 50 intact males were left isolated and undisturbed, except for routine maintenance, for 16 weeks. After 16 weeks of sexual inactivity, weekly sex tests were instituted for these two sexually deprived groups and were continued for 11 more weeks.*

Figure 6.3 shows the results for ejaculation latency for the four groups. Hollow symbols represent castrates; solid symbols intact animals. Contin-

* At that point, the groups were divided and other variables, not relevant to the present report, were manipulated. These results along with details on other behavioral measures will be reported elsewhere.

uous lines signify continuous testing; broken lines indicate those animals subjected to 16 weeks of sexual inactivity. Three major conclusions can be drawn from Figure 6.3. First, for the continuously tested castrates, the difficult period, as seen in Figure 6.2, is again apparent in Figure 6.3 (which also includes an intact control group). Second, castrates subjected to 16 weeks of sexual inactivity did not go through a difficult period when testing resumed. Instead, after the test on the 16th week, their average ejaculation latencies were indistinguishable from those animals, castrates or intacts, who had experienced continuous weekly tests. In other words, the difficult period must depend upon a physiological mechanism that is largely independent of sexual experience during the immediate postcastration weeks. Third, and surprisingly, the group that appeared aberrant were the intact animals who had experienced 16 weeks of sexual inactivity. Their ejaculation latencies for weeks 16–26 were significantly longer than those of any other group.

Figure 6.4 shows similar results for the percentage of tests on which the ejaculatory reflex was observed. The same three general conclusions stated above for ejaculation latency also apply to the measure illustrated in Figure 6.4. The detrimental effect of 16 weeks of sexual inactivity on intact ani-

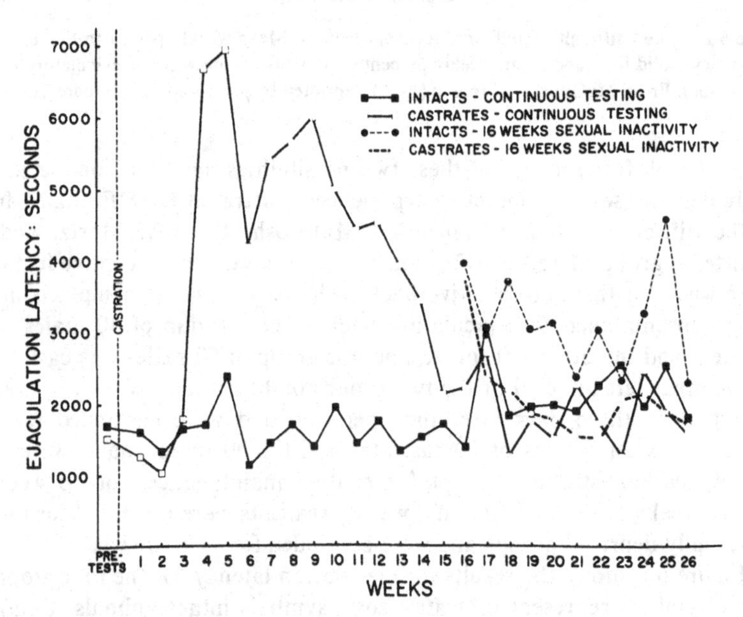

Figure 6.3. Ejaculation latencies of castrated or intact groups of B6D2F$_1$ male house mice either continuously tested (once per week) or subjected to 16 weeks of sexual inactivity before weekly testing resumed.

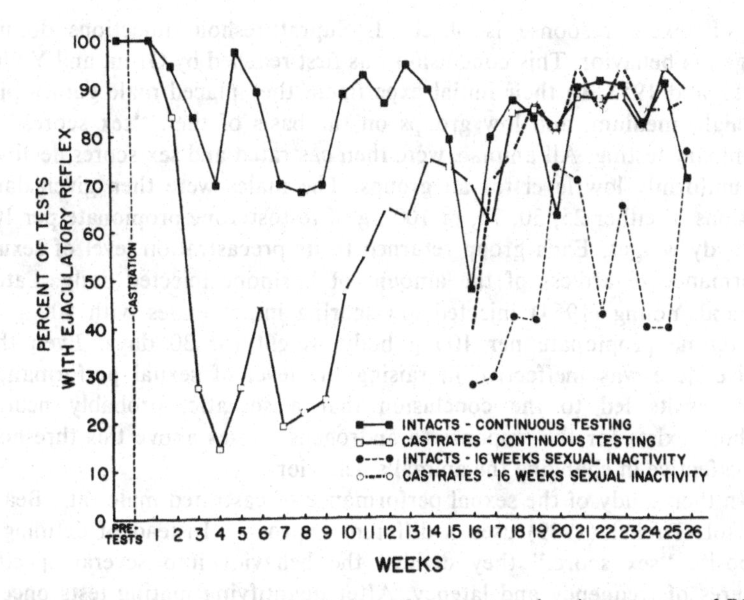

Figure 6.4. Percent of tests with ejaculatory reflex in castrated or intact groups of B6D2F$_1$ male house mice either continuously tested (once per week) or subjected to 16 weeks of sexual inactivity before weekly testing resumed.

mals is particularly apparent in this figure. Intact sexually deprived animals had significantly fewer ejaculatory reflexes than did any of the other three groups. There were no significant differences among the other three groups on this measure.

In sum, the difficult period appears to be independent of sexual experience during the immediate postcastration weeks. And, intact animals are significantly more affected by sexual inactivity than are castrates, as indicated by the measures of ejaculation latency and percentage of tests with ejaculatory reflex. Both of these finding appear to be worthy of further investigation as to the physiology underlying the phenomena.

5. *Response of B6D2F$_1$ and CD2F$_1$ Male Mice to Testosterone Propionate*

As was briefly noted in an earlier section, the response of castrated male guinea pigs to endogenous androgen fits a "threshold," rather than a dose–response, model. That is, if enough androgen is present, a "normal" level of sexual response is observed. Suprathreshold injections do not change the behavior. This conclusion was first reached by Grunt and Young in 1952 and 1953. In their initial experiment they placed male guinea pigs into high, medium, and low groups on the basis of their "sex scores" in preliminary testing. All animals were then castrated and sex scores declined to a uniformly low level for all groups. The males were then given daily injections of either 25, 50, 75, or 100 μg of testosterone propionate per 100 g of body weight. Each group returned to its precastration level of sexual performance regardless of the amount of hormone injected daily. Later, Riss and Young (1954) injected low-scoring intact males with 500μg of testosterone propionate per 100 g body weight for 30 days. Even this massive dose was ineffective in raising the level of sexual performance. These results led to the conclusion that a somatic, probably neural, threshold exists for the response to androgens. Doses above this threshold are ineffective in changing the animal's behavior.

In their study of the sexual performance of castrated male rats, Beach and Holz-Tucker (1949) used a different approach. Instead of defining a composite "sex score," they divided the behavior into several specific measures of frequency and latency. After quantifying mating tests once a week for six consecutive weeks, the males were castrated and placed into groups that were injected daily with testosterone propionate ranging from 1 to 500 μg. While they found several dose-dependent behavioral differences, their conclusions regarding ejaculation latency, defined as the time from the first mount to the ejaculatory reflex, are most important for the present discussion. In general, their results indicated that the larger the dose of testosterone propionate injected daily the more rapidly the animals ejaculated.

Whalen and DeBold (1974) worked with castrated hamsters and examined the effects of increasing doses of testosterone, dihydrotestosterone, and androstenedione. Their general finding was that these androgens produced progressively shortened ejaculation latencies with increases in the amount injected daily.

In contrast to these findings for guinea pigs, hamsters, and rats, McGill, Albelda, Bible, and Williams (1976) have recently reported different results for B6D2F$_1$ castrated male house mice. They found that ejaculation latency in males of this genotype was proportional to the amount of testosterone propionate injected daily in both pre- and postpubertally castrated animals. The results for their postpubertal, sexually experienced castrates and an intact (not castrated) control group are shown in Figure 6.5. In terms of statistical significance, the 500-μg group had longer ejaculation latencies than all other groups; the 150-μg animals' scores were longer than both the 20-μg and intact groups; while the 60-μg males had longer latencies than did the intact animals. In addition, B6D2F$_1$ females neonatally androgenized by the method described in Manning and

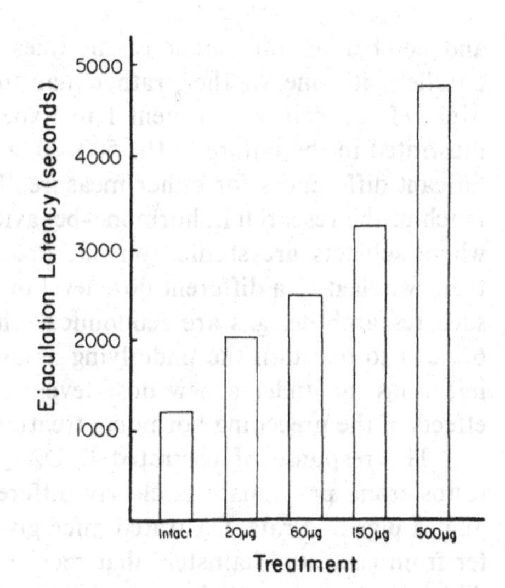

Figure 6.5. Ejaculation latencies of castrated B6D2F$_1$ male house mice with different daily injections of testosterone propionate. Results for an intact (not castrated; not injected) control group are also show. From McGill *et al.* (1976), reprinted by permission of Academic Press, Inc.

McGill (1974) also exhibited significantly increased ejaculation latencies when injected with 100 μg of testosterone propionate daily.

McGill *et al.* (1976) had tested their postpubertally castrated males for eight weeks. In an extension of that experiment Steven M. Albelda, Lawrence J. Kaplan, and I instituted certain dose changes for these animals, allowed them to remain on the new schedule for two weeks, and then tested them again each week for eight weeks while daily injections continued. The results are shown in Figure 6.6.

As Figure 6.6 shows, changing the amount of hormone injected daily in Experiment 2 resulted in changes in ejaculation latency in the directions predicted from the results of Experiment 1. All the dose changes made from Experiment 1 to Experiment 2 resulted in statistically significant alterations in ejaculation latency except for the group that went from 500 to 20 μg.

Figure 6.7 shows the results for number of intromissions preceding ejaculation in the two experiments. Once again all changes were in the predicted direction and significant except for the 500–20-μg group. While the correspondence is not perfect, the results indicate that dose changes are not producing their effects by changing the pacing of sexual behavior in terms of the number of intromissions per unit time. Rather, the measures of number of intromissions and ejaculation latency are correlated and vary in closely similar ways.

One other observation may be made from an examination of Figures 6.6 and 6.7. For example, consider the animals on 60 and 150-μg doses in Experiment 1. Both of these groups were moved to a 1,000-μg dose in Experiment 2. While both groups significantly increased ejaculation latency and number of intromissions, the lines connecting the means tend to run parallel with one another, rather than to converge. This indicates a "carry over" effect from Experiment 1 to Experiment 2. This phenomenon is also illustrated in the failure of the 500–20μg change to produce statistically significant differences for either measure. These results are troublesome since much of the research in hormone–behavior interactions has involved designs where subjects are studied on one dose level, perhaps rested briefly, and then switched to a different dose level or even to a different hormone. While such research designs are economical, the results shown in Figures 6.6 and 6.7 call to question the underlying assumption that a week or two without injections, or under a new dose level or a different hormone, removes the effects of the preceding hormonal treatment.

The response of castrated B6D2F$_1$ house mice to different doses of testosterone propionate is clearly different from the response of castrated guinea pigs and rats. Castrated mice given testosterone propionate also differ from castrated hamsters that received testosterone, androstenedione, or dihydrotestosterone (Whalen and DeBold, 1974). While B6D2F$_1$ male mice exhibited ejaculation latencies in proportion to the amount of hormone injected daily, castrated male rats, guinea pigs, and hamsters tended either

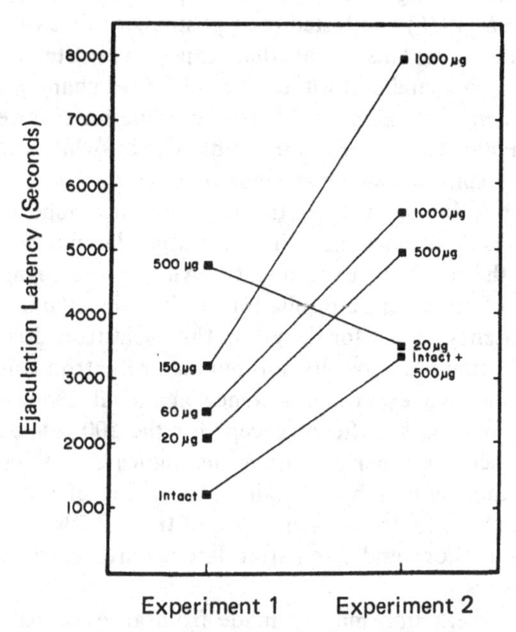

Figure 6.6. The top four groups show alterations in ejaculation latency in castrated B6D2F$_1$ male house mice produced by changing the amount of testosterone propionate injected daily. The last group of males were intact (not castrated) animals that received no exogenous androgen in Experiment 1 and 500 μg of testosterone propionate in Experiment 2.

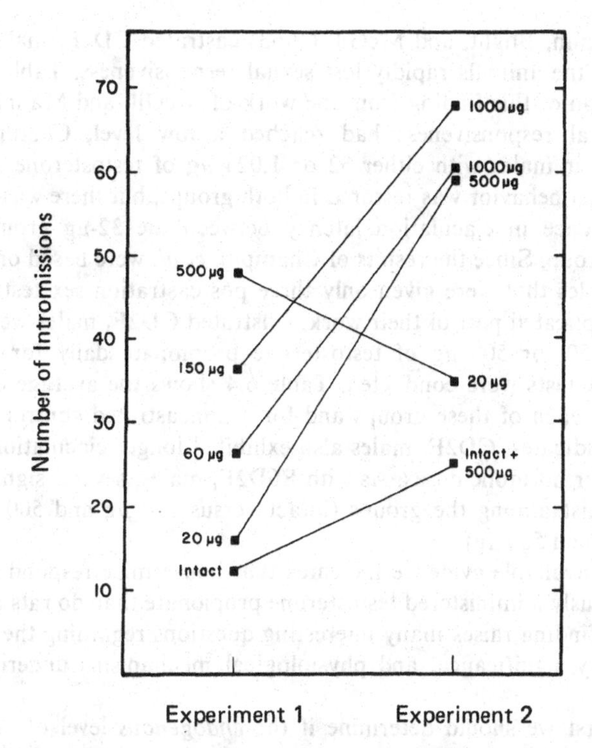

Figure 6.7. The top four groups show alterations in number of intromissions preceding ejaculation in castrated B6D2F$_1$ male house mice produced by changing the amount of testosterone propionate injected daily. The last group of males were intact (not castrated) animals that received no exogenous androgen in Experiment 1 and 500 μg of testosterone propionate in Experiment 2.

to be unaffected or to ejaculate more rapidly with increasing doses of androgen. But in light of Gorzalka and Whalen's results described in the first section of this chapter, we must exercise caution before declaring this to be a "species difference." It is possible that different genotypes of house mice do not respond as B6D2F$_1$ males (and that B6D2F$_1$ castrates respond differently to different androgens). Perhaps the response of B6D2F$_1$ hybrids is somehow related to their remarkable retention of all elements of sexual behavior including the ejaculatory reflex after removal of the gonads, or the gonads and the adrenals (Thompson, McGill, McIntosh, and Manning, 1976). In other words, it is possible that animals that continue to mate in the absence of hormones from the gonads or the adrenal glands respond differently to exogenous androgen than do those genotypes that lose sexual behavior rapidly after castration. And, as a matter of fact, there has been evidence that such might be the case.

Champlin, Blight, and McGill (1963) castrated $CD2F_1$ male mice and found that the animals rapidly lost sexual responsiveness. Table 6.2 shows confirmation of this finding from the work of McGill and Manning (1976). After sexual responsiveness had reached a low level, Champlin *et al.* injected their males with either 32 or 1,024 μg of testosterone propionate daily. Sexual behavior was restored in both groups, but there was no significant difference in ejaculation latency between the 32-μg group and the 1,024-μg group. Since the results of Champlin *et al.* were based on relatively small samples that were given only three postcastration sex tests, we have recently replicated part of their work. Castrated $CD2F_1$ males were injected with 20, 150, or 500 μg of testosterone propionate daily for 15 weeks. Weekly sex tests were conducted. Table 6.4 shows the average ejaculation latency for each of these groups and for a noncastrated control group. As the table indicates, $CD2F_1$ males also exhibited longer ejaculation latencies to the larger hormone doses. As with $B6D2F_1$ males, several significant differences exist among the groups (intact versus 150 μg and 500 μg; 20 μg versus 150 and 500 μg).

Thus, available evidence indicates that house mice respond differently to exogenously administered testosterone propionate than do rats and guinea pigs. This finding raises many interesting questions regarding the functions, evolutionary significance, and physiological mechanisms underlying these differences.

But first we should determine if the *endogenous* level of androgen is correlated with ejaculation latency. There is some evidence that this is the case. Albelda (1975) found that BALB/c males have two and one-half to three times as much endogenous plasma testosterone than do $B6D2F_1$ males. This difference, determined by radioimmunoassay, is also reflected in the ejaculation latencies of the two genotypes. While $B6D2F_1$ males require about 20–25 min to reach the ejaculatory threshold, BALB/c males exhibit ejaculation latencies of approximately 60 min (McGill, 1962, 1970). Other results from radioimmunoassay plasma–testosterone analyses indicate that genotypes similar to $B6D2F_1$ in ejaculation latency (C57BL/6; DBA/2; $D2B6F_1$) have plasma-testosterone levels similar to those found in $B6D2F_1$ males. If this relationship in fact holds, it should be possible to take any heretofore unstudied strain, measure one variable (either ejaculation

Table 6.4. Mean Ejaculation Latency of Intact (Not Castrated; Not Injected) $CD2F_1$ Male House Mice and of Castrated Males Injected Daily with Different Amounts of Testosterone Propionate

	Intact	20 μg	150 μg	500 μg
Ejaculation latency (sec)	1633	1363	2493	3624

latency or plasma testosterone level), and successfully predict the other variable.

6. *Response of B6D2F$_1$ Male Mice to Dihydrotestosterone, Estradiol Benzoate, and Androstenedione*

A second series of experiments has resulted from our findings regarding the response of B6D2F$_1$ castrated males to testosterone propionate. How would castrated males of this genotype respond to different hormones? Is testosterone operating as a "prehormone" and thus having its effects via its metabolites? Testosterone propionate is aromatizable to estradiol and it may also be reduced to 5 a-dihydrotestosterone. Thus the first two hormones that we selected for study were estradiol benzoate and dihydrotestosterone. If testosterone propionate has its effect on ejaculation latency via its conversion to estradiol, then estradiol benzoate might be expected to have similar effects while dihydrotestosterone (not convertible to estradiol) would not, and vice versa.

Groups of 15 sexually experienced B6D2F$_1$ males were castrated and injected daily with either 20, 150, or 500 μg of dihydrotestosterone; or with 0.2, 1.5, or 5.0 μg of estradiol benzoate. A group of 30 males was left intact. Hormone treatments were initiated on the day following castration and continued for 11 weeks. Weekly sex tests began one week following the operation. Figure 6.8 shows the average ejaculation latencies for the groups over the 10 weeks of testing. For the dihydrotestosterone groups, all comparisons revealed statistically significant differences except for 20 and 150 μg versus the intact animals. For the groups receiving estradiol benzoate, the 0.2-μg animals had significantly longer ejaculation latencies than the other three groups, which did not differ among themselves.

On the day following the tenth sex test, the hormone doses for certain groups were changed as shown in Figures 6.9 and 6.10. In addition, the intact control group of 30 animals was divided into three matched groups on the basis of ejaculation latency. One group remained intact and untreated, the second group remained intact and received daily injections of 500 μg of dihydrotestosterone, while the third group remained intact and received daily injections of 5.0 μ of estradiol benzoate. Eight weekly sex tests followed. Then all hormone treatments were withdrawn and the animals were tested for a final 11-week period. Figures 6.9 and 6.10 show the changes that occurred as a result of the experimental manipulations.

For dihydrotestosterone, the changes resulting from shifting from treatment 1 to treatment 2 were in the direction predicted by the results of treatment 1. The results for animals receiving estradiol benzoate are

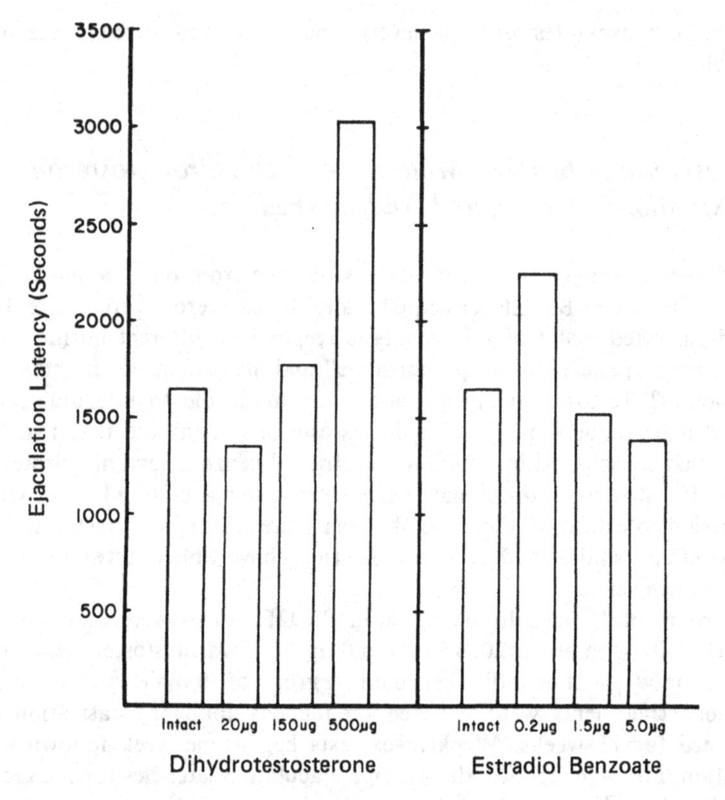

Figure 6.8. Ejaculation latencies of castrated B6D2F$_1$ male house mice with different daily injections of dihydrotestosterone or estradiol benzoate. Results for the intact (not castrated; not injected) control group are shown twice to facilitate comparisons.

considerably more complicated and will require further study. In particular, we need to know what peripheral changes are being produced in the various treatment groups and what the effects of even lower hormone doses might be.

We have recently completed the first phase (treatment 1) of a similar experiment where androstenedione, which may be aromatized to estrone, was the injected hormone. As Table 6.5 indicates, the results for androstenedione are similar to those found with estradiol benzoate. That is, an inverse dose–response relationship occurred such that animals receiving high amounts of the hormone had shorter ejaculation latencies than did animals receiving lower amounts. Statistical analysis revealed significant differences for intact versus 500 μg and for 20 μg versus 150 and 500 μg. All other comparisons just failed to reach significant levels.

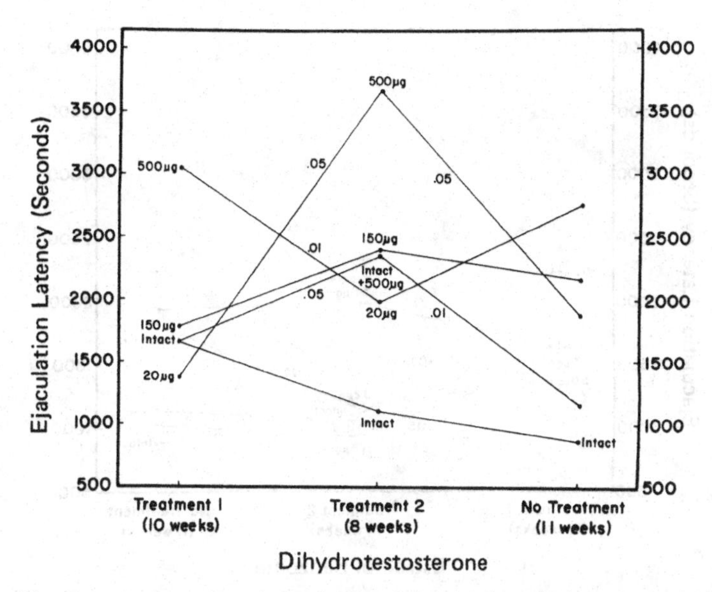

Figure 6.9. Treatment 1 repeats the results for ejaculation latency shown in the left half of Figure 6.8. Treatment 2 shows the results when various changes were made in the amount of dihydrotestosterone injected daily. The final data points show ejaculation latencies after withdrawal of hormone treatment. Significance levels resulting from the various experimental manipulations are also depicted.

In sum, increasing amounts of testosterone propionate and dihydrotestosterone tend to increase ejaculation latency, while the opposite is true for estradiol benzoate and androstenedione. With further research it should be possible to determine whether testosterone propionate and androstenedione are producing their effects directly or through conversion to dihydrotestosterone and estrone.

One further observation should be made regarding the experiments just described. All dose levels of all the hormones studied, testosterone propionate, dihydrotestosterone, estradiol benzoate, and androstenedione, prevented the occurrence of the difficult period described above. The fact

Table 6.5. Mean Ejaculation Latency of Intact (Not Castrated; Not Injected) B6D2F$_1$ Male House Mice and of Castrated Males Injected Daily with Different Amounts of Androstenedione

	Intact	20 µg	150 µg	500 µg
Ejaculation latency (sec)	1273	2017	1052	768

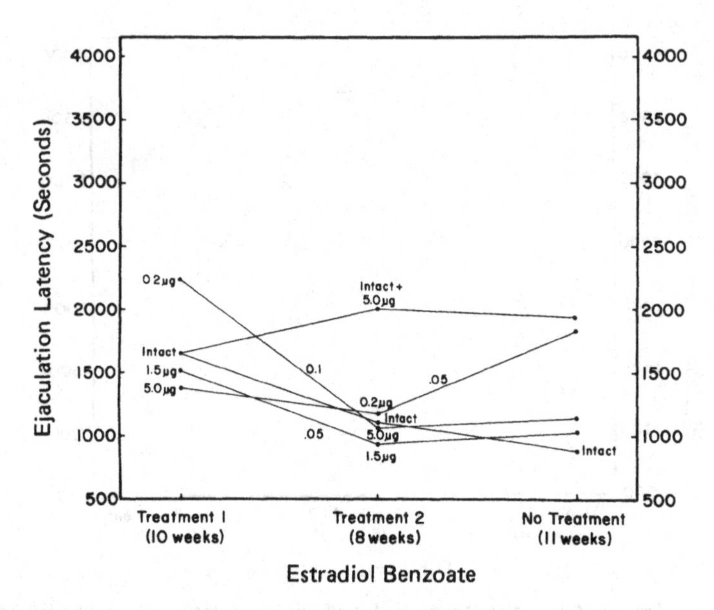

Figure 6.10. Treatment 1 repeats the results for ejaculation latency shown in the right half of Figure 6.8. Treatment 2 shows the results when various changes were made in the amount of estradiol benzoate injected daily. The final data points show ejaculation latencies after withdrawal of hormone treatment. Significance levels resulting from the various experimental manipulations are also depicted.

that all of the steroids studied have this effect may aid in the search for the physiological mechanism underlying the difficult period.

7. Summary and Prospectus

The first sections of this chapter described studies that revealed important genotype–hormone interactions *within* species. In particular, I reviewed Gorzalka and Whalen's experiments showing that Swiss–Webster female house mice are unresponsive to dihydroprogesterone, and research that my colleagues and I have conducted regarding genotypically determined differences in retention of sexual behavior following castration in male house mice. Then the "difficult period" observed in $B6D2F_1$ male mice in the immediate postcastration weeks was described. The difficult period consists of a temporary increase in ejaculation latency and a temporary decrease in the percent of males ejaculating on each test. Further experiments showed that denying sexual experience during the difficult period did not delay the return to "normal" levels of both ejaculation latency and percent of animals exhibiting the ejaculatory reflex during the weekly tests. Furthermore, intact control males were affected by sexual inactivity, while castrated males denied sexual contact during the difficult period were not significantly different from continuously tested intact animals when testing resumed.

Next I described the unusual responses of B6D2F$_1$ and CD2F$_1$ castrated male house mice to injections of testosterone propionate. For both of these genotypes, ejaculation latency was found to increase in proportion to the amount of hormone injected daily. Apparently no other species studied to date exhibits such a response to exogenous androgen. This series of experiments also showed that ejaculation latency could be altered by changing the amount of hormone administered, but that there appeared to be residual effects from the first injection series. It was noted that this latter finding is cause for some concern since experimental designs in hormone-behavior research have frequently used changes in the amount, or even the kind, of hormone administered to the same subject. Further research with castrated B6D2F$_1$ males showed that dihydrotestosterone also produced increased ejaculation latency with increasing dosage. Androstenedione, and to a lesser extent estradiol benzoate, produced the opposite result; larger doses resulted in shorter ejaculation latencies than did smaller doses.

Thus while the early sections of the chapter were concerned with genotype–hormone interactions within species, the emphasis in the later sections was on genotype–hormone interactions as reflected in cross-species comparisons. At several points I suggested the course that research in the immediate future might take. I will not repeat such suggestions here but instead will consider some broader questions.

One such question is rather troublesome. Are species differences in hormone–behavior interactions "real" or are they the result of sampling error? For example, if one selected Swiss–Webster female house mice as representatives of the species, one would conclude that house mice were refractory to dihydroprogesterone. If CD-1 females were selected, the opposite conclusion would be reached. A similar situation exists with respect to retention of sexual responsiveness following gonadectomy in male house mice.

How many "species differences" in response to hormones or hormonal withdrawal are the accidental result of the particular genotype (race, breed, or strain) selected for study? Perhaps many, perhaps only a few. But it should be clear at this point that we cannot generalize about "species differences" when only one or two genotypes within each species have been sampled.

Why is it important to consider species differences, even if they do exist? One answer lies in the consideration of questions of the function (or survival value) and the evolution of the behavioral pattern. For example, in

1974 Hart reviewed the effects of gonadal androgen on the sociosexual behavior of certain male mammals. He noted that whereas cats appear to have more highly developed forebrains than dogs (and thus, theoretically, should be less dependent on gonadal hormones), castrated cats lose sexual responsiveness, aggressiveness, and scent-marking behavior more rapidly and completely than do castrated dogs. Following a suggestion by Dale F. Lott, Hart provided the following speculations regarding these differences:

> Natural selection may have operated on the neural substrate in some species differently than others. The feral relative of the domestic cat, for example, was apparently an asocial species. As such, the maintenance of intense aggressive and scent-marking activities at times other than the breeding season in this species would have had no advantage in terms of maintaining social group structure and in fact might have led to frequent physical injury because of intermale conflicts. A seasonal fluctuation in androgen secretion, as Aronson and Cooper [1968] have suggested occurs, and a responsive nervous system allowing for subsequent rapid changes in aggressive and scent-marking activities would be advantageous. In social animals, such as wolves (the feral ancestor of dogs) and primates, the year-round maintenance of aggressive and scent-marking activity is necessary for the survival of group structure. Even with a seasonal fluctuation in gonadal androgen, as, for example, is evident in some free ranging rhesus monkeys [Vandenbergh, 1969; Vandenbergh and Vessey, 1968], a nervous system that does not respond to these fluctuations in terms of altering aggressive and scent-marking activity is advantageous. Unfortunately too little data is available to carry this notion past the conjecture stage.

It is clear then that "real" species differences in hormone–behavior interactions may lead to interesting conclusions regarding the function and evolution of behavioral patterns as they relate to the responsiveness of neural tissue to hormonal influences. It is also clear that such generalizations are best based on samples from naturally occurring wild populations. Domesticated or laboratory populations may be unsuited for such research. For example, in considering olfactory communication systems, Doty (1974) has written:

> The last example suggests some of the problems in attempting to generalize from inbred laboratory rodents to their wild counterparts, if and when such counterparts exist. A comparison of sexual behaviors of wild (*Cavia opera*), domesticated (*Cavia porcellus*), and F_1 hybrid guinea pigs suggested to Rood [1972] the hypothesis that distinctive male and female odors may have been lost in the process of domestication. Bruce [1968] concluded that the olfactory identities of both wild and outbred laboratory *Mus musculus* are eliminated by inbreeding. Bronson [1971] found that estrous symchrony in female deer mice (*Peromyscus maniculatus bairdi*) could be produced by adding as little as .01 milliliter male urine per day for a two-day period to their bedding, whereas a minimum of 1.0 milliliter per day is required to produce the effect in inbred SJL/J *Mus musculus*, possibly adding further sup-

port to this notion. It does not appear surprising that animals bred for many generations without regard to selection pressures for characters and behaviors related to sexual isolation would not behave as their wild progenitors in relation to heterospecifics. It is clear, however, that more research is needed to determine to what degree domestication and inbreeding have influenced characters related to mate selection and copulation.

As both Hart and Doty implied, there is a paucity of data concerned with hormone–behavior interactions in undomesticated, nonlaboratory animals. More such studies are badly needed if problems concerned with the evolution and function of the hormonal control of behavior are to be solved.

Of what use then are studies of domesticated or laboratory animals? Aside from the fact that they may lead to hypotheses concerning wild populations, their primary use is in answering different sorts of questions: questions of causation or mechanism that can be separated from questions of the function and evolution of behavior. Such mechanistic questions can be posed for species-typical responses, for cross-species comparisons, and for genotypically distinct hormone–behavior differences within species. Several chapters of the present volume illustrate the value of these approaches.

Two important mechanistic questions are, "Where in the body are the hormones acting?" And, "How are they producing their effects?" Considering the first question, Beach (1948) listed four ways by which hormones could produce differences in behavior: (1) by affecting the organism as a whole through such variables as growth and metabolism; (2) by affecting specific structures used in response patterns, for example, growth of the penis during puberty; (3) by affecting peripheral receptor mechanisms, which would alter the organism's tendency to respond to particular stimuli; and (4) by affecting the central nervous system. As Beach predicted at that time, it is the final variable that is proving most important in hormone–behavior interactions. This leads us to the second question; that of the means by which hormones produce their effects.

The question of how hormones operate has been investigated and some of the first steps in the causal chain have been elucidated. Hormones act directly on the genome, or on the immediate products of the genome, in particular cells so that certain biochemical substances are present or absent in the presence or absence of certain hormones (see Whalen, Chapter 7). The identification of these substances, and the means by which they then produce behavioral effects, constitute some of the major unresolved problems in the area.

This is where research on genotypically determined hormone–behavior differences within species may make important contributions. It is easier to determine what the genes are actually doing if one can work with *differences* in biochemistry in genetically different individuals *within* a species.

We may hypothesize that somewhere in the brain biochemical differences exist between female mice that respond to dihydroprogesterone and those that do not. Similarly, there should be biochemical differences between castrated males that continue to copulate and those that do not. The search for such differences might be simplified if research can provide examples of single genes, or at least short chromosome segments, producing such important effects. But even with polygenic inheritance, it should be possible to elucidate the physiological differences underlying the behavioral differences. In sum, then, our understanding of hormone–behavior interactions has much to gain by intensive investigation of genotype–hormone interactions within species.

References

Albelda, S. *Examination of plasma testosterone levels in B6D2F$_1$ male mice before and after non-contact sexual exposure and following intracranial cycloheximide infusions.* Unpublished senior honors thesis, Williams College, 1975.

Aronson, L. R. Hormones and reproductive behavior: Some phylogenetic considerations. In A. Gorbman (Ed.), *Comparative endocrinology.* New York: Wiley, 1959.

Aronson, L. R., and Cooper, M. L. Seasonal variation in mating behavior in cats after desensitization of glans penis. *Science,* 1966, *152,* 226–230.

Beach, F. A. Central nervous mechanisms involved in the reproductive behavior of vertebrates. *Psychological Bulletin,* 1942, *39,* 200–226.

Beach, F. A. Evolutionary changes in the physiological control of mating behavior in mammals. *Psychological Review,* 1947, *54,* 297–315.

Beach, F. A. *Hormones and behavior.* New York: Cooper Square Press, 1948.

Beach, F. A. Locks and beagles. *American Psychologist,* 1969, *24,* 971–989.

Beach, F. A. Hormonal factors controlling the differentiation and display of copulatory behavior in the Ramstergig and related species. In E. Tobach, L. R. Aronson, and E. Shaw (Eds.), *The biopsychology of development.* New York: Academic Press, 1971.

Beach, F. A., and Holz-Tucker, A. M. Effects of different concentrations of androgen upon sexual behavior in castrated male rats. *Journal of Comparative and Physiological Psychology,* 1949, *42,* 433–435.

Bronson, F. H. Rodent pheromes. *Biology of Reproduction,* 1971, *4,* 344–357.

Bruce, H. M. Absence of pregnancy-block in mice when stud and test males belong to an inbred strain. *Journal of Reproduction and Fertility,* 1968, *17,* 407–408.

Champlin, A. K., Blight, W. C., and McGill, T. E. The effects of varying levels of testosterone on the sexual behaviour of the male mouse. *Animal Behaviour,* 1963, *11,* 244–245.

Davidson, J. M. Characteristics of sex behaviour in male rats following castration. *Animal Behaviour,* 1966, *14,* 266–272.

Doty, R. L. A cry for the liberation of the female rodent: Courtship and copulation in *Rodentia. Psychological Bulletin,* 1974, *81,* 159–172.

Gorzalka, B. B., and Whalen, R. E. Effects of genotype on differential behavioral responsiveness to progesterone and 5α-dihydroprogesterone in mice. *Behavior Genetics,* 1976, *6,* 7–15.

Gorzalka, B. B., and Whalen, R. E. Genetic regulation of hormone action: Selective effects of progesterone and dihydroprogesterone (5α-pregnane-3,20-dione) on sexual receptivity in mice. *Steroids*, 1974, *23*, 499–505.

Goy, R. W., and Young, W. C. Strain differences in the behavioral responses of female guinea pigs to alpha-estradiol benzoate and progesterone. *Behaviour*, 1957, *10*, 340–354.

Grunt, J. A., and Young, W. C. Differential reactivity of individuals and the response of the male guinea pig to testosterone propionate. *Endocrinology*, 1952, *51*, 237–248.

Grunt, J. A., and Young, W. C. Consistency of sexual behavior patterns in individual male guinea pigs following castration and androgen therapy. *Journal of Comparative and Physiological Psychology*, 1953, *46*, 138–144.

Hart, B. L. Gonadal androgen and sociosexual behavior of male mammals: A comparative analysis. *Psychological Bulletin*, 1974, *81*, 383–400.

Manning, A., and McGill, T. E. Neonatal androgen and sexual behavior in female house mice. *Hormones and Behavior*, 1974, *5*, 19–31.

McGill, T. E. Sexual behavior in three inbred strains of mice. *Behaviour*, 1962, *19*, 341–350.

McGill, T. E. Genetic analysis of male sexual behavior. In G. Lindzey and D. D. Thiessen (Eds.), *Contributions to behavior-genetic analysis: The mouse as a prototype*. New York: Appleton-Century-Crofts, 1970.

McGill, T. E. Genetic factors influencing the action of hormones on sexual behavior. In J. B. Hutchison (Ed.), *Biological determinants of sexual behaviour*. London: Wiley, 1977.

McGill, T. E., and Haynes, C. M. Heterozygosity and retention of ejaculatory reflex after castration in male mice. *Journal of Comparative and Physiological Psychology*, 1973, *84*, 423–429.

McGill, T. E., and Manning, A. Genotype and retention of the ejaculatory reflex in castrated male mice. *Animal Behaviour*, 1976, *24*, 507–518.

McGill, T. E., and Tucker, G. R. Genotype and sex drive in intact and in castrated male mice. *Science*, 1964, *145*, 514–515.

McGill, T. E., Albelda, S. M., Bible, H. H., and Williams, C. L. Inhibition of the ejaculatory reflex in B6D2F$_1$ mice by testosterone propionate. *Behavioral Biology*, 1976, *16*, 373–378.

Riss, W., and Young, W. C. The failure of large quantities of testosterone propionate to activate low drive male guinea pigs. *Endocrinology*, 1954, *54*, 232–235.

Rood, J. P. Ethological and behavioural comparisons of three genera of Argentine *Cavies*. *Animal Behaviour Monographs*, 1972, *5*, 1–83.

Thompson, M. L., and Edwards, D. A. Experiential and strain determinants of the estrogen–progesterone induction of sexual receptivity in spayed female mice. *Hormones and Behavior*, 1971, *2*, 299–305.

Thompson, M. L., McGill, T. E., McIntosh, S. M.,and Manning, A. The effects of adrenalectomy on the sexual behaviour of castrated and intact BDF$_1$ mice. *Animal Behaviour*, 1976, *24*, 519–522.

Vandenbergh, J. G. Endocrine coordination in monkeys: Male sexual responses to the female. *Physiology and Behavior*, 1969, *4*, 261–264.

Vandenbergh, J. G., and Vessey, S. Seasonal breeding of free-ranging rhesus monkeys and related ecological factors. *Journal of Reproduction and Fertility*, 1968, *15*, 71–79.

Whalen, R. E., and DeBold, J. F. Comparative effectiveness of testosterone, androstenedione and dihydrotestosterone in maintaining mating behavior in the castrated male hamster. *Endocrinology*, 1974, *95*, 1674–1679.

Young, W. C. The hormones and mating behavior. In W. C. Young (Ed.), *Sex and internal secretions*, 3rd ed. Baltimore: Williams and Wilkins, 1961.

7

Molecular Mechanisms of Hormone Action

Richard E. Whalen

Over the past year or two I have had a fine time arguing with Frank Beach about ways one can proceed in the analysis of the effects of hormones upon behavior. I have argued that the wave of the future is the study of the molecular actions of hormones, and immediately Frank responds that our greater need is a finer grained analysis of behavior; he then proceeds to illustrate this point by his work on proceptivity, which is elegant both in concept and data (Beach, 1976). In this presentation, however, I shall attempt to illustrate the other approach, namely a finer grained analysis of the effects of hormones on those neural systems that mediate sexual behavior. In what follows I will focus on the receptive behavior of female rodents.

Indeed, the primary problem of interest to me is behavior. There is no question that gonadal hormones have various effects in the brain, one of which is the control of gonadotropin secretion and one of which is the control of behavior. Ultimately it should be possible to distinguish between these two control systems, although it should be made clear that it is not easy to do so.

A major component of the sexual response of the female rat is the lordotic reflex, a concave arching of the back that exposes the perineum, permitting intravaginal penetration of the male's penis. It has been known for many years that the lordotic reflex is under hormonal control. The

---

Richard E. Whalen • Department of Psychobiology, University of California, Irvine, California. The research was supported by Grant No. HD-00893 from the National Institute of Child Health and Human Development.

cyclic display of lordotic responses disappears following ovariectomy and the display of such responses can be reinstated by the administration of estrogen and progesterone. In 1942 Beach showed that progesterone acted synergistically to induce lordosis in the estrogen-primed rat. Recently, we reexamined the interactive effects of estradiol benzoate (EB) and progesterone (P) by varying the doses of each concurrently (Whalen 1974). Ovariectomized rats were administered 1, 2, 4, or 8 μg of EB, each dose being combined with 10, 25, 50, 100, or 200 μg of progesterone. As is shown in Figure 7.1, there is a strong relationship between the dose of *each steroid* and the probability that a mounting response by a male will elicit lordosis.

Figure 7.1 presents group averages. It is also important to examine individual differences in hormone response. Figure 7.2 shows the probability of lordosis in four females administered varying doses of EB and a single dose (100 μg) of P. Large individual differences are evident. One of the females showed an almost maximal behavioral response to 1 μg of EB while another female showed no response until the dosage reached 8 μg of EB. Thus, one has a large dose range of response even within a relatively homogeneous group of animals.

In addition to dose–response relationships, studies have revealed potentially important relationships between the time of hormone administration and the resultant display of the behavior. For example, Figure 7.3 shows changes in the probability of lordosis as a function of time after the intravenous administration of free estradiol (Green, Luttge, and Whalen, 1970). Progesterone was given four hours before testing. No behavioral

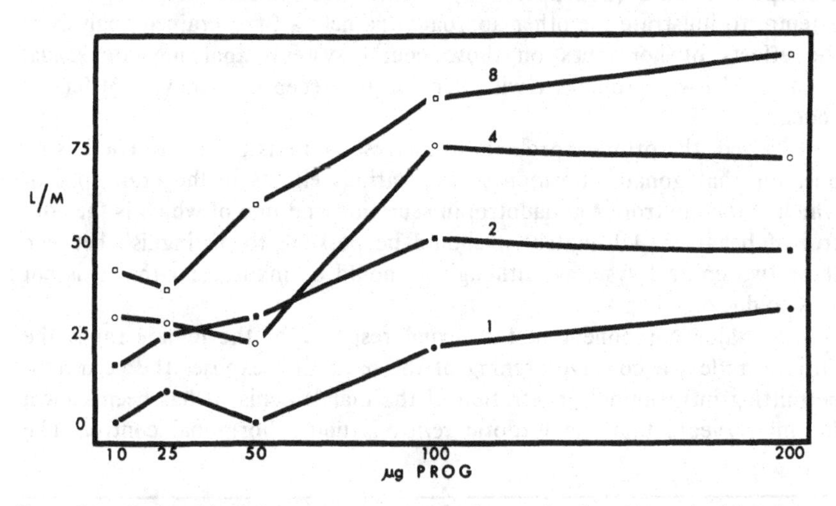

Figure 7.1. Ratio of lordotic responses to mounts by a male of ovariectomized female rats treated with 1, 2, 4, or 8 μg of estradiol benzoate and 10, 25, 50, 100, or 200 μg of progesterone. From Whalen (1974).

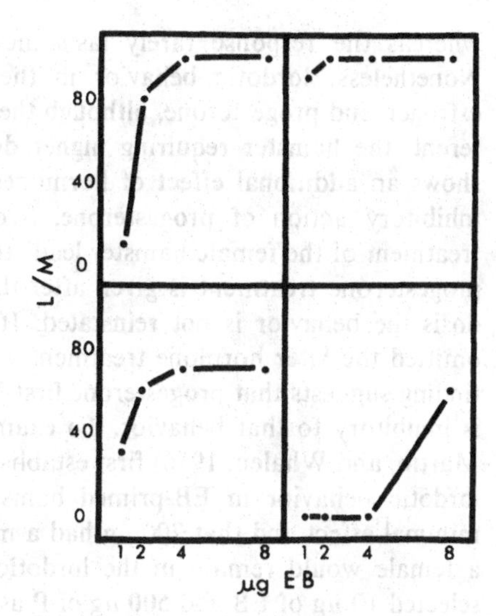

Figure 7.2. Ratio of lordotic responses to mounts by a male of four individual ovariectomized female rats treated with 1, 2, 4, or 8 μg of estradiol benzoate and 100 μg of progesterone. From Whalen (1974).

effects of the estrogen were evident for 15 hr; then between hours 16 and 24 there was an almost linear increase in the intensity with which the behavior was displayed.

Receptive behavior in the hamster differs somewhat from that seen in the rat. The lordotic response is less striking—the back flattens rather than arches; however, the female hamster may hold this response for minutes,

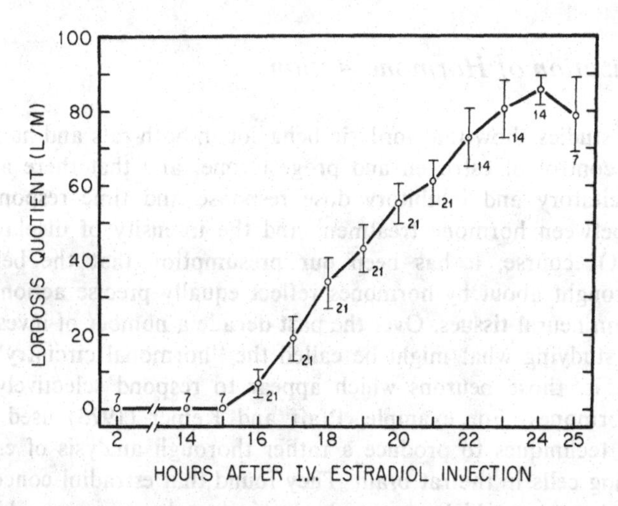

Figure 7.3. Ratio of lordotic responses to mounts by a male of ovariectomized female rats administered free estradiol intravenously as a function of time after estrogen treatment. From Green, Luttge, and Whalen (1970).

whereas the response rarely lasts more than a few seconds in the rat. Nonetheless, lordotic behavior in the hamster is under the control of estrogen and progesterone, although the dose–response relationships are different, the hamster requiring higher doses of estradiol. The hamster also shows an additional effect of hormones upon behavior, namely a clearcut inhibitory action of progesterone. Sequential estrogen and progesterone treatment of the female hamster leads to the display of lordosis. If a second progesterone treatment is given after the animal has ceased displaying lordosis the behavior is not reinstated. If the first progesterone treatment is omitted the later hormone treatment will lead to lordotic responding. This finding suggests that progesterone first facilitates lordosis behavior and then is inhibitory to that behavior. To examine this phenomenon, we (DeBold, Martin, and Whalen, 1976) first established a dose–response relationship for lordotic behavior in EB-primed hamsters. We found that 25 μg had a minimal effect and that 200 μg had a maximal effect on the total time that a female would remain in the lordotic posture during a 10-min test. We selected 10 μg of EB and 500 μg of P as a standard procedure for the induction of lordosis and then interpolated a second treatment with various doses of P. We found that under these conditions, 25 μg of progesterone, a dose which failed to facilitate lordosis, significantly inhibited the EB–P induction of receptivity. Control procedures confirmed that the inhibition was due to the action of the progesterone. A similar inhibitory action of progesterone has been demonstrated in the rat (Nadler, 1970) although it is *much less* dramatic in that species.

1. *Localization of Hormone Action*

These studies show that lordotic behavior in both rats and hamsters is under the control of estrogen and progesterone, and that there are quite precise excitatory and inhibitory dose–response and time–response relationships between hormone treatment and the intensity of display of the behavior. Of course, it has been our presumption that the behavioral changes brought about by hormones reflect equally precise actions of the hormones on neural tissues. Over the past decade a number of investigators have been studying what might be called the "hormonal circuitry" of the brain, that is, those neurons which appear to respond selectively to the gonadal hormones. For example, Pfaff and Keiner (1973) used autoradiographic techniques to produce a rather thorough analysis of estradiol-accumulating cells in the rat brain. They found that estradiol concentrated in a complex system, which transcends classical nuclear groups and includes the medial preoptic region, medial anterior hypothalamus, ventromedial nucleus, arcuate nucleus, ventral premammillary nucleus, amygdala, septum, diagonal band, ventral hippocampus, olfactory tubercle, prepiriform and entorhinal cortex, and even a few cells in the central gray.

In our laboratory we have used liquid scintillation counting techniques to assess the accumulation of radiolabeled steroids by brain tissues. Typically, samples of anterior and posterior hypothalamus, anterior mesencephalon, and cerebral cortical samples are taken. Figure 7.4 shows tissue retention of radioactivity as a function of time after the administration of estradiol or estrone. Following estradiol treatment one finds an anterior to posterior declining concentration gradient. Estrone shows a different concentration pattern with a tendency for more posterior structures to show higher concentrations of the label. For both estrogens, tissue levels of radioactivity decline over time.

When one administers radiolabeled progesterone one finds that the neural accumulation of steroid is quite different from that found with estrogen treatment. The highest concentrations are in mesencephalic rather than in diencephalic structures (Whalen and Luttge, 1971a,b).

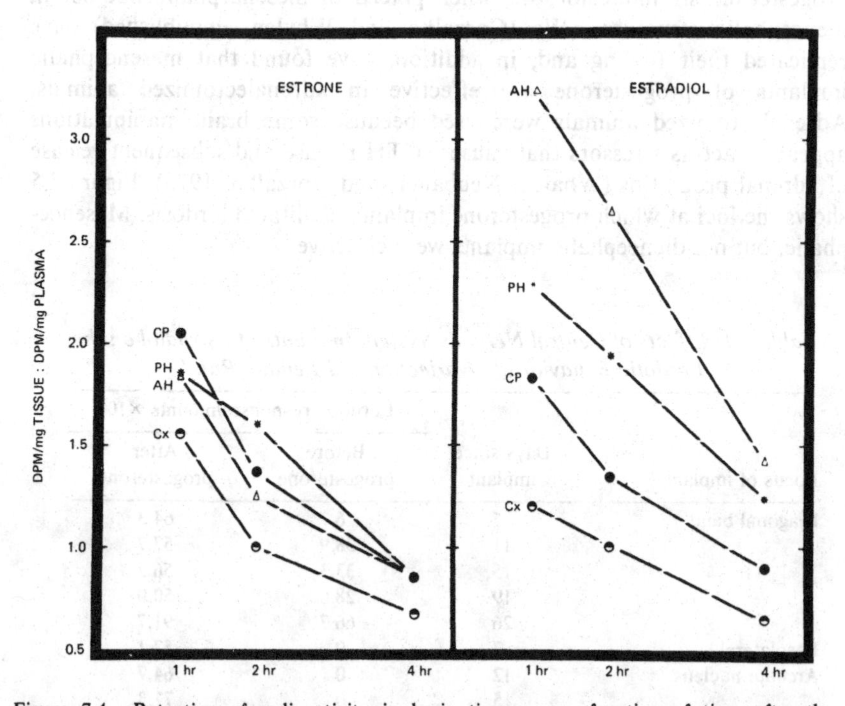

Figure 7.4. Retention of radioactivity in brain tissues as a function of time after the intravenous administration of ^{3}H-estrone or ^{3}H-estradiol. The ratio of disintegrations per minute of tissue samples to dpm of plasma are presented. AH is the anterior hypothalamus; PH the posterior hypothalamus; CP the cerebral peduncle region of anterior mesencephalon; and Cx the parietal cortex.

The findings on the neural localization of radiolabeled hormones might suggest that the effects of estrogen and progesterone on behavior are mediated by different neural structures. Evidence obtained through the local implantation of hormones into the brain indicates that this is the case. In early work Lisk (1962) reported that estradiol was effective in inducing lordotic behavior in the rat when implanted in the region of the suprachiasmatic nucleus. We have found that the sensitive area is more widespread, running from the bed nucleus of the diagonal band anteriorally to the region of the arcuate–ventromedial nucleus, posteriorally. The effects of such implants prior to and following systemic progesterone administration are illustrated in Table 7.1.

It should be noted that implants stimulate lordotic behavior in regions that the autoradiographic and liquid scintillation studies indicate concentrate estradiol. The same appears to be true for progesterone. We found that progesterone tends to accumulate in mesencephalic structures. Ross, Claybaugh, Clemens, and Gorski (1971) found that implants of progesterone stimulate lordosis when placed in mesencephalic but not in diencephalic structures. We (Gorzalka and Whalen, unpublished) have replicated their finding and, in addition, have found that mesencephalic implants of progesterone are effective in adrenalectomized animals. Adrenalectomized animals were used because some brain manipulations appear to act as stressors that cause ACTH release and subsequent release of adrenal progestins (Whalen, Neubauer, and Gorzalka, 1975). Figure 7.5 shows the loci at which progesterone implants facilitated lordosis. Mesencephalic, but not diencephalic implants were effective.

Table 7.1. Effects of Central Nervous System Implants of Estradiol on the Lordotic Behavior of Ovariectomized Female Rats[a]

		Lordotic responses/mounts × 100	
Locus of implant	Days since implant	Before progesterone	After progesterone
Diagonal band	5	6.7	64.3
	11	38.9	57.7
	15	33.3	56.2
	19	28.0	50.0
	26	66.7	91.7
Basolateral	7	0	57.1
Arcuate nucleus	12	0	64.7
	15	0	77.3
	19	0	83.3
	22	0	60.0
	26	0	84.2

[a] On the test day the lordotic behavior was assessed before and three hours after the administration (s.c.) of 500 μg of progesterone.

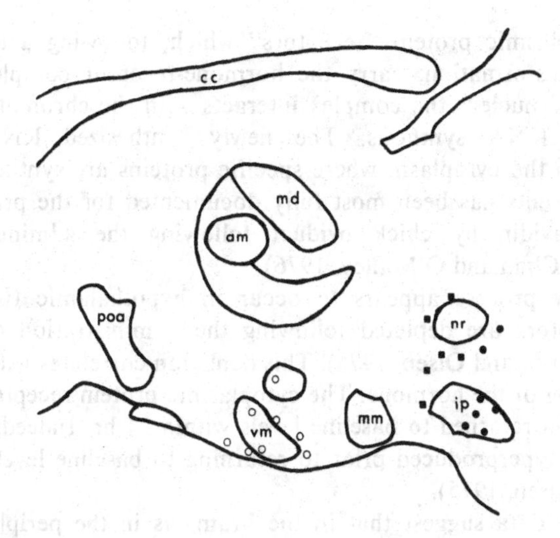

Figure 7.5. Location of progesterone implants in the mesencephalon that stimulated (solid circles and squares) or failed to stimulate (open circles) lordotic responses in ovariectomized (circles) and ovariectomized-adrenalectomized (squares) estrogen-primed female rats.

More recently we have used the implant technique in search of both excitatory and inhibitory loci for progesterone in the hamster. Although several sites have been explored, to date, no implant has facilitated lordosis. Progesterone implants in the posterior hypothalamus and anterior mesencephalon, however, inhibited the estrogen–progesterone induction of receptivity.

2. *Subcellular Studies*

Studies in which we attempted to localize the site of action of gonadal hormones for controlling behavior using both implants and radiolabeled hormone techniques suggest that in the rat and hamster extradiol acts in the diencephalon and progesterone acts in the mesencephalon. It is now possible to take the analysis of hormone action to more molecular levels. For example, Zigmond and McEwen (1970) showed that following the administration of radiolabeled estradiol the hormone comes to be concentrated in the nuclei of hypothalamic, but not of cortical cells. We have replicated this finding (Whalen and Massicci, 1975). In addition we have found that within the nuclei of hypothalamic cells estradiol binds to the chromatin material which is composed of DNA, histones and nonhistone proteins. The kinetics of estradiol binding to brain chromatin are shown in Figure 7.6.

Current theory based on studies of peripheral hormone-sensitive tissues holds that steroid hormones enter cells freely, but that in target tissues they bind to cytoplasmic protein "receptors" which, following a temperature dependent transformation, carry the hormone-receptor complex into the nucleus. In the nucleus the complex interacts with the chromatin material to stimulate RNA synthesis. The newly synthesized RNA is then transported to the cytoplasm where specific proteins are synthesized. This sequence of events has been most fully documented for the production of the protein avidin by chick oviduct following the administration of progesterone (Chan and O'Malley, 1976).

A similar process appears to occur in hypothalamic tissue. Cytoplasmic receptors are depleted following the administration of estradiol (Whalen, Martin, and Olsen, 1975). This depletion correlates with increased nuclear binding of the hormone. The cytoplasmic protein receptors are then replenished and returned to baseline levels within 24 hr. Indeed, these proteins may be hyperproduced prior to returning to baseline levels (Whalen, Martin, and Olsen, 1975).

Thus, the data suggest that in the brain, as in the peripheral target tissues, estradiol binds to a cytoplasmic protein receptor, that the estradiol–receptor complex is translocated to the nucleus where the complex binds to the chromatin, and that receptor protein synthesis is initiated. A similar process of receptor depletion and replenishment is seen in the cytoplasmic compartment of hamster hypothalamic tissue following estrogen administration (DeBold, Martin, and Whalen, 1976, and Figure 7.7.)

To date there has been no demonstration of the existence of cytoplasmic receptors in the brain that selectively respond to progesterone. This is not strong evidence for the absence of such receptors, however, since the receptors could be "fragile" and not survive the extraction procedures needed to demonstrate their existence. Nonetheless, it should be noted that

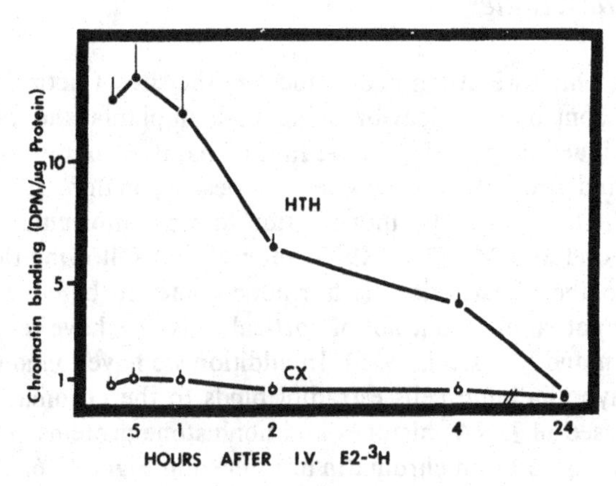

Figure 7.6. Binding of estradiol to nuclear chromatin material from the hypothalamus and cortex of ovariectomized rats administered ^{3}H-estradiol.

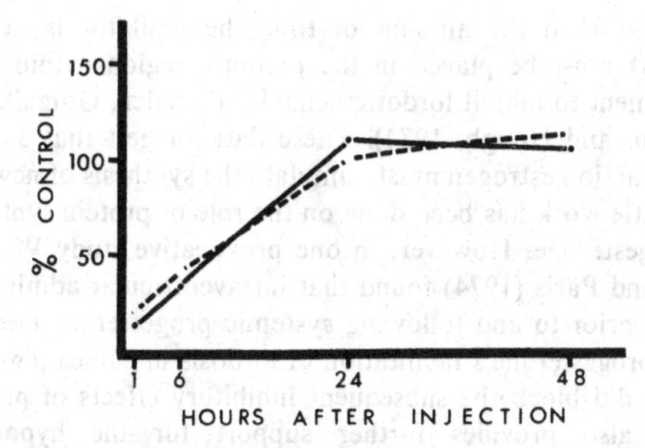

Figure 7.7. Replenishment of cytoplasmic estrogen receptors from the hypothalamus of female hamsters administered either estradiol (solid line) or estradiol plus progesterone (dashed line). One hour after treatment receptors are depleted and return to baseline levels 24 hr after treatment. From DeBold, Martin, and Whalen (1976).

the brain does not show one characteristic of hormone target tissues; namely, a limited capacity in the tissue response to progesterone. Treatment of rats with up to 10 mg of progesterone prior to the administration of a tracer dose of labeled progesterone has no effect upon the concentration of radioactivity found in the brain (Whalen and Gorzalka, 1974). This contrasts strikingly with the inhibition of ^{3}H-estradiol binding by the hypothalamus in rats pretreated with unlabeled estradiol. It is therefore possible that the neural effects of progesterone are not mediated by a limited-capacity cytoplasmic receptor binding system. If this proves to be the case then the models for steroid hormone action generated by studies of peripheral target tissues will have only limited applicability to the brain and new models will have to be developed.

3. *Mechanisms of Action of Estrogen and Progesterone*

As noted above, the brains of rats and hamsters appear to possess cytoplasmic receptors that bind estradiol; the estradiol-receptor complex is translocated to the nucleus where the complex binds to nuclear chromatin. In the uterus and oviduct, chromatin binding of estrogen is followed by the synthesis of specific proteins. To date the hormonal induction in neural tissues of proteins which appear to have specific regulatory funtions has not been demonstrated. However, studies using antibiotic RNA and protein synthesis inhibitors suggest that such must be the case. The intrahypothalamic application of actinomycin-D (Quadagno, Shryne, and Gorski, 1971), or of cycloheximide (Quadagno and Ho, 1975), block estrogen-induced lordosis behavior in rats. A critical feature of the blockade is the time interval between estrogen administration and treatment with the inhibitor, rather than the amount of time the inhibitor is in the brain. Actinomycin-D must be placed in the preoptic region within 6–12 hr of estrogen treatment to inhibit lordotic behavior (Whalen, Gorzalka, DeBold, Quadagno, Ho, and Hough, 1974). These data suggest that as part of its mechanism of action estrogen must stimulate the synthesis of new proteins.

Rather little work has been done on the role of protein synthesis in the action of progesterone. However, in one provocative study Wallen, Goldfoot, Joslyn, and Paris (1974) found that intraventricular administration of cycloheximide prior to and following systemic progesterone treatment had no effects on progesterone's facilitation of lordosis in guinea pigs. However, cycloheximide did block the subsequent inhibitory effects of progesterone. This finding also provides further support for the hypothesis that progesterone's excitatory effects are mediated by a mechanism other than a nuclear-binding–protein-synthesis system.

4. Hormone Metabolism

Investigators of the hormonal control of female mating behavior usually employ the major secretory products of the ovaries, estradiol and progesterone for their studies. These steroids may or may not be the active agents in the control of behavior. It is now known that neural tissue is capable of metabolizing hormones much as the prostate metabolizes testosterone to dihydrotestosterone, the latter being the steroid that is bound to the nucleus of prostatic cells. It is possible therefore that estradiol and progesterone act as prohomones, which are active only following metabolic conversion.

With respect to the estrogen system in the rat, Beyer, Morali, and Vargas (1971) found that at low doses (1 μg/day) estradiol was somewhat more effective than estrone in inducing lordosis, while estriol was ineffective. At a higher dose (4 μg/day) estradiol and estrone were equally effective and estrone was partially effective in stimulating lordotic behavior. These data suggest that estradiol is indeed the active agent, but they also suggest that those cells responding to estrogen are not entirely specific, that is the exact molecular configuration of estradiol is not needed for a response.

If one measures estrogenic metabolites in the female rat brain following the administration of radiolabeled estradiol, one finds that the predominant steroid accumulated is estradiol. Estrone is present, but to a much lesser degree (Kato and Villee, 1967). The estrone that is found is at higher levels in the posterior hypothalamus than in the anterior hypothalamus (Luttge and Whalen, 1970). Following estrone administration 29% of the radioactivity in the anterior hypothalamus is estradiol, while in the posterior hypothalamus only 17% is estradiol. In contrast, 72% of the radioactivity in the pituitary is estradiol following estrone administration

(Luttge and Whalen, 1972). Thus, it would appear that the brain can retain both estradiol and estrone, and that the anterior and posterior hypothalamus differ in the retention patterns of these hormones. It is possible therefore that metabolism may play a role in estrogen action, but the available evidence does not favor that hypothesis.

The situation might be different for progesterone, since it is known that that hormone is readily metabolized by brain (Karavolis and Herf, 1971). We have been studying this problem by comparing the effectiveness of progesterone and its major metabolites in inducing lordosis in estrogen-primed rats and hamsters. In our initial study (Whalen and Gorzalka, 1972) we found that progesterone was the most effective agent followed by its 5α-reduced metabolite dihydroprogesterone (5α-pregnane-3,20-dione); 20α-hydroxyprogesterone and 5α-pregnan-3α-ol-20-one were approximately equivalent and both were less active than dihydroprogesterone. Lordosis was not stimulated by 5α-pregnan-3β-ol-20-one, 5α-pregnan-20α-ol-3-one, or by the oil vehicle used in this study.

Since none of the major metabolities of progesterone was found to be more effective than progesterone itself in eliciting lordosis, we concluded that progesterone need not be metabolized to be active. However, following the completion of this work we became aware of the finding of Kato, Takahashi, and Omori (1971) that dihydroprogesterone (DHP) has a significantly greater anesthetic potency when administered in a Tween-80 (polyoxyethylene sorbitan mono-oleate) vehicle than when administered in an oil vehicle. We (Gorzalka and Whalen, 1977) therefore administered both progesterone and DHP in either oil or Tween-80. With the oil vehicle DHP was again found to be less effective than progesterone. With Tween-80 DHP and progesterone were equally effective and identical to the effect of progesterone in oil. Since DHP was no more effective than progesterone we again concluded that progesterone is not simply a prohormone.

In our most recent work we have been administering progestins intravenously rather than subcutaneously (Kubli-Garfias and Whalen, in press) to estogen-primed rats. Figure 7.8 compares the effects of progesterone and 20α-hydroxyprogesterone. This latter steroid, which we found to have low potency when administered subcutaneously, was more effective than progesterone when given intravenously. In this study we also found that DHP was completely inactive when given intravenously in propyleneglycol, and was only moderately effective in Tween-80. Thus, not only the vehicle, but also the route of administration are critical factors that determine the behavioral effectiveness of progestins. These factors become serious road blocks to the determination of the active agent and role of metabolism in hormone action.

We have also looked for both the excitatory and inhibitory actions of various progestins in the hamster (Martin and Whalen, unpublished) and have found that only progesterone is behaviorally active. Pregnenolone,

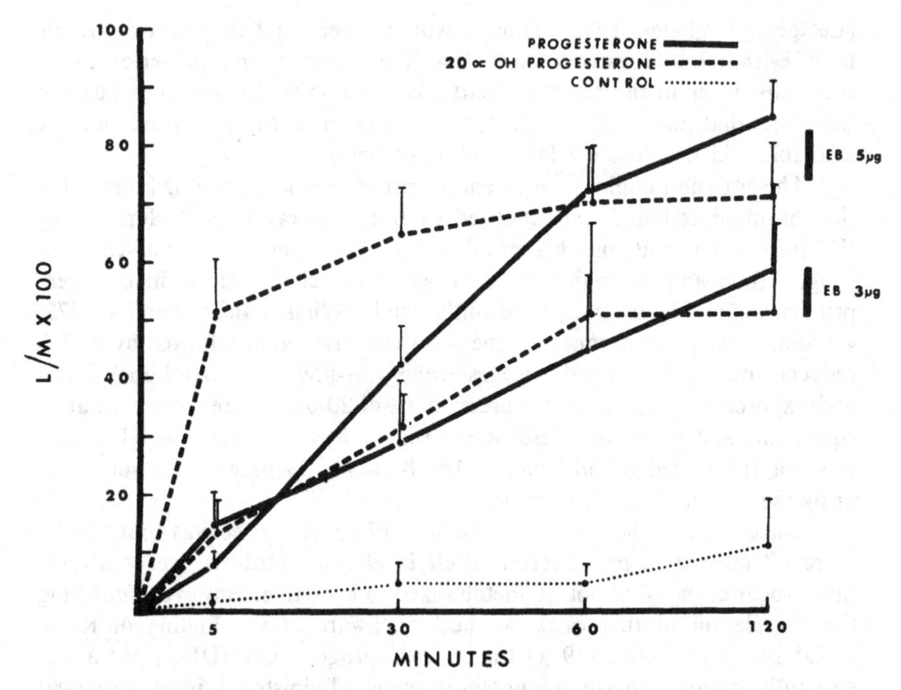

Figure 7.8. Ratio of lordotic responses to mounts by a male of ovariectomized female rats primed with either 3 or 5 μg of estradiol benzoate and administered progesterone, 20α-hydroxyprogesterone, or the control vehicle as a function of time after progestin administration.

20α-hydroxyprogesterone, and dihydroprogesterone exerted no significant behavioral effects, either excitatory or inhibitory. While these data are preliminary, they do suggest major species differences in the neural response to progesterone and its metabolites. Unraveling the biochemical basis of progestin action may indeed be a difficult task.

5. The Future

This brief review has only touched upon some approaches we have taken during the past ten years toward gaining an understanding of how hormones act at the level of the central nervous system to control sexual behavior. A great deal has been learned since 1966 about which steroids can be active in modulating behavior. The concept of "prohormones" has developed and has led to some exciting biochemical and behavioral research. We now know that behavior can be mediated by metabolites of the primary hormones secreted by the gonads and we know that neural tissue is capable of effecting steroid metabolism. We still do not know, however, whether the prohormone concept has *physiological* meaning with respect to the control of behavior. I expect that we will be much closer to an answer to that question a decade from now.

In the past ten years we have learned much about where gonadal steroids work in the central nervous system to mediate behavior. Nonetheless, a careful analysis of the studies employing microimplantation of steroids in the brain reveals a number of inconsistencies; for example, some investigators report that progesterone is active in the diencephalon while others find only mesencephalic implants to be effective. Certainly this controversy should be resolved within the next few years.

Knowing which steroids are physiologically active, and knowing in which cell groups they exert their action may provide us with opportunities for gaining further insight into the biochemistry of hormone action at the cellular level. Studies of the molecular actions of hormones at peripheral target tissues have provided some models for how hormones may be working. I predict, however, that these models will prove inadequate for a detailed understanding of how gonadal steroids act upon brain tissue with its unique intercellular communication system and its lack of propensity for growth and cell division.

The decades to come will reveal the precise biochemical and electrophysiological basis for the synergistic interaction between estrogen and progesterone on the induction of lordotic behavior. These decades will also be characterized by a more precise analysis of the subtle features of behavior, an analysis that Beach has suggested is equally critical for our understanding of hormone–behavior interactions.

Acknowledgments

This account of the research we have been pursuing for several years owes an indebtedness to Frank A. Beach. No attempt has been made to cite all of the relevant sources that have helped to stimulate the research reviewed here. However, I would like to acknowledge the efforts of my students, W. G. Luttge, B. B. Gorzalka, J. F. DeBold, K. L. Olsen, and J. V. Martin, and of my colleague, P. I. Yahr.

References

Beach, F. A. Importance of progesterone to induction of sexual receptivity in spayed female rats. *Proceedings of the Society for Experimental Biology and Medicine,* 1942, *51,* 369–371.

Beach, F. A. Sexual attractivity, proceptivity and receptivity in female mammals. *Hormones and Behavior*, 1976, *7*, 105–138.

Beyer, C., Morali, G., and Vargas, R. Effect of diverse estrogens on estrous behavior and genital tract development in ovariectomized rats. *Hormones and Behavior*, 1971, *2*, 273–278.

Chan, L. and O'Malley, B. W. Mechanism of action of the sex steroid hormones. *New England Journal of Medicine*, 1976, *294*, 1322–1328.

DeBold, J. F., Martin, J. V., and Whalen, R. E. The excitation and inhibition of sexual receptivity in female hamsters by progesterone: Time and dose relationships, neural localization and mechanisms of action. *Endocrinology*, 1976, *99*, 1519–1527.

Gorzalka, B. B., and Whalen, R. E. The effects of progestins, mineralcorticoids, glucocorticoids and steroid solubility on the induction of sexual receptivity in rats. *Hormones and Behavior*, 1977, *8*, 94–99.

Green, R., Luttge, W. G., and Whalen, R. E. Induction of receptivity in ovariectomized female rats by a single intravenous injection of estradiol-17β. *Physiology and Behavior*, 1970, *5*, 137–141.

Karavolis, H. J., and Herf, S. M. Conversion of progesterone by rat medial basal hypothalamic tissue to 5α-pregnane-3,20-dione. *Endocrinology*, 1971, *89*, 940–942.

Kato, J., and Villee, C. A. Preferential uptake of estradiol by the anterior hypothalamus of the rat. *Endocrinology*, 1967, *80*, 567–575.

Kato, R., Takahashi, A., and Omori, Y. The mechanism of sex differences in the anesthetic action of progesterone in rats. *European Journal of Pharmacology*, 1971, *13*, 141–149.

Kubli-Garfias, C. and Whalen, R. E. Induction of lordosis behavior in female rats by intravenous administration of progestins. *Hormones and Behavior* in press.

Lisk, R. D. Diencephalic placement of estradiol and sexual receptivity in the female rat. *American Journal of Physiology*, 1962, *203*, 493–496.

Luttge, W. G., and Whalen, R. E. The accumulation retention and interaction of oestradiol and oestrone in central neural and peripheral tissues of gonadectomized female rats. *Journal of Endocrinology.*, 1972, *52*, 379–395.

Nadler, R. D. A biphasic influence of progesterone on sexual receptivity of spayed female rats. *Physiology and Behavior*, 1970, *5*, 95–97.

Pfaff, D., and Keiner, M. Atlas of estradiol-concentrating cells in the central nervous system of the female rat. *Journal of Comparative Neurology*, 1973, *151*, 121–158.

Quadagno, D. M., and Ho, G. K. W. The reversible inhibition of steroid-induced sexual behavior by intracranial cycloheximide. *Hormones and Behavior*, 1975, *61*, 19–26.

Quadagno, D. M., Shryne, J., and Gorski, R. A. Inhibition of steroid-induced sexual behavior by intrahypothalamic actinomycin-D. *Hormones and Behavior*, 1971, *2*, 1–10.

Ross, J., Claybaugh, C., Clemens, L. G., and Gorski, R. A. Short latency induction of estrous behavior with intracerebral gonadal hormones in ovariectomized rats. *Endocrinology*, 1971, *89*, 32–38.

Wallen, K., Goldfoot, D. A., Joslyn, W. D., and Paris, C. A. Modification of behavioral estrus in the guinea pig following intracranial cycloheximide. *Physiology and Behavior*, 1972, *8*, 221–223.

Whalen, R. E. Estrogen–progesterone induction of mating in female rats. *Hormones and Behavior*, 1974, *5*, 157–162.

Whalen, R. E., and Luttge, W. G. Role of the adrenal in the preferential accumulation of progestin by mesencephalic structures. *Steroids*, 1971, *18*, 141–146.

Whalen, R. E., and Luttge, W. G. Differential localization of progesterone uptake in the brain; role of sex, estrogen pretreatment and adrenalectomy. *Brain Research*, 1971, *33*, 147–155.

Whalen, R. E., and Gorzalka, B. B. The effects of progesterone and its metabolites on the induction of sexual receptivity in rats. *Hormones and Behavior*, 1972, *3*, 221–226.

Whalen, R. E., and Gorzalka, B. B. Estrogen–progesterone interactions in uterus and brain of intact and adrenalectomized immature and adult rats. *Endocrinology,* 1974, *94,* 214–223.

Whalen, R. E., Gorzalka, B. B., DeBold, J. F., Quadagno, D. M., Ho, G., and Hough, J. C., Jr. Studies on the effects of intracerebral actinomycin-D implants on estrogen-induced receptivity in rats. *Hormones and Behavior,* 1974, *5,* 337–343.

Whalen, R. E., Neubauer, B. L., and Gorzalka, B. B. Potassium chloride induced lordosis behavior is mediated by the adrenal glands. *Science,* 1975, *187,* 857–858.

Whalen, R. E., and Massicci, J. Subcellular analysis of the accumulation of estrogen by the brain of male and female rats. *Brain Research,* 1975, *89,* 255–264.

Whalen, R. E., Martin, J. V., and Olsen, K. L. Effect of oestrogen antagonist on hypothalamic oestrogen receptors. *Nature* (London), 1975, *258,* 742–743.

Zigmond, R. E., and McEwen, B. S. Selective retention of oestradiol in specific brain regions of the ovariectomized rat. *Journal of Neurochemistry,* 1970, *17,* 889–899.

8

Reflexive Mechanisms in Copulatory Behavior

Benjamin L. Hart

1. Introduction

Reflexive mechanisms involve some of the most basic considerations in the study of sexual behavior. In female mammals, exclusive of primates, the behavioral signs of receptivity often include spinal reflexes as do the important responses of erection and ejaculation in males of several mammalian species.* Studies of the temporal aspects of sexual activity, including sexual refractoriness, must take into account the role of basic reflexive mechanisms. Finally, the degree to which endogenous and exogenous gonadal steroids influence sexual reflexes of males and females is obviously important in understanding the total picture of hormonal control of sexual behavior.

Sexual reflexes are among the most complex of all visceral and somatic reflexes. In the male, for example, one stimulus can evoke rhythmic contractions in one striated penile muscle, tonic contraction in another penile muscle, vascular engorgement of the erectile tissue of the penis, peristaltic contraction of smooth muscle in the genital tract (resulting in

* Neurologists working with human paraplegics sustaining relatively complete transection of the spinal cord have been aware of the embarrassment suffered by patients from accidental triggering of penile erection. Ejaculation can also be evoked from many human paraplegic patients (Munro, Horne, and Paull, 1948; Talbot, 1955; Zeitlin, Cottrell, and Lloyd, 1957.

Benjamin L. Hart • School of Veterinary Medicine, University of California, Davis, California. Research in the studies by the author has been supported by Grant No. MH 12003 from the National Institute of Mental Health, US Public Health Service.

seminal expulsion), and movements of the hind limbs and trunk. This type of visceral-somatic integration is not commonly seen in other reflexes.

A survey of experimental studies of reflexive mechanisms in sexual behavior of male and female vertebrates has been presented in a comprehensive review by Beach (1967). That review documents that species-specific copulatory patterns consist in part of reflexes mediated by the spinal and myelencephalic areas which are capable of functioning after separation from more rostral parts of the central nervous system. The intent of the present chapter is to survey the research on sexual reflexes in mammals that has occurred since the earlier review by Beach and particularly to emphasize the influence that gonadal hormones apparently have on these reflexes.

Two historical lines of research are appropriate to mention. The relatively simple copulatory pattern of male frogs and toads, in which the male clasps the female (amplexus) throughout the period of oviposition and fertilizes the eggs as they emerge from the cloaca, has provided many investigators with a convenient animal model of sexual reflexes. The early work of Steinach (1910) and Aronson and Noble (1945), and the more recent studies by Hutchinson (1964, 1967), and Hutchison and Poynton (1963) in locating a myelencephalic region that mediates amplexus have provided an excellent model of the role of cephalic disinhibition in the display of sexual reflexes.

The work of the famous investigator, Charles Sherrington, on reflexive mechanisms is widely recognized. However, Sherrington apparently gave little attention to sexual reflexes *per se,* although they obviously caught his interest. In his contribution to a textbook (1900) he refers to an unpublished observation made on spinal male dogs in his laboratory. He noticed that touching the preputial skin over the penis evoked extension of the legs and that stimulating the portion of the penis proximal to the glans would result in a ventral curvature causing the penile bone to be thrust forward. Sherrington apparently did not study erection and ejaculation but did note that the trunk movements suggested themselves as belonging to the act of copulation. These observations by Sherrington and the documentation that erection and ejaculation can be evoked from human males with a completely transected spinal cord (Munro *et al.*, 1948; Talbot, 1955; Zeitlin *et al.*, 1957) provided the initial basis for much of the research I will discuss in this review.

In the discussion that follows, studies of spinal sexual reflexes in which I have been involved are presented first for female and then for male mammals; within these categories studies on various species are grouped together. The chronological order of the investigations was, of course, not this well planned. Initially, I set out to replicate the observations of Sherrington on male dogs and to extend the analysis to include the responses of erection and ejaculation. Later, male rats were included to broaden the species representation. Since then, the work has switched back and forth between males and females and among different species with an orientation towards examining the influence of hormones on reflexive mechanisms. In the conclusion, I will discuss the conceptual issues involved in interpreting the role of reflexive mechanisms in copulatory behavior.

2. *Spinal Reflexes in General*

Before proceeding to a specific discussion of sexual reflexes, I would like to consider briefly experimental work on reflexes in general. The extensive studies by Sherrington (1910, 1947) on the reflexes of decerebrate dogs and cats laid much of the conceptual foundation for our views of reflexive behavior. According to Sherrington, reflexes were not just meaningless, immutable movements in response to specific stimuli. His spinal animals would sometimes scratch parts of the body that had been tickled, withdraw the foot from a needle that was stuck into the foot, swallow if milk was placed into the mouth but reject substances such as acids, and even walk or run according to the speed of a treadmill. Thus, Sherrington stated that such reflexes seemed to have obvious functional meaning. Furthermore, reflexes were not identical each time they occurred, and a response to a weak stimulus would often be weak and to a strong stimulus more intense or longer lasting.

Although neurophysiologists do not agree on an acceptable anatomical or physiological definition of a reflex, I shall, for the purposes of this paper, define reflexes as repeatable, stereotyped responses that are evoked by relatively specific stimuli and that do not require conscious mediation or even cerebral awareness. Reflexes are mediated in presumably localized areas of the brain stem and spinal cord. Since all sexual reflexes are polysynaptic, and there are many interneurons involved in the central mediation of the responses, there is a great deal of opportunity for complex neural processing and integration to occur within the reflex system.

The reason spinal reflexes are studied so much more frequently than those of the brain stem is that the lower part of the spinal cord is particularly easy to isolate surgically and chronic spinal animals are easily and humanely maintained.

2.1. *Spinal Shock*

In isolating a section of the nervous system such as the spinal cord, there is a sudden deprivation of tonic facilitative and inhibitory descending influences (Chambers, Lui, and McCouch, 1973). For a period of time after surgical transection of the spinal cord, reflexes are not evokable or at least are depressed, presumably because the isolated neurons are suffering from a transient or even permanent type of transneuronal degeneration. This depression of reflex activity in the isolated cord is referred to as spinal shock. Most spinal neurons seem to recover from the effects of transection and continue to function in a relatively normal manner. Some activities of the isolated spinal cord may even appear to be hyperreflexive and the development of denervation hypersensitivity, or even the sprouting of new collateral fibers from parent axons have been offered as explanations by which some reflexes become hyperactive (Stavraky, 1961).

Simple reflexes such as the knee jerk and the flexion reflex appear to recover from spinal shock and to gain normal strength within a few hours afer spinal transection. More complex reflexes such as those involved in copulatory responses do not appear normal for weeks. The degree of encephalization of the species is apparently related to the duration of spinal shock. In dogs and cats reflexes recover more rapidly than in primates, and in rats spinal reflexes recover even sooner.

2.2. *Reflexes and Behavior*

A number of reflexes of the rat, dog, cat, and monkey have been examined for their role in behavior (Hart, 1973a; Kellogg, Deese, and Pronko, 1946; Sherrington, 1906, 1947; Stelzner, Ershler, and Weber, 1975). The most intensively studied is the flexion reflex, which is activated by nociceptive stimulation of the limb. Others are the crossed extension, knee jerk, scratching, and defecation reflexes. Standing and walking represent more complicated reflective activity involving complex interactions of opposing sets of flexor and extensor muscles. Sherrington observed that a dog in which the cerebral hemispheres were removed was capable of walking or standing when placed on a treadmill. Work by Shurrager and Dykman (1951) has revealed that cats and dogs subjected to spinal transection at a very young age show a surprising degree of walking ability in the back legs, including jumping and running for short distances. Some dogs transected in adulthood will also display walking and this reflex is facilitated by the administration of strychnine, a drug that blocks postsynaptic inhibition. This indicates that there is a degree of intrinsic spinal inhibition normally associated with the walking reflex (Hart, 1971a).

Rats that are transected in adulthood virtually never show any standing or walking reflexes even if administered a subconvulsive dose of strychnine (Hart, unpublished). The legs are simply dragged about. In a recent study, however, Stelzner *et al.* (1975) found that when rats were transected before 12 days of age the animals later developed standing and walking reflexes that even included placing responses. It appears that in rats, as Shurrager and Dykman found in cats and dogs, early transection allows a more complex level of reflexive activity to develop within the spinal cord.

2.3. Habituation and Conditioning

Another line of research that should be touched upon in regard to spinal reflexes and that may have some relevance of the understanding of sexual reflexes, is the expanding work on habituation of spinal reflexes and classical conditioning in spinal animals. Quantitative relationships between various stimulus parameters and the degree of habituation of spinal reflexes has been studied in spinal cats. Habituation of the flexion reflex in spinal cats shares all the characteristics of habituation of other responses that have thus far been evident in intact animals (Groves and Thompson, 1970; Thompson and Spencer, 1966). Although initially a controversial issue, it now seems to be accepted that a type of classical conditioning can even occur at the spinal level (Patterson, Cegawski, and Thompson, 1973; Patterson, 1977).

2.4. Level of Transection with Regard to Sexual Reflexes

In the following discussion of sexual reflexes of females and males, I will limit my comments to reflexive activity mediated below the midthoracic level of the spinal cord, because all of the studies have involved spinal subjects with a transection at this level. It is probably true that the more complex sexual reflexes involve movement of the head, neck, back and all four limbs as well as somatic and visceral reactions of the genitalia. Thus, subjects with a midthoracic transection show only a limited form of the total reflex system because upper spinal and brain stem areas are not activated by stimulation of the genitalia. Also, afferent input from the head, neck, and upper trunk to reflex elements in the lumber spinal region are blocked by transection. This may be important in the elicitation of some aspects of female receptive postures.

Finally, none of the studies have included a transection of the sympathetic trunk. From the anatomical standpoint, a midthoracic spinal transection is apparently anterior to the central connections between visceral afferents and autonomic innervation of the pelvic genitalia. However, there is the possibility that some visceral afferent nerves may be relaying sensory information from the pelvic genital organs cranially past the level of transection or that some sympathetic innervation originating above the level of transection (possibly activated by forebrain structures) could influence visceral responses in pelvic genital structures. However, there has never been any obvious indication of awareness of spinal subjects of genital stimulation or any sign of supraspinally influenced responses in genital organs of spinal subjects (Hart, unpublished observations).

3. Spinal Sexual Reflexes of Females

3.1. Sensory Mechanisms

The involvement of various sensory regions in the afferent arm makes spinal sexual reflexes somewhat difficult to interpret in females. There are several different tactile sensory fields that probably contribute to the activation of a receptive posture including the neck, back, flanks, surface of the vulva and surrounding perineum, clitoris, and walls of the vagina and posterior cervix. The limited work that has been done suggests a rather confusing picture of the relative importance of these different sensory areas. In spinal female dogs, for example, stimulation of the clitoris evokes rhythmic contractions of the posterior wall of the vagina (Hart, 1970a). In spinal cats stimulation of the clitoris and vaginal wall produces no visible reaction (Hart, 1971b). Yet, in the intact cat, probing the vagina with a glass rod evokes the copulatory cry and even postcoital rubbing and rolling. These responses of the intact cat are eliminated or attenuated by denervation of the vagina but not by denervation of the clitoris (Diakow, 1971).

3.2. Effector Mechanisms

In regard to the effector side of sexual reflexes in females, movements of the limbs and postural adjustments such as leg movement, lordosis, or lateral curvature of the spine have received the most attention. Other possible types of responses, which have not been studied in spinal females, include contraction of smooth muscle in the ovarian stroma, walls of the oviduct, uterus, and vagina. Theoretically, smooth muscle of the female genital tract could be modulated by reflexive processes just as the smooth muscle of the male genital tract lining the vas deferens and urethra is activated during elicitation of the ejaculatory reflex. Modified erectile tissue of the wall of the vagina and clitoris may also become engorged, as does erectile tissue in the penis of males when the genitalia are stimulated. The reason such responses have not been studied is obvious; whereas dilation of blood vessels and contraction of genital tract smooth muscle in males lead to visible penile erection and seminal expulsion, comparable responses in the female would be difficult to observe.

In this summary of experimental work on sexual reflexes of females, we shall consider the cat, dog, and rat. The question of estrogenic facilitation of these reflexes will be discussed when the reflexes are described for each species. The use of autoradiography to identify estrogen-concentrating neurons has now revealed that such neurons do occur in the spinal cord of the rat (Keefer, Stumpf, and Sar, 1973), a finding that certainly suggests a possible estrogenic facilitation of sexual reflexes through direct action on spinal neurons.

3.3. Female Cats

3.3.1. Early Experiments

The first detailed report of spinal sexual reflexes in any female mammal was a publication by Maes in 1939, which was later to become quite controversial because in this publication Maes indicated that these reflexes were altered by estrogen levels. Maes reported that in female cats with a thoracic transection, stimulation of the perineal region evoked dorsal flexion of the pelvis, treading or stepping of the back legs, and lateral deviation of the tail. These are all responses that are very common in the receptive behavior of estrous female cats. Maes reported that these reflexes could only be evoked in his spinal cats if the female was in natural estrus or had been injected with estrogen when the transection was performed.

Shortly after this report by Maes, Bard (1940) stated that in an experiment he conducted with Bromiley, the responses of pelvic elevation, treading, and tail deviation could be evoked from spinal female cats regardless of their hormonal state. Both Maes and Bard left out essential methodological details regarding their experimental procedures, and apparently neither Maes nor Bard and Bromiley maintained their spinal subjects until reflex function had stabilized after recovery from spinal shock. Therefore, they did not report a comparison of the same subjects before and after administration of estrogen.

3.3.2. Reevaluation of Estrogenic Influence

To reexamine this problem, female cats were first spayed and later injected with estrogen to determine how readily treading and tail deviation were evoked by male cats and by manual stimulation by the experimenter (Hart, 1971b). Spinal transection was then performed on these animals and they were maintained with no hormonal stimulation until recovery from spinal shock. The reflexes evoked by perineal stimulation were evaluated in a series of tests before and after estrogen administration. It was found that both treading and lateral tail deviation in response to tactile stimulation of the perineal region were displayed in most spinal subjects after, but not before, administration of estrogen. Pelvic elevation, which was reported by both Maes and Bard, was not seen, probably because the midthoracic transection was not high enough to allow the entire set of back muscles to respond as a unit.

An important observation in the above experiment was that not all spinal animals responded to hormonal treatment in the same manner. For example, in tests on the nine subjects before estrogen administration, one cat displayed treading but no tail deviation and another tail deviation but no treading; the seven other subjects exhibited no responses to tactile stimulation. Following estrogen administration, both treading and tail deviation were seen in seven subjects including the two subjects that showed partial responses before estrogen administration. However, in two subjects no reflexes were evoked either before or after prolonged estrogen administration; somatic reflexes such as leg withdrawal, crossed extension, and knee jerk were active in these subjects. These two animals exhibited strong receptive behavior prior to transection. Thus, it was clear that facilitation from more rostral areas of the spinal cord or brain was probably important for the display of the sexual reflexes in the rear quarters of these animals.

In the spinal cat, probing of the vagina and cervix with a glass rod evoked no response. But, as mentioned above, in intact female cats glass rod stimulation of the vagina and cervix induces an afterreaction characterized by arousal and excitement followed by rolling and rubbing on the floor.

3.4. Female Dogs

3.4.1. Description of Reflexive Responses

The study of spinal female dogs offered an opportunity to examine not only reflexes involved in the receptive posture but also the contractions of muscles surrounding the vagina that can be evoked by stimulation of the clitoris (Hart, 1970a). The signs of receptivity of the intact female dog under conditions of both natural and induced estrus include lateral curvature of the rear quarters (towards the male), looping or arching of the tail to the side and vertical movement of the genital organs when the male begins pelvic thrusting (Figure 8.1). Following complete intromission by the male and engorgement of the glans penis, resulting in a genital lock, most female dogs exhibit a vigorous twisting and turning reaction, which lasts 5–30 sec.

In spinal female dogs it was found that lateral curvature of the rear quarters could be evoked by stimulation of the vulva or perineum (Figure 8.2). It was also possible to monitor, electromyographically, the contraction of the constrictor muscle (constrictor vestibuli) of the posterior vagina and vulva. Spinal female dogs, previously spayed, were tested before and after

Figure 8.1. Receptive postures in a female dog that represent a spinal reflex. Top: Lateral curvature and tail deviation evoked by stimulation of the vulva by the male. Bottom: Lateral curvature and tail deviation is presented to a male who cannot quite reach the female; in this case the receptive posture is displayed in the absence of tactile stimulation of the vulva. Compare these responses to those in Figure 8.2.

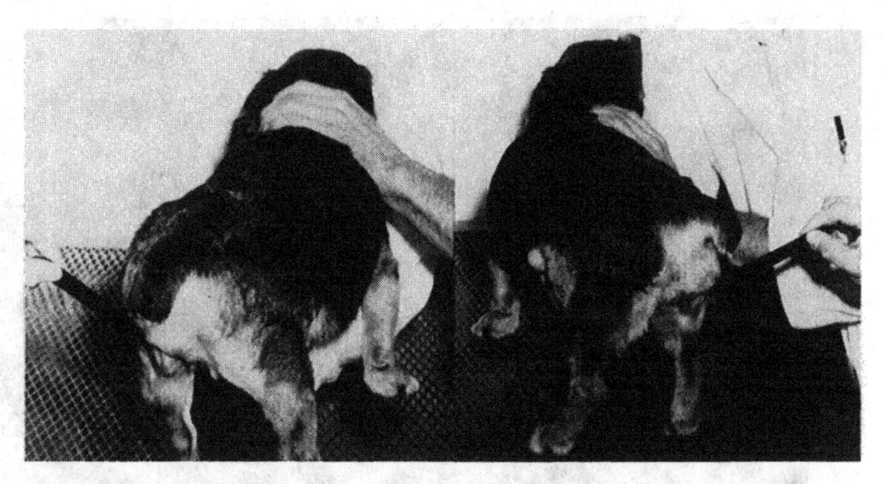

Figure. 8.2. Lateral curvature in an estrogen-treated spinal female dog evoked by stroking the skin on the side of the vulva (with a black tongue depressor) while the subject is restrained in a standing position. Stimulation is on the left side in the left photograph and on the right side in the right photograph. From Hart (1970a,b). Reprinted by permission of Academic Press, Inc.

estrogen administration. Before estrogen administration two of the seven subjects showed a lateral curvature of the rear quarters toward the side of perineal stimulation. Five subjects showed this response following estrogen administration, and the animals showing lateral curvature prior to hormonal treatment exhibited an increase in intensity of the response after estrogen. Lateral curvature was accompanied by tonic unilateral contraction of the constrictor muscle, as illustrated in Figure 8.3. Both the lateral curvature and unilateral contraction of the constrictor muscle are probably activated in the intact female dog when the male rubs his penis against the perineal region. This results in orientation of the genital region of the female toward the male's penis.

Rhythmic contractions of the constrictor muscle occurring at about 1 per sec were evoked when the clitoris was stimulated by digital palpation. This is illustrated in Figure 8.3 (b and c). The duration of these contractions varied from 6 sec to over 2 min. Ventral flexion of the pelvis and leg stepping also occurred during the time of the rhythmic contractions. In contrast to the apparent facilitation of the lateral curvature reflex by estrogen, neither the latency to the onset of the rhythmic contractions of the constrictor muscle nor the duration of the contractions was influenced by estrogen administration. The primary evidence that a reflex consisting of rhythmic contraction of the constrictor muscle does occur in intact female dogs during copulation is that in the first minute or so of the genital lock, rhythmic contractions of the anal sphincter muscle often are apparent.

Since the anal sphincter muscle and the constrictor muscle tend to contract as a unit, the constrictor muscle is probably contracting as well.

3.4.2. Homologous Responses in Females and Males

The constrictor muscle of the female dog is homologous to the bulbocavernosus (bulbospongiosus) muscle of males. In spinal male dogs, rhythmic contraction of the bulbocavernosus muscle, at the rate of about 1 per sec, is characteristic of the intense ejaculatory reaction that is evoked by penile stimulation. Thus, females and males have a similar response to stimulation of homologous genital organs.

One aspect of the study of female sexual reflexes involved six adult spinal male dogs. These subjects were examined for the "female" lateral curvature response by stroking the skin over the scrotal region. In a series of tests conducted before and after injections of estrogen, one dog showed lateral curvature during one test prior to estrogen treatment, whereas five of six dogs showed this response after estrogen treatment.

It seems evident that in dogs, females have the ability to display a response characteristic of males (rhythmic contraction of the striated genital tract musculature) and males can, with the appropriate hormonal stimulation, display a response characteristic of females (lateral curvature of rear quarters).

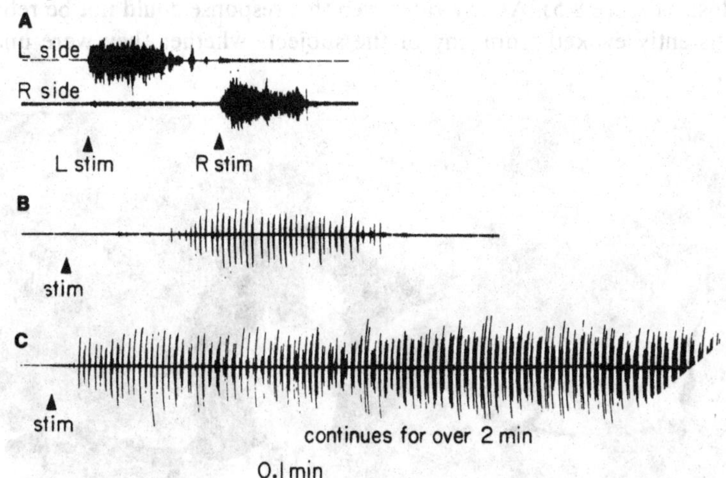

Figure 8.3. Reflexive responses in the constrictor vestibuli muscle of spinal female dogs as monitored electromyographically. (a) Tonic contraction in response to stimulation of the vulva on the left and right sides (note unilateral contraction of the muscle). (b) and (c) Rhythmic contractions evoked by clitoral stimulation from the same female on two different days. Arrows indicate onset of stimulation. From Hart (1970a,b). Reprinted by permission of Academic Press, Inc.

3.5. Female Rats

Rodents are a logical group of animals in which to study reflexes related to the receptive posture because of the frequent reference to the lordotic posture as a lordotic "reflex" (Figure 8.4).

3.5.1. Females Transected in Adulthood

In an attempt to determine if a lordotic reflex could be evoked from spinal female rats, ovariectomized females that had been transected in the midthoracic region were maintained for about 20 days for postsurgical recovery before they were tested (Hart, 1969a). Two procedures were used in an attempt to evoke the lordosis. One was to manually palpate the female in the perineal and flank regions. The other was to allow sexually active male rats to mount the spinal females. The females were given a series of tests prior to any hormonal treatment and after injections of estrogen and progesterone in amounts sufficient to induce receptivity in nonspinal females.

Seven out of nine subjects in one experiment and six out of eight in another showed a slight or moderate arching of the back on at least one test before or after hormonal treatment. With the paralysis of the back legs, it was difficult to evaluate whether this occasional arching represented a true lordosis (Figure 8.5). At any rate, even this response could not be reliably or consistently evoked from any of the subjects whether they were under the

Figure 8.4. Lordosis in a female rat that is often referred to as the "lordotic reflex." From Hart (1969a,b). Reprinted by permission of Academic Press, Inc.

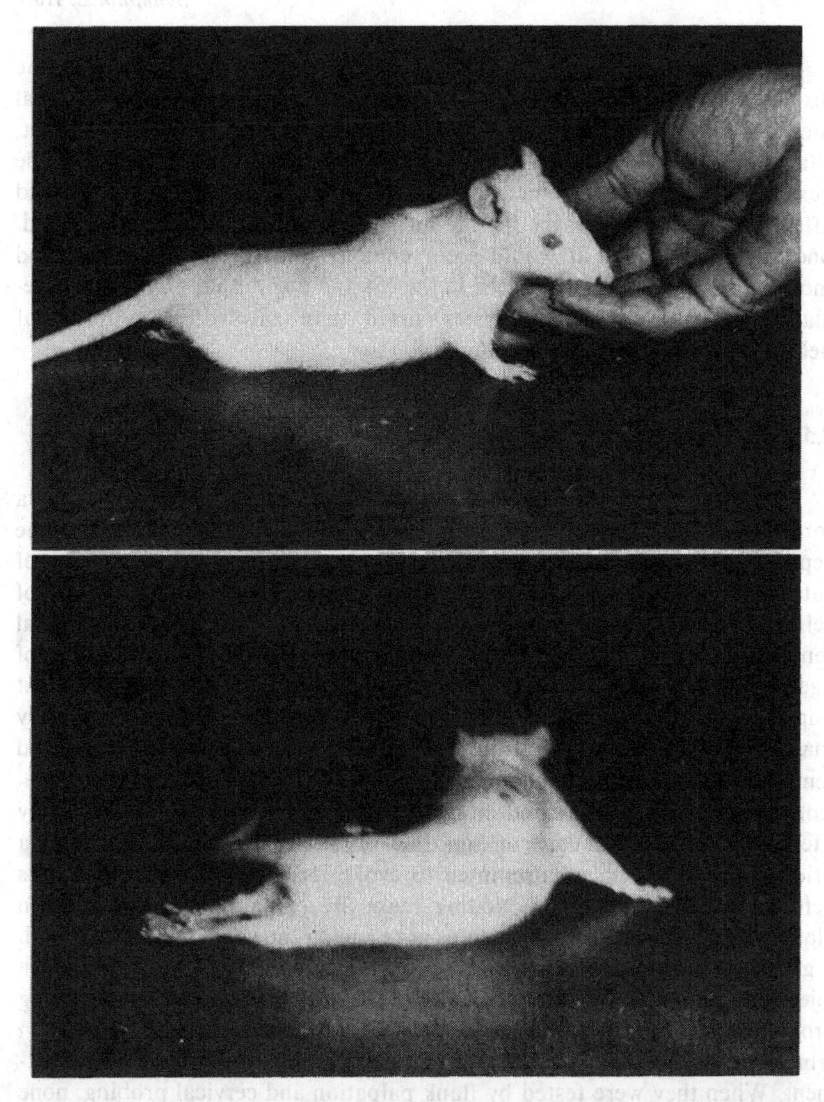

Figure 8.5. This is the closest response to the "lordotic reflex" that has been observed in a spinal female rat. Top: Normal posture. Bottom: Slight arching of the back maintained for a few seconds immediately after manual palpation of the flank region; the body is also curved. From Hart (1969a,b). Reprinted by permission of Academic Press, Inc.

influence of ovarian hormones or not. Neither estrogen alone, estrogen plus progesterone, nor even testosterone (unpublished observation) influenced either the intensity or frequency of this arching. Essentially, the same results were subsequently found by Pfaff, Lewis, Diakow, and Keiner (1972).

One conclusion of the above work on rats is that, for some reason, the display of the lordotic reflex in rats requires the interaction of supraspinal mechanisms. This is the conclusion drawn by Pfaff *et al.* (1972) and, in fact, Pfaff and co-workers, and Kow and Pfaff (1976) have outlined the peripheral sensory requirements and some of the apparent descending and afferent pathways necessary for the display of lordosis. On the other hand, one might argue that it would seem unlikely that receptive postures and movements are spinally mediated in the cat and dog but not in the rat, especially since the rat is less encephalized than carnivores, in terms of behavioral function.

3.5.2. Females Transected Neonatally

In more recent unpublished work, we again attempted to evoke a lordotic response from spinal female rats. The approach was based on the report by Stelzner, Ershler, and Weber (1975) that neonatal transection of rats (prior to day 10) leads to a more complete and integrated type of reflexive behavior than is found in rats transected in adulthood. Neonatal female rats were transected in the midthoracic region at 3–4 or 8–9 days of age and returned to the nest for recovery and care by the mother. Most pups survived and gained weight. The rats were weaned in the customary manner and maintained until they were adults. In our neonatally transected females, the walking and placing activities of the back legs were much superior to those of rats transected in adulthood, confirming the observations by Stelzner *et al.* At 100 days of age they were ovariectomized, and after a brief recovery period we attempted to evoke lordosis in these female rats before hormonal treatment. Neither manually palpating the flank region along with touching of the perineum, nor probing the vagina and cervix with a glass rod evoked any response resembling lordosis. They were then given injections of 40-μg estradiol benzoate followed 72 hr later by a 500-μg progesterone, a hormonal regimen that usually induces a full receptivity in female rats. The subjects were tested 4–6 hr after the progesterone treatment. When they were tested by flank palpation and cervical probing, none of the females exhibited a lordotic response. Even when placed with sexually active males, which were allowed to mount them, the subjects did not display lordosis. The hormonal treatment and subsequent testing was repeated at weekly intervals for several weeks, with consistently negative results.

Whether spinal female rats will ever be found to display a lordotic response under any treatment is probably an open question. It is conceivable that the appropriate methodological techniques such as a higher level of transection may allow the reflex to be displayed by spinal subjects.

4. Spinal Sexual Reflexes of Males

4.1. Sensory Mechanisms

In contrast to the relatively large peripheral sensory field presumably involved in sexual reflexes of females, the sensory field related to sexual reflexes of males seems to be restricted to specific receptors on the penis. Outside of the fact that sectioning the dorsal nerve of the penis appears to prevent erection and ejaculation during mating tests in cats (Aronson and Cooper, 1968), rhesus monkeys (Herbert, 1973), and rats (Larsson and Sōdersten, 1973), there is little known about the actual penile receptors that trigger or evoke erectile and ejaculatory reflexes. In fact, such specific receptors have never been histologically described for any species. In addition to the presumed existence of such genital receptors there are, of course, other receptors located in the penis related to thermal and tactile reception as well as nociception.

The glans penis of the rat contains prominent cornified papillae (Phoenix, Copenhaver and Brenner, 1976) and the encapsulated nerve endings beneath the papillae (Beach and Levinson, 1950) are sometimes thought possibly to represent specialized genital receptors. These receptors are undoubtedly affected by topical anesthetization and intromission is blocked by surface anesthetization of the glans penis of the rat as it is by sectioning the dorsal nerve (Adler and Bermant, 1966; Carlson and Larsson, 1964; Sachs and Barfield, 1970). However, the receptors involved in activating penile movement, erection, and ejaculation are probably not associated with these cornified papillae, but are rather deep pressure receptors located on the body of the penis, proximal to the glans. In rats (Hart, 1968a) and dogs (Hart, 1967a), for example, only deep pressure stimulation to the body of the penis, not stimulation of the glans, will evoke erection and/or ejaculation. Topical anesthetization of the glans penis of male rats attenuates but does not block the elicitation of sexual reflexes (Hart, 1972a). In fact, male rats with the glans penis removed are still capable of showing all the copulatory patterns including ejaculation (Spaulding and Peck, 1974). The superficially located receptors of the glans would appear to function in the male's ability to detect the vaginal orifice rather than activate the responses of erection, ejaculation, or penile movement.

4.2. Effector Mechanisms

On the efferent (or response) side, penile engorgement, seminal expulsion, and movements of the trunk and limbs can all be activated by the same stimulus, indicating that the ejaculatory reflex constitutes a complex of autonomic and somatic motor activities.

Penile erection is apparently initiated by dilation of the helicene arteries in the erectile bodies. Thus, the blood pressure in the erectile tissue reaches that of systemic blood pressure and the erectile bodies fill with blood. However, in the penis of the goat, bull, and horse, pressure in one erectile body (corpus cavernosum penis) exceeds, by several fold, that of systemic blood pressure (Beckett *et al.*, 1972, 1973, 1974). This pressure is caused by the contraction of the striated bulbocavernosus (bulbospongiosus) and ischiocavernosus muscles surrounding this erectile tissue. It is not known whether striated muscle plays a role in the erection of the rat penis, but certainly the rapid engorgement and detumescence would suggest involvement of such striated muscle contractions. It is important to realize that these extrinsic penile muscles are probably androgen sensitive and atrophy in the absence of androgen and undergo recrudescence in its presence.

The response of ejaculation involves (1) the movement of seminal fluid along the ductus deferens, (2) emptying of secretions of various accessory sexual organs into the genital tract (by contraction of smooth muscle in the walls), (3) contraction of the external sphincter muscle of the bladder, (4) peristaltic contraction of the smooth muscle surrounding the urethra, and (5) rhythmic contraction of the striated bulbospongiosus muscle, which surrounds the body of the penis as it emerges from the pelvic cavity.

4.3. *Theoretical Considerations*

There are some theoretical considerations involved in an analysis of the role that spinal mechanisms play in postural adjustments, copulatory movements, penile erection, and ejaculation. One consideration is the degree to which supraspinal facilitation and inhibition normally modulate sexual reflexes. Animal husbandrymen, in attempting to evoke ejaculation for artificial insemination in domestic animals are aware of the suppressive effects of strange environments or threatening situations on ejaculation and semen quality. It is sometimes necessary to "tease" the male with the presence of an estrous female prior to manual penile stimulation to get a full yield of semen. Experimental work has shown that manual stimulation of the penis of normal, intact male rats and dogs in the laboratory situation rarely evokes the same intensity of penile erection and limb movements as that seen in intact animals during copulation (Hart, 1967a; 1968a). This difference could be due to lack of necessary supraspinal facilitation or pronounced supraspinal inhibition of the reflex activity. At least in male dogs, sexual reflexes of spinal subjects match the intensity of comparable responses in intact dogs during copulation. This suggests that sexual reflexes of the intact animal during copulation are displayed under conditions of reduction of supraspinal inhibition rather than an increase in supraspinal facilitation.

Another consideration is whether the postejaculatory refractory period is partially a function of spinal cord refractoriness. It has been proposed that the postejaculatory refractory period should be subdivided into an absolute refractory phase during which no stimulation can elicit copulation (intromission or ejaculation) and a relative refractory phase during which copulation may occur with heightened sexual excitation (Beach and Holz-Tucker, 1949; Larsson, 1959). As we will see, there is reason to believe that at least part of the absolute refractory phase is a reflection of spinal cord refractoriness.

The following summary of experimental work on sexual reflexes of male dogs and rats deals with the role that the reflexes may have in copulatory behavior. The influence of gonadal androgen on these reflexes will also be considered. In this regard, it is important to note that the autoradiographic technique has revealed that spinal motoneurons of males do concentrate androgen (Sar and Stumpf, 1977). The different ways in which androgen may modulate sexual reflexes of males will be dealt with more extensively in the discussion.

4.4. Male Dogs

4.4.1. Sherrington's Observations

The observation of sexual reflexes in male dogs goes back to a report by Sherrington in 1900 in which he describes his observations on a spinal male dog. He states,

> In the dog, after spinal transection above the lumbar region, movements due to the skeletal musculature are easily excited from certain genital regions. Touching the preputial skin evokes bilateral extension at the knees and ankles, and to a less extent at the hip-joints; accompanying this there is depression of the tail. If through the skin more posteriorly to the glans penis be pressed, the posterior end of the body is curved downwards, pushing the penile bone forward. These reflex movements suggest themselves as belonging to the act of copulation, and the suggestion is strengthened by the failure to obtain similar movement on touching analogous parts of the spinal bitch.

As I shall outline below, the spinal male dog displays even more reflexive activity related to copulation than observed by Sherrington.

4.4.2. Mating Behavior and Reflexive Responses

Before describing the reflexive behavior it would be well to briefly review the copulatory behavior of the male dog (Figure 8.6). When placed

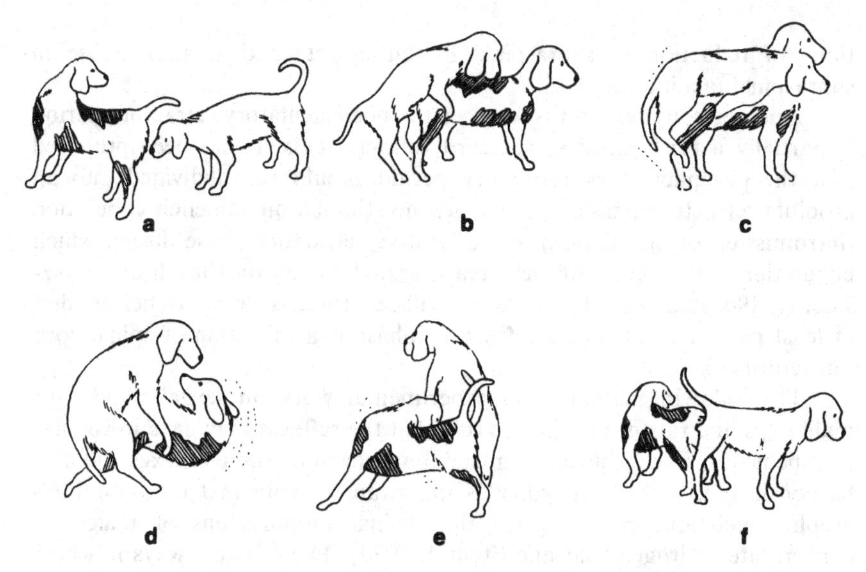

Figure 8.6. A number of reflexive responses in the male and female dog are apparently displayed during mating. In (a) and (b) lateral curvature and tail deviation of the female occurs when the male licks and investigates the female genitalia and during mounting and thrusting by the male. These movements by the female as well as some reflexive movement of the vulva probably aid in intromission by the male. In (c) the intense pelvic movement by the male along with leg stepping, rapid engorgement of the penis and seminal expulsion which occur just after intromission, characterize the reflex referred to as the intense ejaculatory reaction. In (d) twisting and turning of the female, which occurs just after complete engorgement of the penis, may be a response to clitoral stimulation; reflexive rhythmic contractions of the constrictor vestibuli muscle probably occur at this time in response to clitoral stimulation. In (e) and (f) the genital lock is maintained because the erect bulb of the glans penis of the male is too large to be withdrawn from the vaginal orifice. This prolongation of penile engorgement in the male, which is accompanied by some seminal expulsion, is also a reflex. From Hart (1970a). Reprinted by permission of Academic Press, Inc.

with a receptive female, the male usually first investigates the genital region and sometimes makes a few attempts at "playing" with the female. Mounts without thrusting are usually exhibited once or twice before pelvic thrusting occurs along with mounting. Intromission seems to be achieved as a result of "trial and error" pelvic thrusting, and intromission may not occur until the third or fourth mount that includes pelvic thrusting. Just following intromission there is the onset of a highly stereotyped response, in which the front legs are pulled posteriorly, the tail deflected downwards and stepping activity of the rear legs along with pelvic movement initiated. The first stage of ejaculation occurs also occurs at this time. This behavior, which has been referred to as behavior of intromission by Beach (1970) appears to be the manifestation of one complex spinal reflex, which I will refer to as the intense ejaculatory reaction here. The duration of this reaction is usually 15–30 sec; after the termination of the stereotyped leg stepping and pelvic movement, intromission is continued into the genital lock. Most available experimental evidence (Hart, 1970b; 1972b) indicates that the vascular and muscular mechanisms, which keep the bulb of the penis swollen to the extent that it cannot be withdrawn from the vaginal orifice during the lock, are reflexive responses of the male rather than responses of the female dog in restraining or holding the penis within the vagina. During the genital lock the two sexual partners remain tied in the tail-to-tail position for usually 10–30 min. Although both dogs remain quite passive throughout the lock there may occur some twisting and turning by the female at the beginning and possibly some exploration of the surroundings by the animals while locked.

Intermittent rhythmic contractions of the external anal sphincter muscle may be seen, during the lock, and by careful palpation it is evident that these contractions may occur simultaneously with the contraction of the bulbospongiosus (bulbocavernosus) muscle. These rhytjmic contractions often subside when the dogs are standing quietly and are stimulated when the male or female moves. Following separation from the lock, the male licks the genital area and the exposed penis. If the female licks the penis, the male often shows a lordosis or arching of the back, this lordosis also occurs as the penis is withdrawn into the sheath.

One major problem in comparing the reactions of spinal subjects with those of intact animals is that the paralyzed condition of the spinal subjects eliminates general supportive movement of the back legs. The use of electromyography (EMG) for monitoring reactions, such as the contractions of the striated penile muscles, is one procedure that allows for some degree of quantitative comparison between intact and spinal subjects. In work on dogs, we have utilized EMG monitoring of three penile muscles (Hart, 1967a,b; Hart and Kitchell, 1966). In Figure 8.7, contradictions of the ischiourethral, ischiocavernosus, and bulbospongiosus muscles are shown during (a) copulation in an intact male dog and (b) elicitation of an ejaculatory response in a spinal dog.

4.4.3. Evaluation of Four Different Reflexes

Based on contraction patterns of the striated penile muscles, erection, seminal expulsion, and movements of the posterior trunk and hind legs, one can discern four different sexual reflexes in spinal male dogs (Hart, 1967a). One of the reflexes is a response involving shallow thrusting along with partial erection, and is elicited by rubbing the preputial sheath over the glans of the penis. This reflex does not include any expulsion of seminal fluid. A second reflex, of a different type, is characterized by detumescence

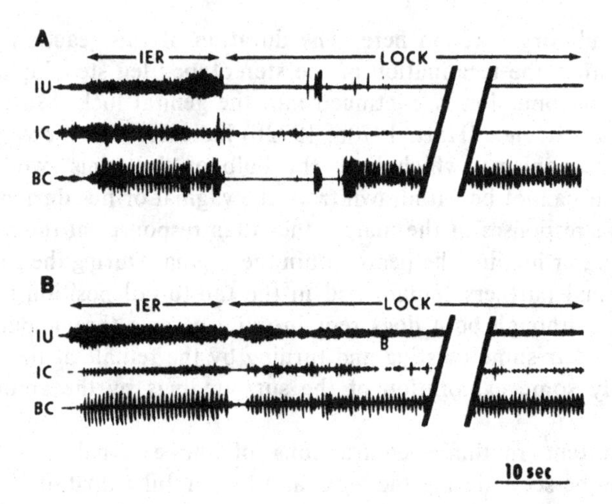

Figure 8.7. Electromyographic recordings from the ischiourethral (IU), ischiocavernosus (IC) and bulbospongiosus (bulbocavernosus) (BC) muscles of male dogs. (a) Recordings during copulation in an intact dog during an intense ejaculatory reaction (IER) and initial part of the copulatory lock. From Hart (1972b). Reprinted by permission of the Wistar Press. (b) Recordings from a spinal male dog during elicitation of an IER and simulated lock. The break in the record indicates a long section has been removed. From Hart (1967a). Copyright 1967 by the American Psychological Association. Reprinted by permission.

of the penis if it is erect at the time of stimulation. The reflex is elicited by touching the coronal area of the glans and is characterized by lordosis of the back, strong extension of the rear legs and penile detumescence.

The two reflexes just discussed appear to play important roles in copulation. The reflex characterized by shallow pelvic thrusting probably contributes to the male dog's ability to achieve complete intromission once partial intromission is obtained. The reflex characterized by lordosis and rapid detumescence may be important in inducing detumescence if penile erection has progressed too far during the male's mounting and thrusting for him to achieve intromission, since complete engorgement of the bulb of the glans would prevent intromission.

Two other distinct reflexes involve erection and ejaculation. The intense ejaculatory reaction is evoked by applying pressure and rubbing the body of the penis just proximal to the glans (Figure 8.7). The duration of this reaction is 15–30 sec in the spinal subject, as it is in the intact dog, and the response is characterized by pelvic movement, alternate stepping of the back legs, plus rapid penile engorgement and the expulsion of seminal fluid. This is the only reflex that seems to have a rather short definite duration with an abrupt ending. Thus, even if stimulation (pressure and rubbing) of the body of the penis is continued, the reflex does not continue or recur, presumably because of a refractory period. This is the best evidence from spinal animals for the existence of a postejaculatory refractory interval that occurs at the spinal level. In a series of subjects in which attempts were made to evoke the intense ejaculatory reaction after different intervals of time subsequent to the elicitation of an initial reaction, it was found that the refractory interval in spinal animals ranged from 5–30 min (Hart, 1967a). In a series of intact dogs, it was found that the postejaculatory refractory interval among several subjects ranged from 1 hr to 24 hr (Hart, 1967a).

If, after the subsidence of the intense ejaculatory reaction in spinal dogs, the penis is still held proximal to the glans, and pressure is applied to the penile tip as it is turned posteriorly, a fourth reflex, referred to as the simulated lock, is evoked. No leg movements occur during this reflex, but full erection of the penis can be maintained for 10–30 min (Figure 8.7). Seminal expulsion continues on an intermittent basis during the reflex. The simulated lock seems to "run down" after 10–30 min, as rhythmic contractions of the bulbospongiosus muscles subside and detumescence of the penis occurs. There is a type of refractoriness that follows this reaction but since it is impossible to evoke the simulated lock without first eliciting the intense ejaculatory reaction, the most that can be said is that the simulated lock probably adds to the refractoriness that immediately follows the intense ejaculatory reaction.

4.4.4. *Influence of Testosterone*

The work on spinal dogs described above was performed on subjects that were castrated prior to transection but maintained on daily injections of replacement testosterone. It was a logical question to ask whether testosterone withdrawal and subsequent administration would effect the intensity and duration of the reflexes. Behavioral studies have shown that while many nonspinal male dogs maintain their ability to copulate following castration, there is an impairment of the responses involved in intromission. The most evident of these impairments is a pronounced decline in the duration of the genital lock (Beach, 1970; Hart, 1968b). There is some indication as well of a reduction in the duration of the intense ejaculatory reaction (Beach, 1970). Since the intense ejaculatory reaction and the genital lock appear to be manifestations of spinal reflexes, spinal dogs were studied to determine if the sexual reflexes were altered by testosterone.

Subjects were castrated prior to transection and then immediately following transection half the subjects were maintained on testosterone replacement while the other half were given no androgen (Hart, 1968c). After a series of tests for all four sexual reflexes following recovery from spinal shock (60 days after transection), the hormone injection schedules were reversed for the two groups and another series of tests conducted 60 days later.

The most obvious effect of the withdrawal of testosterone was a marked reduction in the duration of the simulated lock (Figure 8.8). For example, dogs that initially recieved no testosterone had a mean lock duration of 2 min, which was extended to 10.5 min after 60 days of testosterone treatment. By contrast the dogs initially given testosterone injections and then taken off the hormone had a simulated lock mean duration reduced from 11.6 to 4.2 min, after 60 days. The mean duration of the intense ejaculatory reaction as measured by the duration of rhythmic contractions in the bulbospongiosus muscle was also affected by testosterone. Since there were two groups of dogs maintained on a parallel basis in terms of days recovery from the spinal transection, but in which hormone treatment schedules were reversed, we can assume that these changes in the sexual reflexes were not a

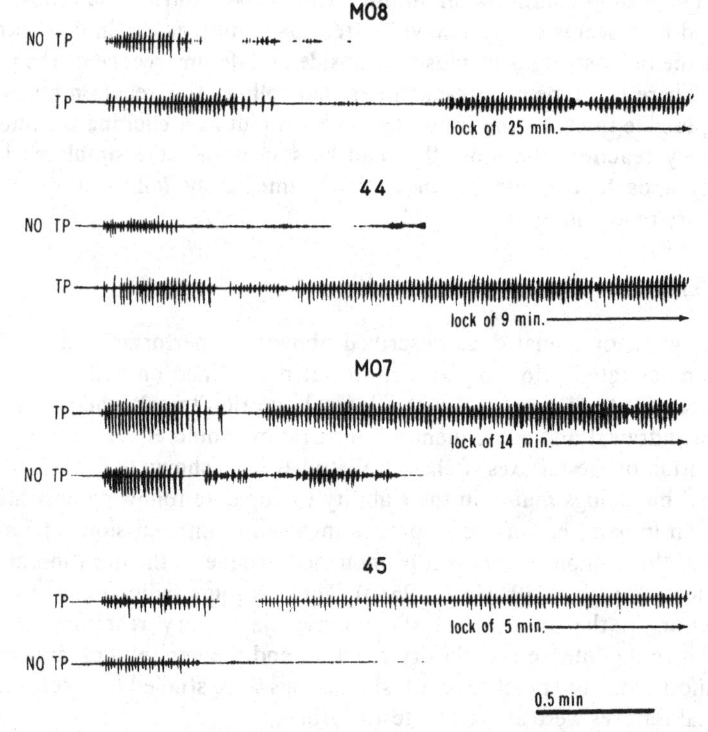

Figure 8.8. Electromyographic recordings from the bulbospongiosus (bulbocavernosus) muscle of 4 spinal male dogs during elicitation of an intense ejaculatory reaction (initial series of contractions on each recording) and simulated lock (onset of rhythmic contractions after a brief pause). Recordings were taken after 60 days of testosterone administration (TP) and 60 days after withdrawal of testosterone (no TP). From Hart (1968c). Copyright 1968 by the American Psychological Association. Reprinted by permisssion.

function of further recovery from spinal shock or the development of hypersensitivity of neurons.

One other effect of testosterone withdrawal was an apparent elevated threshold for the elicitation of the intense ejaculatory reaction; this reflex could not be evoked from some subjects in which testosterone injections were no longer given. Importantly, there was no evident change in the post-ejaculatory refractory period as a function of testosterone withdrawal or administration. Furthermore, there was no change in the response pattern or intensity of skeletal movements involved in either pelvic thrusting or leg stepping. This experiment on the effects of testosterone on sexual reflexes of dogs did not involve a direct examination of the question of whether the effect of hormones was primarily on peripheral sensory receptors, effector mechanisms or on spinal neurons. This question will be dealt with in the following section on spinal reflexes of the male rat and in the general discussion of hormonal influences on spinal neurons.

4.5. *Male Rats*

The above work on sexual reflexes was conducted on dogs because Sherrington's observations had indicated that with further testing a strong correlation could probably be found between copulatory behavior of the intact dog and responses of spinal animals. Indeed, the dog serves as an excellent example of the species specificity of reflexive mechanisms in copulatory behavior. When it became apparent that analysis of the effects of hormones on sexual reflexes was the next logical step, attention was turned primarily to the male rat because copulatory behavior had been studied more extensively in this species, and behavioral studies on the effects of castration revealed that we could expect a more rapid change in sexual reflexes as a function of alteration of androgen than in the dog (Davidson, 1966a).

Because sexual behavior in the male rat and male dog is very different, we were also interested in assessing the degree to which differences in copulatory behavior between dogs and rats might be represented by species-specific differences in sexual reflexes. The male rat displays a series of brief copulatory intromissions before ejaculation and has no genital lock.

4.5.1. *Description of Reflexive Activity*

Spinal male rats that are the typical subjects of the experiments described below have been transected in the midthoracic region and allowed approximately 20 days to recover from spinal shock. By this time the rats are grooming the paralyzed hind quarters, have developed automatic emptying of the urinary bladder, and move about and care for themselves very well.

A rather surprising finding in the initial experiment on spinal male rats was that, unlike dogs, erection and ejaculation could not be evoked by simply touching or rubbing part of the penis (Hart, 1968a). The only body movement that could be evoked in this manner was some kicking of the back legs. When the preputial sheath was held behind the glans, there occurred a series or cluster of genital responses every 2–3 min (Figure 8.9). These responses could not be elicited more frequently by any other type of genital stimulation.

A standard type of test was developed in which the preputial sheath was held behind the glans for 30 min (Hart, 1968a). This puts some pressure on the body of the penis and allows for virtually no stimulation of the glans itself. Clusters of genital responses that are elicited by this type of stimulation begin with three or four partial or complete erections. During some erections the penis becomes so engorged that it resembles a flared cup. The

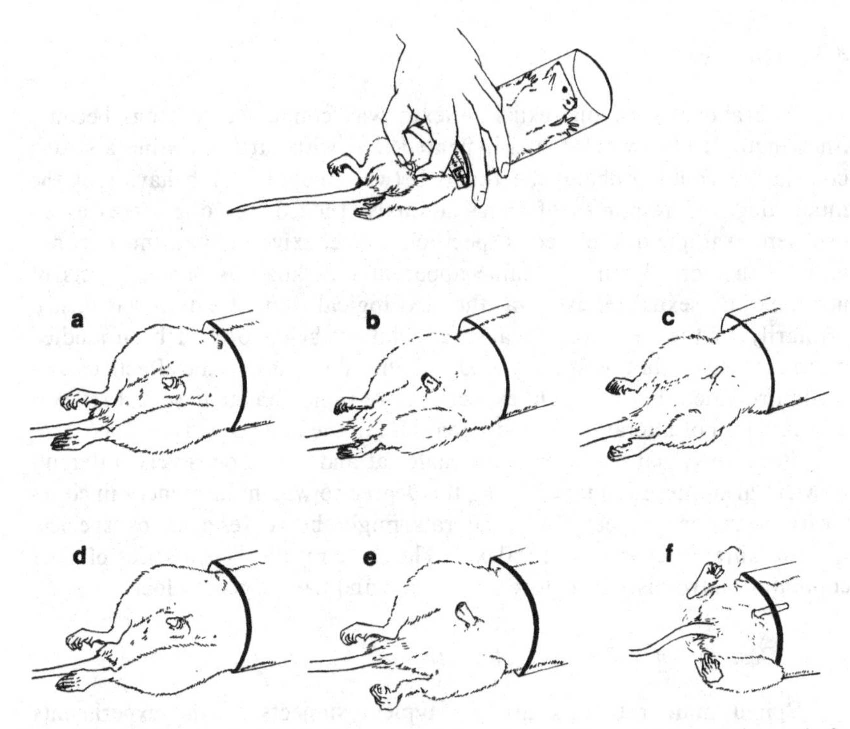

Figure 8.9. Responses in a typical response cluster of a test for sexual reflexes in a spinal male rat showing (a) quiescent stage before onset of response cluster, (b) response cluster beginning with a brief penile erection, (c) a quick flip, which generally followed several erections, (d) brief quiescence stage within a response cluster, (e) another erection within the response cluster, and (f) a long flip, which generally occurred near the end of a response cluster. From Hart (1968a). Copyright 1968 by the American Psychological Association. Reprinted by permission.

tumescence and detumescence of each erection takes about 1 sec, and some erections are followed immediately by a quick dorsal flip of the glans. These are referred to as quick flips and have a duration of less than 1 sec. The quick flip is accompanied by no body or pelvic movement. After 1–3 quick flips there usually occurs a third response consisting of erection of the glans followed immediately by a long extended flip of the glans. This response has a duration of 1–2 sec and is referred to as a long flip. The long flip is accompanied by strong ventral flexion of the pelvis. When additional pressure is applied to the sides of the penis during the long flip, its duration and intensity are often increased and the glans remains briefly engorged in the flip position.

4.5.2. Relation to Copulatory Behavior

Perhaps the major problem with the work on genital responses in spinal male rats is that these responses do not clearly or logically relate to different elements of sexual behavior. The mating pattern of the male rat is characterized by a series of approximately 5–15 intromissions which are followed by ejaculation. Intromissions are usually spaced 30–60 sec apart but may occur as frequently as every 5 or 10 sec. The occurrence of genital responses in spinal animals as clusters appears to have little resemblance to the copulatory responses displayed during mating. It appears likely that the long flip is most closely related to the ejaculatory response, but even this is not certain, since in spinal animals, seminal expulsion has rarely been observed. Furthermore, the type of genital stimulation needed to evoke sexual reflexes in spinal rats does not seem to be comparable to the genital stimulation achieved by the male rat during copulation.

Nonetheless, these sexual reflexes undoubtedly relate in some way to copulatory behavior, and in fact, erections, quick flips, and long flips have been observed in intact male rats during copulation. With genital grooming prevented by fitting the rats with a plastic collar, brief erections of the glans penis were observed after intromissions. Following separation of the male from the female after ejaculation there occurred erections, quick flips, and long flips identical to those seen in our spinal subjects (Hart and Haugen, 1971). These observations gave rise to the notion that in some way the genital grooming that normally follows ejaculations and intromissions prevents or suppresses genital responses such as flipping. Thus, if oral genital grooming is prevented, the responses are overt and quite apparent.

4.5.3. Influence of Testosterone

The first experiment directed toward an analysis of the effects of testosterone on sexual reflexes of male rats involved initially castrating the subjects and subsequently administering daily injections of testosterone propionate (Hart, 1967b). A few days later, a spinal transection was performed and the subjects were allowed to recover from spinal shock. All subjects were then given baseline tests for sexual reflexes while still receiving testosterone injections. Subjects were then separated into two groups, so that one group could be taken off testosterone while the other group was maintained on the androgen. Later, when the initial group was again given testosterone, the second group was taken off. This procedure of counterbalancing two groups, which is illustrated in Figure 8.10, was utilized to control for any possible influences of the effects of further recovery from spinal shock on the intensity or frequency of the sexual reflexes.

The effects of testosterone withdrawal and administration are quite apparent in Figure 8.10. All of the genital responses, erections, quick flips, and long flips, were markedly altered by testosterone, and the effects were very rapid. Within 4 days after withdrawal of testosterone, and within 2 days after testosterone replacement, there were significant changes in the frequency of these genital reponses. Interestingly, the timing mechanism related to the onset of the genital response clusters was not influenced.

This experiment indicated that the decline in ejaculatory and intromission responses of intact rats following castration (Davidson, 1966a; Young, 1961), and the reappearance of these responses following androgen replacement, may be due, in part, to the influence of gonadal androgen at the spinal level. Assuming that the long flip responses of spinal male rats are related to the ejaculatory and possibly intromission responses of intact animals, then the marked decline in long flip following androgen withdrawal is probably related to the weakening of intromission and ejaculatory responses in male rats following castration.

4.5.4. Central Versus Peripheral Action of Testosterone

Because testosterone was administered subcutaneously in the above experiment, it could not be determined whether the effect was primarily one of influence on peripheral receptors and effectors or on neural tissue within the spinal cord. Our initial approach to this question was rather direct. The experiment was based on an approach used by Davidson (1966b) on castrated male rats in which he found activation of sexual behavior following the implantation of crystalline testosterone propionate into preoptic and anterior hypothalamic areas of the brain. The localized effect of the androgen on this part of the brain was evident because examination of accessory sexual organs showed that the increases in sexual activity could not be attributed to elevated systemic levels of androgen as a result of vascular absorption from the implantation site.

Our experiment involved castrated spinal male rats that were maintained without testosterone replacement until recovery from spinal shock

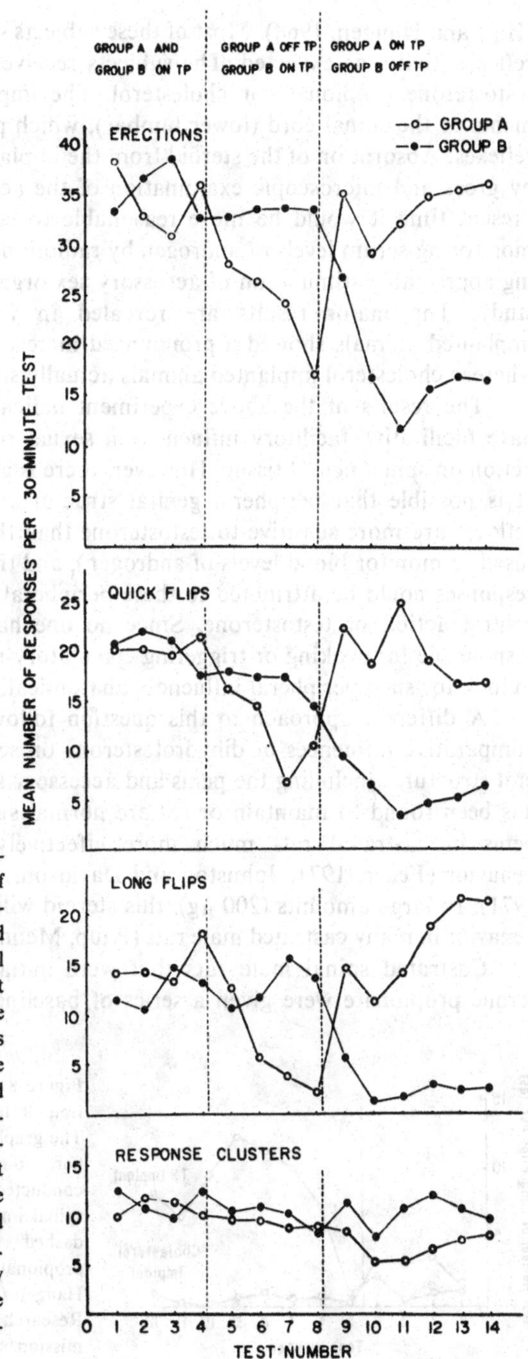

Figure 8.10. Influence of withdrawal and administration of testosterone on number of erections, quick flips, long flips, and response clusters per test for spinal male rats. Tests were conducted at 2-day intervals. When hormone withdrawal is indicated there was no injection given on the day of the last test which was conducted while the animals were on testosterone (TP). When readministration of TP is indicated, the first injection was given 48 hr before the first test which was conducted while the animals were on TP. From Hart (1967b). Copyright by the American Association for the Advancement of Science. Reprinted by permission.

(Hart and Haugen, 1968). Most of these subjects showed a low level of sexual reflex activity, as expected. The subjects received spinal implants of either testosterone propionate or cholesterol. The implant was placed into segments of the spinal cord (lower lumbar), which presumably mediate sexual reflexes. Absorption of the steroid from the implantation site was monitored by gross and microscopic examination of the accessory sex organs (at the present time it would be more reasonable to conduct this experiment by monitoring serum levels of androgen by radioimmunoassay). Subjects showing appreciable stimulation of accessory sex organs were excluded from the study. The major results are revealed in Figure 8.11. Testosterone-implanted animals showed a pronounced increase in numbers of long flips, whereas cholesterol implanted animals actually showed a decline.

The results of the above experiment indicated that testosterone does have facilitative facilitory influence on sexual reflexes by means of direct action on spinal neural tissue. However, there is a problem in interpretation. It is possible that peripheral genital structures involved in triggering the reflexes are more sensitive to testosterone than the accessory sexual organs (used to monitor blood levels of androgen), and that the increase in long flip responses could be attributed to these peripheral influences rather than the central action of testosterone. Since no one has identified the receptors responsible for evoking or triggering ejaculatory responses, it is not possible to look for such peripheral influences anatomically.

A different approach to this question followed some research on the comparative influences of dihydrotestosterone on sexual behavior and peripheral structures including the penis and accessory sexual organs. This steroid has been found to maintain or restore normal size and morphology of the penis in castrated rats much more effectively than it restores sexual behavior (Feder, 1971; Johnston and Davidson, 1973; Whalen and Luttge, 1971). In large amounts (200 μg), this steroid will, however, activate sexual behavior in many castrated male rats (Paup, Mennin, and Gorski, 1975).

Castrated spinal male rats that were initially maintained on testosterone propionate were given a series of baseline tests for sexual reflexes.

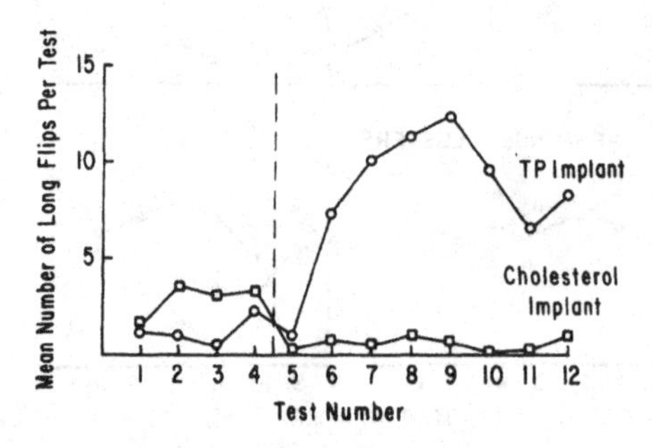

Figure 8.11. Effects of spinal implantation of testosterone in spinal male rats. The graph shows the number of long flips per 30-min test for sexual reflexes conducted every 2 days before and after spinal implantation (indicated by vertical dashed line) of either testosterone propionate or cholesterol. From Hart and Haugen (1968). Copyright 1968 by Brain Research Publications. Reprinted by permission.

Then some of the subjects were given no steroid, others were switched to a low level (25 μg) of dihydrotestosterone, and still others were left on testosterone propionate injections (Hart, 1973b). Those switched to the low level of dihydrotestosterone showed a significant decline in genital responses paralleling the decline of subjects receiving no steroid after the initial testosterone treatment. Those continuing to receive testosterone injections showed no decline in responses. The penile papillae and length and weight of the penile shaft in the dihydrotestosterone-treated subjects were markedly greater than in subjects taken off testosterone. Since, with this low level of dihydrotestosterone sexual reflexes were not maintained at previous levels but peripheral genital morphology was maintained, we can conclude that the changes in sexual reflexes could not be completely accounted for by androgenic influences on peripheral tissues. Thus, this experiment was also consistent with the notion that androgen may directly activate spinal neural elements.

4.5.5. Influence of Other Steroids

Until a few years ago, work on hormones and sexual behavior in male rats had indicated that, whereas testosterone propionate could maintain or restore normal copulatory behavior in castrates, neither dihydrotestosterone nor estrogen alone could restore normal copulatory behavior. However, small amounts of estradiol benzoate given together with moderate amounts of dihydrotestosterone propionate would maintain or restore normal copulatory behavior of castrates. These findings led to the notion that estrogen acts in some way upon the brain and dihydrotestosterone on the penis and that the combined effect of the two is sufficient to activate normal sexual activity. The most recent work has shown that high levels of estradiol benzoate (50–100 μg) administered over a prolonged period of time (about 30 days) can restore the normal copulatory pattern in male rats (Södersten, 1973; Paup, Mennin, and Gorski, 1975), and a high level of dihydrotestosterone (200 μg) is sufficient to restore sexual activity in at least some male rats (Paup, Mennin, and Gorski, 1975).

Since we are assuming that the spinal cord, peripheral sensory receptors and effector structures, and the brain all play a role in hormonal activation of copulatory behavior, it is of interest to determine what effects different steroids have on sexual reflexes of spinal male rats when given in amounts sufficient to evoke copulatory behavior of castrated (nonspinal) male rats. In an experiment still in progress, male rats were first castrated and then subjected to spinal transection a few days later. About 20 days after transection they were given baseline tests for sexual reflexes, and, as expected, the number of genital responses per 15-min test was low. They were then given injections of either testosterone propionate (200 μg), dihydrotestosterone propionate (200 μg), or estradiol benzoate (100 μg progressing to 200 μg 30 days later). Dihydrotestosterone propionate in the high dose, as well as testosterone propionate, markedly stimulated sexual reflexes, but high levels of estrogen had no apparent influence. The estrogen-treated subjects, when later given testosterone propionate, increased their reflex activity. Thus, we are faced with the contradiction in these preliminary data that estradiol benzoate can evoke the complete normal pattern of male sexual behavior in previously castrated, sexually inactive male rats, but that estrogen does not have the same effect as testosterone on sexual reflexes of spinal male rats. It is possible that estrogen and testosterone facilitate different neural mechanisms when they activate sexual activity in male rats.

5. Discussion of Conceptual Issues

5.1. Species-Specific Reflexive Mechanisms

One of the major points emphasized by Beach in his 1967 review of reflexive mechanisms in copulatory behavior was that species-specific copulatory patterns are, in part, comprised of reflexes mediated by spinal or myelencephalic regions that are capable of functioning after separation from other parts of the brain. Certainly our observations on the receptive postures of spinal female dogs and cats and the various reflexes of spinal male dogs support this proposition. This concept can, if fact, be expanded in terms of the possible kinds of behavioral responses included. Work on the dog has indicated that the postejaculatory refractory period is partially a function of spinal refractoriness. The intrinsic timing mechanism that controls the occurrence of genital response clusters in spinal male rats may also relate in some way to a sexual refractory period.

Beach did not deal with the issue of sex-specific reflexive mechanisms, but a basic question might be: Since both males and females can display copulatory responses typical of the opposite sex under certain hormonal conditions, do both sexes possess the reflexive mechanisms characteristic of the opposite sex? This question has been explored to only a very limited degree. In spinal dogs we have seen that females can display a reaction somewhat resembling an ejaculatory response and estrogen-primed males will show the feminine lateral curvature reaction to stimulation of the scrotal skin.

Other experiments I have reviewed here have revealed that there are not, in all species, copulatory reflexes that clearly relate to some aspects of sexual behavior. The lordotic response of rodents, which has been repeatedly referred to as a reflex by numerous workers, has been displayed to only a questionable degree on a few occasions by spinal female rats. No investigator has yet reported the reliable occurrence of a strong lordosis in a spinal or myelencephalic female rodent, regardless of hormonal treatment or type of peripheral stimulation.

Another difficulty lies with evaluating sexual reflexes of male rats. Although erections, quick flips, and long flips are readily evoked, none of these responses has been clearly related to any aspect of copulatory behavior. This is especially puzzling because according to the conventional neuroanatomical perspective the rat is a less encaphalized species than the dog. Yet, in terms of spinal reflexive mechanisms, the dog appears to have a more functionally complete reflexive system. Transecting male rats as neonates, which has allowed various somatic reflex systems to develop more completely than seen in rats transected in adulthood, has not allowed for any enhancement or alteration of sexual reflexes in either male or female rats (Hart, unpublished observation).

5.2. *Effects of Gonadal Hormones on Spinal Reflexes*

It is an accepted belief that gonadal hormones affect copulatory behavior by acting upon various substrates including, of course, the nervous system. Certainly, parts of the brain are altered by such hormones, and we have strong evidence that spinal reflexive mechanisms are modulated by gonadal hormones. At the spinal level, hormones may facilitate sexual reflexes by acting peripherally on receptor or effector mechanisms. Such peripheral action would probably enhance the ease with which a reflex could be evoked or increase the intensity of the response. The central action of hormones on spinal neural elements could have the same effect. It is entirely possible that hormonal modulation of sexual reflexes in some animals occurs both peripherally and centrally. These possibilities will receive more attention below.

5.2.1. *Effects of Peripheral Action of Hormones*

In terms of the peripheral influence of hormones, the most frequently cited examples in males are the changes in penile papillae or spines on the surface of the glans penis of the rat (Beach and Levinson, 1950; Phoenix, Copenhaver, and Brenner, 1976) and cat (Aronson and Cooper, 1967). which undergo atrophy or recrudescence as the animals are castrated or given androgen replacement, respectively. In the rat several investigators have noted changes in weight and size of the entire penis as a function of circulating androgen levels. Aside from any additional effects that androgen may have on sensory receptors *per se*, it is hard to imagine that changes in penile size or structure (papillae or spines) would be without influence on general copulatory behavior and copulatory reflexes. However, long-term castrates can occasionally show the ejaculatory patterns, and even male rats with the glans penis removed are capable of ejaculating (Spaulding and Peck, 1973).

As noted earlier, the receptors that seem to be involved in triggering or releasing ejaculatory responses would appear to be proximal to the glans penis in all of the animals that have been studied. The precise location and structure of these receptors are yet to be determined. Electrophysiologically recording neural activity in nerves supplying the penis would indicate the type of influences peripherally acting hormones may have on neural input from the penis. One such study has been conducted with cats. It was found that sensitivity threshold and pattern of neuronal discharge in sensory nerves supplying the penis did not differ among intact and castrated male cats (Cooper and Aronson, 1974).

For females, there is some recent work that is very suggestive of hormone-induced changes in the peripheral sensory field. Receptors, which are involved in the elicitation of a receptive posture and movements, are undoubtedly tactile in nature and exist over a fairly wide region. The peripheral genital sensory field supplied by the pudendal nerve has been mapped in female rats, and it is now clear that this field is enlarged by the administration of estrogen (Komisurak, Adler, and Hutchison, 1972; Kow and Pfaff, 1973). Although the study of sexual reflexes in female rats has been disappointing, we have seen that for most female cats and dogs there is an apparent estrogen facilitation of a sexual reflex. Part of this facilitation could be accounted for by a type of enlargement of the peripheral genital sensory field by estrogen, allowing a greater summation of sensory stimuli to arrive at spinal neurons that mediate reflexive responses.

Some attention should also be given to the possibility of changes in effector elements of sexual responses. The striated muscles of the penis of the rat (also the levator ani muscle) are influenced by androgen (Hayes, 1965); presumably penile muscle size is altered in other species as well. However, in terms of effects on sexual reflexes, one would expect this to alter the strength of muscle contractions, and possibly the degree of erection (to the extent that full erection is dependent upon the contraction of striated penile muscles), but not necessarily the pattern or frequency of muscle contractions.

5.2.2. *Possible Central Action of Hormones*

The direct activation of spinal neurons by gonadal hormones is a concept that should be viewed in light of the criteria that have established that certain specific neural elements in the brain are activated by these hormones. First of all, it is known that particular areas of the hypothalamus, preoptic region, amygdala, and other limbic areas concentrate estradiol (Pfaff and Keiner, 1973; Stumpf, 1968, 1970; Stumpf and Sar, 1971, 1976). Testosterone is concentrated by neurons in several areas, including parts of the hypothalamus, preoptic nuclei, and limbic structures (Pfaff, 1968; Sar and Stumpf, 1973). Second, studies have shown that female sexual behavior may be activated by implanting estrogenic substances into parts of the hypothalamus of cats (Harris and Michael, 1964), rabbits (Palka and Sawyer, 1966), and rats (Barfield and Chen, 1977). Copulatory activity in male rats can be activated by the implantation of testosterone into the preoptic and other hypothalamic areas (Davidson, 1966b; Johnston and Davidson, 1973; Kierniesky and Gerall, 1973).

According to conventional conceptualization, the spinal cord functions primarily as a relay station between the limbs and trunk and the brain. There is a tendency to overlook the complexity of behavioral mediation that can occur in the spinal cord. It has been well documented now that the complex neural processes of habituation and classical conditioning occur at the spinal level (Groves and Thompson, 1970; Patterson, 1977; Thompson and Spencer, 1966). Perhaps the neural events that occur during the post-ejaculatory refractory period of spinal male dogs, and the timing of genital clusters of spinal male rats, relate to some of these complex processes. In terms of capability for complex neural processing, spinal neurons, including those involved in sexual reflexes, would appear to be qualitatively no different than those located in the basal forebrain or brain stem.

It is evident that alcohol, which is usually thought to affect higher neural structures more than lower centers, may have a greater effect on reflexive mechanisms involved in copulatory behavior than on brain structures involved in sexual motivation. A certain dose of alcohol will impair the ability of a male dog or rat to copulate even though they appear to be strongly "interested" in copulating with females (Hart, 1968d, 1969b). This dose of alcohol, when given to spinal male dogs and rats, markedly impairs the genital reflexive mechanisms, suggesting that the loss of sexual potency in the inebriated individual is the result of the lability of spinal reflexive mechanisms to blood-borne chemicals including alcohol.

One of the compelling reasons for seriously entertaining the possibility that spinal neurons are directly affected by gonadal hormones is the recent evidence for selective uptake of both estrogen and androgen by spinal neurons. Using autoradiography, estrogen-concentrating neurons have been found in laminas 5 and 6 of the dorsal horn at several segmental levels in the female rat spinal cord (Keefer, Stumpf, and Sar, 1973; Stumpf and Sar, 1976). More recently, Sar and Stumpf (1977) have observed the accumulation of H^3-dihydrotestosterone in nuclei of motor neurons of the ventral horn of the cervical, thoracic and lumbar regions of the spinal cord of the male rat. It is probably safe to conclude that testosterone would also be concentrated by such neurons. The uptake of hormones by all regions of the spinal cord suggests that perhaps more responses than those seen in animals with midthoracic transections are actually influenced. The procedure of studying spinal animals with the midthoracic transection precludes observations of the action of hormones on neck and front leg movements, but the receptive postures of females and indeed copulatory movements of males do involve these regions of the body as well.

Although it is not the most parsimonious view, it is probably safest, considering available evidence, to assume that gonadal hormones alter sexual behavior at the spinal level as well as by acting on the brain. The modulation of sexual reflexes in turn involves perhaps the integration of hormonal effects on both peripheral receptors and effectors as well as central neural elements.

ACKNOWLEDGMENTS

I would like to thank Benjamin D. Sachs and Donald A. Dewsbury for their reading of the manuscript and helpful suggestions.

References

Adler, N., and Bermant, G. Sexual behavior of male rats: Effects of reduced sensory feedback. *Journal of Comparative and Physiological Psychology,* 1966, *61,* 240–243.

Aronson, L. R., and Cooper, M. L. Penile spines of the domestic cat: Their endocrine-behavior relations. *The Anatomical Record,* 1967, *157,* 71–78.

Aronson, L. R., and Cooper, M. L. Desensitization of the glans penis and sexual behavior in cats. In M. Diamond (Ed.), *Perspectives in reproduction and sexual behavior.* Bloomington: University of Indiana Press, 1968.

Aronson, L. R., and Noble, G. K. The sexual behavior of Anura. 2. Neural mechanisms controlling mating in the male leopard frog, *Rana pipiens. Bulletin of the American Museum of Natural History,* 1945, *86,* 87–135.

Bard, P. The hypothalamus and sexual behavior. *Research Publications of the Association of Nervous and Mental Disorders,* 1940, *20,* 551–579.

Barfield, R. J., and Chen, J. J. Activation of estrous behavior in ovariectomized rats by intracerebral implants of estradiol benzoate. *Endocrinology,* 1977,

Beach, F. A. Cerebral and hormonal control of reflexive mechanisms involved in copulatory behavior. *Physiological Review,* 1967, *47,* 289–316.

Beach, F. A. Coital behavior in dogs: VI. Long-term effects of castration upon mating in the male. *Journal of Comparative and Physiological Psychology,* 1970, *70*(3), pt. 2.

Beach, F. A., and Holz-Tucker, A. M. Effects of different concentrations of androgen upon sexual behavior in castrated male rats. *Journal of Comparative and Physiological Psychology,* 1949, *42,* 433–453.

Beach, F. A., and Levinson, G. Effects of androgen on the glans penis and mating behavior of castrated male rats. *Journal of Experimental Zoology,* 1950, *144,* 159–171.

Beckett, S. D., Hudson, R. S., Walker, D. F., Vachon, R. I., and Reynolds, T. M. Corpus cavernosum penis pressure and external penile muscle activity during erection in the goat. *Biology of Reproduction,* 1972, *7,* 359–364.

Beckett, S. D., Hudson, R. S., Walker, D. F., Reynolds, T. M., and Vachon, R. I. Blood pressure and penile muscle activity in the stallion during coitus. *American Journal of Physiology,* 1973, *225,* 1072–1075.

Beckett, S. D., Walker, D. F., Hudson, R. S., Reynolds, T. M., and Vachon, R. I. Corpus cavernosum penis pressure and penile muscle activity in the bull during coitus. *American Journal of Veterinary Research,* 1974, *35,* 761–764.

Carlson, S. G., and Larsson, K. Mating in male rats after local anesthetization of the glans penis. *Zeitschrift Tierpsychology,* 1964, *21,* 854–856.

Chambers, W. W., Lui, C. N., and McCouch, G. P. Anatomical and physiological correlates of plasticity in the central nervous system. *Brain, Behavior and Evolution,* 1973, *8,* 5–27.

Cooper, K. K., and Aronson, L. R. Effects of castration on neural afferent responses from the penis of the domestic cat. *Physiology and Behavior,* 1974, *12,* 93–107.

Davidson, J. M. Characteristics of sex behaviour in male rats following castration. *Animal Behaviour,* 1966a, *14,* 266–272.

Davidson, J. M. Activation of the male rat's sexual behavior by intracerebral implantation of androgen. *Endocrinology,* 1966b, *79,* 783–794.

Diakow, C. Effects of genital desensitization on mating behavior and ovulation in the female cat. *Physiology and Behavior,* 1971, *7,* 47–54.

Feder, H. H. The comparative actions of testosterone propionate and 5α-androstran-17βol-3-one propionate on the reproductive behaviour, physiology and morphology of male rats. *Journal of Endocrinology,* 1971, *51,* 241–252.

Groves, P. M., and Thompson, R. F. Habituation: A dual-process theory. *Physiological Review,* 1970, *173,* 16–43.

Harris, G. W., and Michael, R. P. The activation of sexual behaviour by hypothalamic implants of oestrogen. *Journal of Physiology,* 1964, *171,* 275–301.

Hart, B. L. Sexual reflexes and mating behavior in the male dog. *Journal of Comparative and Physiological Psychology,* 1967a, *64,* 388–399.

Hart, B. L. Testosterone regulation of sexual reflexes in spinal male rats. *Science,* 1967b, *155,* 1282–1284.

Hart, B. L. Sexual reflexes and mating behavior in the male cat. *Journal of Comparative and Physiological Psychology,* 1968a, *65,* 453–460.

Hart, B. L. Role of prior experience in the effects of castration on sexual behavior of male dogs. *Journal of Comparative and Physiological Psychology,* 1968b, *66,* 719–725.

Hart, B. L. Alteration of quantitative aspects of sexual reflexes in spinal male dogs by testosterone. *Journal of Comparative and Physiological Psychology,* 1968c, *66,* 726–730.

Hart, B. L. Effects of alcohol on sexual reflexes and mating behavior in the male dog. *Quarterly Journal of Studies on Alcohol,* 1968d, *29,* 839–844.

Hart, B. L. Gonadal hormones and sexual reflexes in the female rat. *Hormones and Behavior,* 1969a, *1,* 65–71.

Hart, B. L. Effects of alcohol on sexual reflexes and mating behavior in the male rat. *Psychopharmacologia,* 1969b, *14,* 377–382.

Hart, B. L. Mating behavior in the female dog and the effects of estrogen on sexual reflexes. *Hormones and Behavior,* 1970a, *2,* 93–104.

Hart, B. L. Reproductive system. Male. In A. C. Andersen (Ed.), *The beagle as an experimental dog.* Ames: Iowa State University Press, 1970b.

Hart, B. L. Facilitation by strychnine of reflex walking in spinal dogs. *Physiology and Behavior,* 1971a, *6,* 627–628.

Hart, B. L. Facilitation by estrogen of sexual reflexes in female cats. *Physiology and Behavior,* 1971b, *7,* 675–678.

Hart, B. L. Sexual reflexes in the male rat after anesthetization of the glans penis. *Behavioral Biology,* 1972a, *7,* 127–130.

Hart, B. L. The action of extrinsic penile muscles during copulation in the male dog. *The Anatomical Record,* 1972b, *173,* 1–6.

Hart, B. L. Reflexive behavior. In G. Berman (Ed.), *Perspectives on animal behavior.* Glenview, Ill.: Scott, Foresman, 1973a.

Hart, B. L. Effects of testosterone propionate and dihydrotestosterone on penile morphology and sexual reflexes of spinal male rats. *Hormones and Behavior,* 1973b, *4,* 239–246.

Hart, B. L., and Haugen, C. M. Activation of sexual reflexes in male rats by spinal implantation of testosterone. *Physiology and Behavior,* 1968, *3,* 735–738.

Hart, B. L., and Haugen, C. M. Prevention of genital grooming in mating behaviour of male rats (*Rattus norvegicus*). *Animal Behaviour,* 1971, *19,* 230–232.

Hart, B. L., and Kitchell, R. L. Penile erection and contraction of penile muscles in the spinal and intact dog. *The American Journal of Physiology,* 1966, *210,* 257–261.

Hayes, K. J. The so-called "levator ani" of the rat. *Acta Endocrinologica,* 1965, *48,* 337–347.

Herbert, J. The role of the dorsal nerves of the penis in the sexual behaviour of the male rhesus monkey. *Physiology and Behavior,* 1973, *10,* 293–300.

Hutchison, J. B. Investigation on the neural control of clasping and feeding in *Xenopus laenis* (Daudin). *Behaviour,* 1964, *24,* 47–66.

Hutchison, J. B. A study of the neural control of sexual clasping behaviour in *Rana angolensis* bocage and *Bufo regularis* reuss with a consideration of self-regulatory hindbrain systems in the anura. *Behaviour,* 1967, *28,* 1–57.

Hutchison, J. B., and Poynton, J. C. A neurological study of the clasp reflex in *Xenopus laenis* (Daudin). *Behaviour,* 1963, *22,* 41–63.

Johnston, P., and Davidson, J. M. Intracerebral androgens and sexual behavior in the male rat. *Hormones and Behavior,* 1973, *3,* 345–357.

Keefer, D. A., Stumpf, W. E., and Sar, M. Topographical localization of estrogen-concentrating cells in the rat spinal cord following ^{3}H-estradiol administration. *Proceedings of the Society for Experimental Biology and Medicine,* 1973, *143,* 414–417.

Kellogg, W. N., Deese, J., and Pronko, W. N. On the behavior of the lumbo-spinal dog. *Journal of Experimental Psychology,* 1946, *36,* 503–511.

Kierniesky, N., and Gerall, A. A. Effects of testosterone propionate implanted in the brain on the sexual behavior and peripheral tissue of the male rat. *Physiology and Behavior,* 1973, *11,* 633–640.

Komisaruk, B. R., Adler, N. T., and Hutchison, J. Genital sensory field: Enlargement by estrogen treatment of female rats. *Science,* 1972, *178,* 1295–1298.

Kow, L. M., and Pfaff, D. W. Effects of estrogen treatment on the size of receptive field and response threshold of pudendal nerve in the female rat. *Neuroendocrinology,* 1973, *13,* 299–313.

Kow, L. M., and Pfaff, D. W. Sensory requirements for the lordosis reflex in female rats. *Brain Research,* 1976, *101,* 47–66.

Larsson, K. Effects of prolonged postejaculatory intervals in the mating behavior of the male rat. *Zeitschrift für Tierpsychologie,* 1959, *16,* 628–632.

Larsson, K., and Södersten, P. Mating in male rats after section of the dorsal penile nerve. *Physiology and Behavior,* 1973, *10,* 567–571.

Maes, J. P. Neural mechanisms of sexual behaviour in the female cat. *Nature* (*London*), 1939, *144,* 598–599.

Munro, D. H., Horne, H. W., and Paull, D. P. The effect of injury to the spinal cord and

cauda equina on the sexual potency of men. *New England Journal of Medicine*, 1948, *239*, 903–911.
Palka, Y. S., and Sawyer, C. H. The effects of hypothalamic implants of ovarian steroids on oestrus behavior in rabbits. *Journal of Physiology*, 1966, *185*, 251–267.
Patterson, M. M. Mechanisms of classical conditioning and fixation in spinal mammals. In A. Riesen and R. F. Thompson (Eds.), *Advances in psychobiology*, Vol. 3. New York: Wiley, 1977.
Patterson, M. M., Cegawski, C. F., and Thompson, R. F. Effects of a classical conditioning paradigm on hind-limb flexor nerve response in immobilized spinal cats. *Journal of Comparative and Physiological Psychology*, 1973, *84*, 88–97.
Paup, D. C., Mennin, S. P., and Gorski, R. A. Androgen- and estrogen-induced copulatory behavior and inhibition of luteinizing hormone (LH) secretion in the male rat. *Hormones and Behavior*, 1975, *6*, 35–46.
Pfaff, D. Autoradiographic localization of radioactivity in rat brain after injection of tritiated sex hormones. *Science*, 1968, *161*, 1355–1356.
Pfaff, D. W., and Keiner, M. Atlas of estradiol-concentrating cells in the central nervous system of the female rat. *Journal of Comparative Neurology*, 1973, *151*, 121–158.
Pfaff, D., Lewis, C., Diakow, C., and Keiner, M. Neurophysiological analysis of mating behavior responses as hormone-sensitive reflexes. In E. Stellar and J. Sprague (Eds.), *Progress in physiological psychology*, Vol 5. New York: Academic Press, 1972.
Phoenix, C. L., Copenhaver, K. H., and Brenner, R. M. Scanning electron microscopy of penile papillae in intact and castrated rats. *Hormones and Behavior*, 1976, *7*, 217–228.
Sachs, B. D., and Barfield, R. J. Temporal patterning of sexual behavior in the male rat. *Journal of Comparative and Physiological Psychology*, 1970, *73*, 359–364.
Sar, M., and Stumpf, W. E. Autoradiographic localization of radioactivity in the rat brain after the injection of 1,2-^{3}H-testosterone. *Endocrinology*, 1973, *92*, 251–256.
Sar, M., and Stumpf, W. E. Androgen concentration in motor neurons of cranial nerves and spinal cord. *Science*, 1977, *197*, 77–79.
Sherrington, C. S. The spinal cord. In E. A. Sharpey-Schafer (Ed.), *Text-book of physiology*. Edinburgh: Scotland, 1900.
Sherrington, C. S. Flexion-reflex of the limb, crossed extension-reflex, and reflex stepping and standing. *Journal of Physiology*, 1910, *40*, 28–121.
Sherrington, C. S. *The integrative action of the nervous system*. New Haven: Yale University Press, 1947.
Shurrager, P. S., and Dykman, R. A. Walking spinal carnivores. *Journal of Comparative and Physiological Psychology*, 1951, *44*, 252–262.
Södersten, P. Estrogen-activated sexual behavior in male rats. *Hormones and Behavior*, 1973, *4*, 247–256.
Spaulding, W. S., and Peck, C. K. Sexual behavior of male rats following removal of the glans penis at weaning. *Developmental Psychobiology*, 1974, *7*, 43–46.
Stavraky, G. W. *Supersensitivity following lesions of the nervous system*. Toronto: University of Toronto Press, 1961.
Steinach, E. Geschlechtstrieb und echt sekundare Geschlechtsmerkale als Falge der imersekretorischen Funktion der Keimdrüsen. *Zentralblatt fuer Physiologie*, 1910, *24*, 551–566.
Stelzner, D. J., Ershler, W. B., and Weber, E. D. Effects of spinal transection in neonatal and weanling rats: Survival of function. *Experimental Neurology*, 1975, *46*, 156–177.
Stumpf, W. E. Estradiol-concentrating neurons. Topography in the hypothalamus by dry-mount autoradiography. *Science*, 1968, *162*, 1001–1003.
Stumpf, W. E. Estrogen-neurons and estrogen-neuron systems in the periventricular brain. *The American Journal of Anatomy*, 1970, *129*, 207–218.

Stumpf, W. E., and Sar, M. Estradiol concentrating neurons in the amygdala. *Proceedings of the Society for Experimental Biology and Medicine*, 1971, *136*, 102–106.
Stumpf, W. E., and Sar, M. Autoradiographic localization of estrogen, androgen, progestin and glucocorticosteroid in "target tissues" and "non-target tissues." In J. Pasqualini (Ed.), *Receptors and mechanism of action of steroid hormones*. New York: Marcel Dekker, 1976.
Talbot, H. S. The sexual function in paraplegia. *The Journal of Urology*, 1955, *73*, 91–100.
Thompson, R. F., and Spencer, W. A. Habituation: A model phenomenon for the study of neuronal substrates of behavior. *Psychological Review*, 1966, *73*, 16–43.
Whalen, R. E., and Luttge, W. G. Testosterone, androstenedione and dihydrotestosterone: Effects on mating behavior of male rats. *Hormones and Behavior*, 1971, *2*, 117–125.
Young, W. C. The hormones and mating behavior. In W. C. Young (Ed.), *Sex and internal secretions*. Baltimore: Williams and Wilkins, 1961.
Zeitlin, A. B., Cottrell, T. L., and Lloyd, F. A. Sexology of the paraplegic male. *Fertility and Sterility*, 1957, *8*, 337–344.

9

Neural Plasticity and Feminine Sexual Behavior in the Rat

Lynwood G. Clemens

Long ago in a land called Science there lived a clever toymaker who was called Able. Now Able had many customers and, therefore, he had to work long hours to finish all the toys. So one day he decided to build a machine to help with the work. After many months, clever Able finally completed a robot, and it came to pass that this child of Able's did many things for him. Able loved his robot as a parent would love a child and Able gave the robot a name: Hypothesis. Hypothesis kept Able's shop tidy and neat. Able also found that if he gave Hypothesis the parts to a new toy, Hypothesis could put them together. Because Hypothesis was a capable robot, Able gained a certain noteriety in the land, and this pleased Able.

But times in the land of Science change and with this change came the need for new and different toys. Clever Able adapted easily to this change, going from wooden dolls and mechanical toys to more elaborate electronic gadgets and toys involving mathematics and chemistry. However, as so often happens to busy toymakers, Able did not take the time to make necessary changes in Hypothesis and so the robot just plodded along putting things together in the only way he could. The result was that, although the toys which Able put together worked very well, the ones his Hypothesis assembled had little resemblance to Able's work. But Able loved Hypothesis and to change him would be like destroying a true friend, like destroying a part of himself. Able had not the heart to give up his friend even though the wooden stereoscopes Hypothesis put inside the toy televisions looked silly. And so what if the electronic tennis game had wooden wheels and airplane wings. If this was the way Hypothesis was going to put new parts together, then Able felt obliged to overlook his friend's shortcomings.

Lynwood G. Clemens • Department of Zoology, Michigan State University, East Lansing, Michigan. This research was supported by PHS Grant No. HDO6760.

In time, however, Able did change his robot's name to commemorate their long career together and from that time on Hypothesis was known as Postulate. His toys? Oh, Postulate continued to make his strange toys and people bought them because they knew that Able was a good toymaker. Besides, it is easier to buy a ready made toy than to make one.

Where is Postulate now? After Able died, no one paid much attention to his Postulate and in the land of Science, Postulates cannot live very long without a lot of love. I have heard, though, there is a new toymaker now and they say he has a bionic assistant working for him.

From an unwritten work "Robots I Have Known and Loved" by L. Clemens

1. Introduction

Female mating behavior in the rat is normally under the control of estrogen and progesterone. These ovarian hormones act upon the brain to bring about sexual receptivity. One concept often used to explain how these hormones facilitate mating behavior suggests that lordosis is under some form of tonic neural inhibition and that estrogen and progesterone decrease this inhibition. Since an inhibitory concept is only one of several alternative explanations, and since there is considerable evidence that contradicts an inhibition hypothesis in its general form, it is instructive to review the "crucial" studies for this concept before trying to outline alternative interpretations and hypotheses. In this endeavor I will restrict my comments to the laboratory rat.

In its most general form what I shall refer to as the "inhibition hypothesis" can be summarized as follows: *There exist in the brain, systems that inhibit or prevent the occurrence of lordosis when the female (or male) rat is mounted by another. Sexual receptivity occurs when these inhibitory systems are themselves inhibited by ovarian hormones.* Each author states this hypothesis in a slightly different way. The following is one of the more explicit formulations:

> Estrogen enters the target cells in the preoptic, septal and anterior hypothalamic regions and binds first with cytosolic and then nuclear complex receptors. . . . The estrogen–nuclear complex could then stimulate the production of specific *m*-RNA sequences which in turn could stimulate the production of specific proteins. . . . The newly produced estrogen-specific protein could then facilitate the induction of sexual receptivity by somehow suppressing the tonic inhibitory actions of the preoptic, septal and anterior hypothalamic regions. Lesions in these regions could facilitate receptivity . . . by removing this tonic inhibition, while electrical stimulation in these regions could suppress receptivity . . . by potentiating the tonic inhibitory actions. (Luttge, 1976; pp. 687–688).

Evidence offered in support of inhibitory control of lordosis can be summarized in three propositions:

1. Certain brain lesions increase the probability of lordosis.
2. Electrical stimulation of certain brain areas decreases the probability of lordosis.
3. Compounds that increase levels of brain serotonin reduce the probability of lordosis and compounds that decrease levels or action of serotonin increase the probability of lordosis.

2. Brain Lesions and Inhibition of Lordosis

For the most part, the assumptions that particular investigators make about the effect of a brain lesion are implied rather than explicit. The most frequently implied assumption is that the effects of a lesion are due directly to loss of tissue function. That is, if lordosis frequency increases following a particular lesion, then it is assumed that the removed tissue contained or mediated something that is now lost. With that assumption the conclusion that lesions which facilitate lordosis removed some form of inhibition follows logically. On the other hand, if one assumes that the brain undergoes some form of reorganization after a lesion (Jackson, 1898), then an increase in lordosis following a lesion need not be interpreted as the removal of inhibition but instead may be viewed as product of a new set of neural relationships which may or may not involve inhibition. In brief, interpretation of outcomes is determined by assumptions and these guiding assumptions generally remain unstated. For the lesion studies reviewed below the assumption of "loss of tissue-contained function" seems to be the most prevalent.

For purposes of evaluating the influence of lesions upon sexual receptivity, it is convenient to group lesion studies into three categories: (a) lesions in female rats receiving no gonadal hormones; (b) lesions in females primed with estrogen or estrogen and progesterone; (c) lesions in males primed with estrogen.

2.1. Lesions in Female Rats in the Absence of Ovarian Hormones

Under this condition we look for studies that have reported an increase in lordosis frequency in the absence of detectable levels of estrogen; such studies would constitute the strongest evidence for an inhibitory mechanism, given of course, the appropriate assumption.

In fact, there are no reports where lesions increased sexual receptivity in female rats in the absence of gonadal hormones.* The only rodent study to report a facilitation of lordosis, where we can be reasonably certain of an absence of gonadal hormones, is in the guinea pig (Goy and Phoenix, 1963). In this study 2 of 27 spayed females with hypothalamic lesions were lordotic every time they were tested. Continuous testing for three weeks produced lordosis at every test. Examination of the uteri at autopsy provided no evidence for estrogenic stimulation from endogenous sources. Unfortunately, the lesions were made with stainless steel electrodes and this leaves open the possibility that the lesion had actually become an irritative stimulation site.

Sometimes cited as evidence for an increase in lordosis response in the absence of estrogen is the study by Law and Meagher (1958). In this study of lordosis frequency, animals with lesions in the preoptic region and in the hypothalamus were treated weekly with 500-μg diethylstilbestrol and 1-mg progesterone. Daily lordosis measures were reported in relation to vaginal smears. In some cases lesioned females continued to show high lordosis frequencies even in the presence of vaginal diestrus. The authors suggest that the lesion removed some form of inhibitory control. More recent studies indicate that a frequent consequence of diencephalic lesions is that they lower the threshold required for estrogen to induce lordosis and extend the period of time that a single injection of estrogen will facilitate lordosis frequency (Powers and Valenstein, 1972; Nance, Shryne, and Gorski, 1974, 1975a; Rodgers and Schwartz, 1976). Since Law and Meagher (1958) used a very high weekly dose of synthetic estrogen in an oil solution, we cannot be certain whether the enhancement they saw was removal of inhibition, an increased sensitivity to the estrogen, or both. Their results are particularly difficult to interpret when we consider that a single injection of 2 μg/kg of estradiol benzoate continues to influence lordosis for up to six days after administration in the intact female and up to twenty days in females with lesions in the preoptic area (Powers and Valenstein, 1972). The picture is further complicated by the report that estrogen is lost more slowly in lesioned ovariectomized animals than in nonlesioned ones (Rodgers and Schwartz, 1976). Thus, with the lesioned female we may be seeing receptivity scores that reflect greater exposure to estrogen as a result of decreased ability to shed estrogen following a lesion. Utilization of diestrous smears by Law and Meagher to ascertain exposure to active estrogen levels may not be too meaningful since the experimental treatment of lesioning appears

* Komisaruk (1974) reports the occurrence of lordosis in ovariectomized females in response to vaginal and cervical stimulation. The frequency of such responses is increased by septal lesions in the absence of gonadal hormone treatment. However, these females do not show lordosis to the male rat in the standard test situation. Hence, the relevance of lordosis in this experimental situation to sexual receptivity remains unclear and will not be dealt with in this review. For the present discussion, I am concerned with how hormones bring the female rat to a point of responding to the male.

to modify neural sensitivity of estrogen (Smith and Lawton, 1972). Assumptions about peripheral tissues being more or less sensitive to estrogen than brain tissue are thereby open to question in the lesioned animal.

In summary then, there are no clearcut cases in which lesions enhance sexual receptivity in the rat in the absence of gonadal hormones, and in fact the only evidence for rodents rests with two guinea pigs. In contrast to this, however, considerable evidence is available to suggest that brain lesions can lower the threshold required for estrogen to induce lordosis.

2.2. Lesions in Female Rats Treated with Ovarian Hormones

As noted in the previous section, some lesions in the preoptic region increased responsiveness to estrogen (Powers and Valenstein, 1972). That is, for a given dose of estrogen, preoptic lesions increased the probability of lordosis. Lesioned females also showed lordosis in response to progesterone for a longer interval after the estrogen treatment than did controls. Lesions in the lateral septum also increased the probability of lordosis (Nance, Shryne, and Gorski, 1974, 1975a,b), as did lesions in the olfactory bulb of the estrogen-primed female (Nance, McGinnis and Gorski, 1976), and the estrogen–progesterone-treated female (Moss, 1971; Edwards and Warner, 1972). A conclusion that the increased probability of lordosis in these studies is due to removal of inhibition is not obligatory. For example, these studies do not provide criteria that would allow us to decide whether the change in lordosis frequency may not also be due to enhanced levels of serum estrogen or an increased sensitivity to estrogen, which has nothing to do with neural inhibition. The inhibition concept appears to arise more from assumptions about the effects of the lesion than any specific behavioral criteria.

In a review of cerebral inhibition of sexual responses, Beach (1967) summarized a number of studies on spinal cats and dogs which indicated supraspinal inhibition of certain elements of the female coital response. These studies, along with the work of Hart (1971), indicate that some aspects of the feline coital response may be organized in the spinal cord and appear sensitive to estrogen. This does not appear to be the case in the rat, however, where midthoracic spinal lesions reduced the frequency of lordosis-like responses and estrogen appeared to have little effect upon the probability of copulatory movement in the spinal female rat (Hart, 1969 and this volume).

Lesions of the neocortex (Beach, 1967) result in an enhancement of lordotic responding, which is somewhat different from that described for septal lesions. In the decorticate female, lordosis is often shown to stimuli that do not ordinarily elicit the response, i.e., vaginal penetration in the diestrus female. In addition, the lordosis itself is often exaggerated. These exaggerated responses in intact females suggested to Beach that some form of inhibitory control had been removed by the lesion. Unfortunately, the necessary hormonal controls were not available and we are left with the problem of possible changes in hormone sensitivity.

In a separate series of experiments directed toward neocortical function and lordosis, we found that application of KCl or strong electrical current to the neocortex would induce an increase in lordosis frequencies in estrogen-primed female and male rats (Clemens, Wallen, and Gorski, 1967; Clemens, 1972). Since KCl and electrical stimulation are known to produce spreading depression, we interpreted our results as a facilitation resulting from loss of neocortical inhibition. However, this interpretation must be reevaluated since Whalen and his colleagues have shown that KCl does not facilitate lordosis in the adrenalectomized animal or in the estrogen-primed female treated with dexamethosone, a treatment known to inhibit release of ACTH from the pituitary (Whalen, Neubauer, and Gorzalka, 1975). These experiments are not readily interpreted. It has been suggested that KCl brings about a release of ACTH, which in turn causes a release of adrenal progesterone, and this progesterone then increases the probability of lordosis (Whalen *et al*, 1975). Adrenalectomy removes the progesterone and hence blocks the enhanced lordosis response. If this were the case, one would not expect to see a significant enhancement of lordosis in the male rat since males are normally not responsive to progesterone (Clemens, Shryne, and Gorski, 1970: Davidson and Levine, 1969). Yet, males do respond to KCl (Merari, Frenk, Hirurg, and Ginton, 1975) and to electrical shock to the neocortex (Clemens, 1972) with enhanced lordosis frequencies. Thus, it is unclear by what mechanism these brain treatments facilitate lordosis and equally unclear how adrenalectomy blocks the effect. Consequently, these studies offer little support for a general tonic inhibition of lordosis at the present time. It is intriguing to consider the possibility that spreading depression may enhance sensitivity to progesterone.

2.3. Lesions in Male Rats Treated with Ovarian Hormones

A facilitation of lordosis in the male occurs not only to functional decortication (spreading depression), but can also be observed in males with lateral septal lesions (Nance, Shryne, and Gorski, 1975b). These studies add a new dimension to the problem of how lesions bring about changes in sexual behavior. When septal lesions in the male rat were followed by several weeks of continuous estrogen treatment, the males continued to show an increased responsiveness to estrogen. This enhanced response to estrogen continued for months after the postlesion estrogen treatment was terminated. Males with septal lesions, but not treated postoperatively with estrogen, do not retain the increased lordosis response to estrogen seen shortly after septal lesioning. By providing estrogen during the postlesion recovery period, the males sustained a permanent alteration in responsiveness to estrogen. These postlesion changes in hormone responsiveness appear to reflect an estrogen-related process of reorganization during recovery from lesioning.

The net finding of the lesion studies reviewed in this section is that certain brain lesions result in an increased probability of lordosis in response to estrogen. Four distinct, though not necessarily mutually exclusive, interpretations are available to explain this increase in lordosis: lesions remove inhibition; lesions increase sensitivity to gonadal hormones; lesions decrease ability to remove estrogen from the system; and/or lesions result in the brain undergoing compensatory reorganization.

3. *Lordosis and Electrical Stimulation of the Brain*

High-frequency electrical stimulation (100 Hz) of the preoptic or habenula region before or during mating of estrogen–progesterone-treated females suppressed lordosis frequency (Moss, Paloutzian, and Law, 1974). Similar stimulation of the septal region in the hamster also inhibited lordosis (Zasorin, Malsbury, and Pfaff, 1975). Stimulation of the caudate putamen suppressed lordosis in the rat when the stimulation was applied immediately before mating; stimulation during the mating test produced a slight facilitation of lordosis frequency (Moss *et al.*, 1974). Stimulation of the olfactory bulbs produced a marked suppression of lordosis if the stimulation was applied before mating, but if the female achieved several lordotic responses before stimulation began, the stimulation had no effect. Electrical stimulation of the anterior hypothalamus premamillary region and medial forebrain bundle produced inconsistent effects (Moss *et al.*, 1974).

Moss and his colleague have suggested that certain regions of the brain when activated may be considered inhibitory. In a simplistic model one would expect that electrical stimulation would have results opposite to those of lesions, but this is clearly not the case. While olfactory bulb lesions increase the probability of lordosis, the effect of electrical stimulation is contingent upon whether the female has started to show lordosis. Habenula lesions decrease lordosis (Rodgers and Law, 1967; Modianos, Hitt, and Popolow, 1975) and electrical stimulation has the same net effect upon lordosis frequency. Stimulation of the preoptic region has effects opposite to some lesions (Powers and Valenstein, 1972) but similar to others (Law and Meagher, 1958).

Even less clear is the relevance of the electrical stimulation parameters to ongoing physiological events. Does the high-frequency stimulation used in these experiments activate the structures involved or does it jam the system with noise? For example, in a recent study of theta rhythm and lordosis, Kurtz (1975) found a close correlation between theta frequency and sexual receptivity. Just prior to lordosis, theta was recorded at 8.0–8.7 Hz; during a lordosis with intromission or ejaculation the theta range was between 6.1 and 7.5 Hz. Since a pacemaker for theta appears to lie in the medial septal region (Stumpf, 1965; Gray, 1972) and since this region has important connections with the areas in which Moss *et al.* and Zasorin *et al.* find their effects, it is unclear what the effects of high-frequency stimulation would be in an area whose normal activity during sexual behavior may be vastly different. Studies of other complex behavioral phenomena have shown that electrical stimulation in this region can have a variety of effects depending entirely upon the stimulation parameters (Gray, 1972). Disruption of sexual behavior by one stimulation parameter is a very narrow base upon which to place an inhibitory concept.

4. *Lordosis and Neurotransmitter Processes*

Another variant of the sexual inhibitory hypothesis is one based upon pharmacological findings. This hypothesis proposes the existence of a serotonergic brain system that maintains a tonic inhibition of lordosis (Meyerson, 1964a). As formulated, this hypothesis has two parts to it. One is the assertion that increases in brain serotonin (5-HT) will reduce the probability of lordosis and the second is that decreases in brain serotonin will increase the probability of lordosis. While it might be psychologically comforting, it is not a logical necessity that the validity of one assertion be dependent or predictive of the validity of the other. Therefore we can deal with each prediction separately.

4.1. *Increased Levels of Brain Serotonin Inhibit Lordosis*

This hypothesis is supported by the findings that pharmacological agents that increase brain serotonin levels, e.g., 5-hydroxytryptophan (a precursor of serotonin), reduced the probability of lordosis in the estrogen–progesterone-treated female (Meyerson, 1964). A second line of support comes from reports that compounds that prevent the inactivation of serotonin (MAO inhibitors), and hence increase brain serotonin levels, also decrease lordosis (Meyerson), 1964; Everitt *et al.*, 1975). Thus, treatments leading to increased brain serotonergic levels reduce the probability of lordosis as does the serotonergic agonist α-methyltryptamine (Espino, Sano, and Wade, 1975). The conclusion that increased brain 5-HT inhibits lordosis, of course, requires the assumptions that the causal action of these pharmacological compounds upon lordosis is achieved by alterations of brain serotonin and that these compounds possess no other actions that influence lordosis directly or indirectly. If we accept those assumptions then we can ask "Does the suppression of lordosis by pharmacological enhancement of brain serotonin therefore mean that lordosis is normally under inhibition by a serotonergic system?" This is certainly not the only possibility. In females treated with estrogen and progesterone there may be systems that are important for lordosis, and increased serotonin levels may disrupt those systems. For example, high levels of serotonin may activate the sleep process or other processes and leave the animal too disoriented to respond appropriately to specific sexual stimuli. That the female does not remain totally stuporous after a drug treatment does not therefore mean she has all her faculties about her.

In regard to a nonspecific alteration in behavioral potential following changes in 5-HT it is important to note that pharmacological treatments that decrease serotonin may also reduce the probability of lordosis. Treatment with parachlorophenylalanine (PCPA), a compound that reduces brain serotonin (Koe and Weissman, 1966), lowered the probability of lordosis in females treated with estrogen and progesterone (Segal and Whalen, 1970; Singer, 1972; Gorzalka and Whalen, 1975). Although Meyerson (1964a) reports that reserpine and tetrabenazine, also purported to reduce serotonin levels, did not reduce lordosis in estrogen–progesterone animals, his criterion for receptivity (occurrence of one lordosis) was far less rigorous than the criteria of either Segal and Whalen or Singer.

Thus, if we accept the assumption that PCPA influences sexual behavior via influences upon serotonin brain systems, then decreased serotonin also reduces the probability of lordosis in estrogen–progesterone-primed females in both California and Florida. Geographical specificity aside, these findings suggest that any marked change in serotonin will reduce lordosis in the estrogen–progesterone-treated female.

4.2. *Decreased Levels of Brain Serotonin Will Increase the Probability of Lordosis*

Notwithstanding the contradictory evidence just cited, we will refer here to estrogen-treated females. Two sources of evidence support the notion of increased lordosis frequency consequent to decreased serotonin. Drugs, which reduce brain serotonin, in some cases increase the frequency of lordosis in estrogen-primed rats (Meyerson, 1964; Meyerson and Lewander, 1970; Zemlan *et al.*, 1973; Ward *et. al.*, 1975; Everitt *et al.*, 1975). Second, drugs that block serotonin receptors elevate lordosis (Zemlan *et al.* 1973; Ward *et al.* 1975).

However, considerable evidence is available to show that reductions in brain serotonin do not always facilitate lordosis. For example, Meyerson and Lewander (1970) found that of two compounds that achieved a 70% reduction in brain serotonin, only one, PCPA, facilitated lordosis. Everitt *et al.* (1975) found that for the other compound, H 22/54, much larger doses were required before it also facilitated receptivity. This poses a problem for the serotonin hypothesis: If a reduction in serotonin is the mechanism by which lordosis is being facilitated, then why is not lordosis related to the percent reduction in serotonin rather than drug dosage? These data are an unambiguous contradiction of the serotonergic inhibitory hypothesis as currently stated.

Ahlenius *et al.* (1972) also found independent variation in measures of lordosis and brain serotonin. They reported a reduction in brain serotonin beginning several hours after PCPA treatment; serotonin remained low until the end of the experiment at 26 hr. Although lordosis frequency increased during the early portion of the experiment, while serotonin was decreasing, lordosis frequency had returned to the low control levels by 26 hr, while serotonin was still depressed. These data, like those of Meyerson and Lewander (1970) are exactly opposite to the concept of serotonin inhibition. Everitt *et al.* (1975) have subsequently shown that tests 26 hr after PCPA administration did result in high levels of lordosis. While the reason for this discrepancy is not clear, it is interesting to note that Everitt *et al.* used twice the amount of estrogen as did Ahlenius (2 μg/kg daily as compared to 1 μg/kg daily). But Everitt's positive finding does not detract from the fact that some investigators find no consistent relationship between brain serotonin levels and lordosis frequency.

In a study of the effects of reserpine on lordosis, Paris *et al.* (1971) found that adrenalectomy abolished the facilitative effects of reserpine on estrogen-primed females. Eriksson and Sodersten (1973) were unable to demonstrate a facilitative effect of PCPA on adrenalectomized rats even though in both studies the investigators replicated the facilitative effects of reserpine and PCPA in ovariectomized estrogen-primed females. These findings have suggested that the experimental compounds may be bringing about a release of ACTH, which then results in progesterone secretion from the adrenal gland. This progesterone then facilitates lordosis independent of any brain serotonin system. Several experiments in mice also indicate that the facilitative effects of reserpine and PCPA are eliminated by adrenalectomy (Uphouse *et al.*, 1970; Hansult *et al.*, 1972).

In contrast to these studies, which show a negative effect of adrenalectomy, several studies report that reserpine (Meyerson, 1964a) and PCPA (Zemlan *et al.*, 1973) facilitate lordosis in the estrogen-primed adrenalectomized rat. Everitt *et al.* (1975) also report that PCPA was effective in facilitating lordosis in estrogen-primed adrenalectomized rats but indicated that lordosis was severely reduced and that had they used more rigorous criteria for measuring lordosis ". . . a considerable although not total decrease in lordosis/mount ratio would have resulted and thus a different conclusion as to the effects of adrenalectomy" (Everitt *et al.*, 1975). The implication here seems to be that adrenalectomy only abolishes the effect of PCPA when rigorous behavioral scoring is used. The alternative option is that a good portion of the drug effect is mediated by adrenal steroids.

If estrogen and progesterone facilitate lordosis by reducing serotonin activity, then we would expect to see either decreased levels of serotonin in the brain following treatment with E and P, or altered turnover rates for serotonin during periods of sexual receptivity. Although Meyerson, (Meyerson, 1964; Meyerson and Lewander, 1970) was unable to demonstrate an effect of estrogen and progesterone upon measures of brain serotonin, Everitt *et al.* (1975) found that estrogen had a slight (7%) elevating effect on brain serotonin levels, whereas progesterone decreased serotonin levels by 20%. However, since a reduction of up to 70% does not always facilitate lordosis (Meyerson and Lewander, 1970), this 20% reduction by progesterone remains less than convincing support for the 5-HT reduction idea.

The data for serotonin turnover are more complex. Serotonin metabolism shows a daily fluctuation, and the effect of estrogen and progesterone treatment on turnover depends upon whether these hormones are given at the beginning of the light portion of the day/night cycle or at the beginning of the dark portion (Everitt *et al.*, 1975). When given at the beginning of the dark phase, estrogen increased serotonin turnover above the level of ovariectomized controls. Treatment of the estrogen-primed female with progesterone resulted in a return of turnover rates to the control level. Thus serotonin turnover rates in the estrogen–progesterone-treated female in the dark phase did not differ from ovariectomized control, yet only the hormone-treated females displayed lordosis. Everitt *et al.* (1975, p. 568) chose to explain these findings by using the serotonin inhibitory hypothesis as an explanatory model: "Since a preponderance of the evidence points to an inhibitory role of 5-HT in sexual behavior, and it is well established that progesterone activates estrous behavior in estrogen-primed rats, the effects of progesterone on 5-HT turnover are consistent with the view that the changes in sexual behavior caused by this hormone are mediated by an effective decrease in 5-HT transmission." In short, the hypothesis under test is elevated to a postulate, then the data are explained by the postulate.

But the data on turnover become more complex. When estrogen and progesterone were given during the day, the effect of estrogen was to reduce 5-HT turnover; progesterone elevated 5-HT turnover back to the control level in females treated with estrogen. Again the turnover rate of females treated with estrogen and progesterone was not different from the control. According to a simplistic serotonergic inhibition hypotheses these females should not be receptive. Also, according to the interpretation just quoted from Everitt *et al.* (1975) the elevation of 5-HT by progesterone would eliminate its effect upon receptivity. This author has tested females by giving them estrogen and progesterone during the day and finds high levels of receptivity (Clemens, unpublished data). This outcome is contrary to a serotonin inhibition hypothesis.

The finding that turnover in the nonreceptive and in the estrogen-progesterone-receptive female are not different suggests that 5-HT may play little role in lordosis, and that the decrease in lordosis achieved by increased 5-HT levels (Meyerson, 1964) or by decreased 5-HT levels (Gorzalka *et al.*, 1975; Segal and Whalen, 1970; Singer, 1972) simply reflects the disruption of a neurochemical system that may or may not be directly related to lordosis.

4.3. Localized Changes in Serotonin

Much of the work up to now has involved analysis of 5-HT in the entire brain. Since specific brain regions may differ in their response to a particular compound and since these changes might not be reflected in analyses of the brain as a single unit, several recent studies have focused upon specific brain regions that are important for sexual receptivity. Kow, Malsbury, and Pfaff (1974) carefully laid out a series of neural relationships that could incorporate an inhibitory role for the serotoninergic fibers in the septum and midbrain raphe nucleus.

> The dynamics of the raphe–septal loop normally maintain a continued excitatory state of its neurons. The net effect of descending projections from the loop, the most important of which may be serotonergic fibers from the raphe, is to inhibit responses to somatosensory input, including inhibition of lordosis. . . . Interruption of function in the loop facilitates responses to somatosensory input, including facilitation of lordosis. . . . Therefore, if progesterone decreases the activity and responsiveness of neurons in this loop, it should facilitate lordosis. . . .
>
> A clear prediction from our present model is that lesion of the midbrain raphe nuclei should facilitate lordosis in estrogen-primed ovariectomized female rats. (Kow *et al.*, 1974, pp. 184–185).

Because of its clarity and because the authors insisted upon defining what criteria must be met in order to prove or disprove the hypothesis, it was possible to eliminate the hypothesis within two years after it was published.

> Severe deficits in receptivity were produced by lesions aimed at the dorsal and ventral norepinephrine pathways and by lesions intended to destroy meso-limbic (A-10) dopamine-producing cells. Significant but

less profound impairments were seen following lesions aimed at the serotonin-producing raphe nucleus or the nigro–striatal dopamine cells. (Herndon, 1976, p. 143.)

Lesions of the raphe nucleus reduced the probability of lordosis. This finding is directly opposed to the prediction of the serotonin inhibition hypothesis as interpreted and spelled out so nicely by Kow *et al.* (1974).

Ward and her colleagues (Ward *et al.*, 1975) implanted 5-HT receptor-blocking agents in the anterior hypothalamus and found an increase in the probability of lordosis within 30 min. These data suggest a localized rôle for 5-HT in sexual receptivity if we assume that the only influence of these agents is to block 5-HT receptors. Additional support for 5-HT involvement within specific brain areas comes from the selective depletion of 5-HT by intracerebral injection of 5, 7-dihydroxytryptamine (Everitt, Fuxe, and Jonsson, 1975). Injection of 5, 7-dihydroxytryptamine (5,7-DHT) into the medial ascending 5-HT pathways from B7 and B8 raphe cell groups (Fuxe and Jonsson, 1974) reduced H^3 5-HT uptake in the anterior hypothalamus and increased the probability of lordosis in estrogen-treated females. The time course of this effect, i.e., percentage of animals exceeding a 40% lordosis/mount ratio over 1–20 day time period, appeared to parallel the reductions in hypothalamic 5-HT. Given the appropriate assumptions relating lesions to loss of function and the specificity of 5, 7-DHT to destruction of 5-HT neurons, these data lend support to an inhibition concept.

In summary, there appears a considerable body of conflicting and contradictory evidence relating to the idea that serotonin inhibits lordosis. On the other hand there are data to suggest that some localized reductions in 5-HT activity may facilitate lordosis.

4.4. Serotonin Stimulation of the Preoptic Region

If serotonin inhibits lordosis by action upon estrogen-sensitive neurons in the preoptic anterior hypothalamus then implants of serotonin in this region should reduce lordosis in the estrogen–progesterone-treated female. To test this possibility one of my graduate students, Raymond Humphrys, introduced 5-HT into the POA of estrogen–progesterone-treated females. In control tests, on alternate weeks, females were implanted with methysergide, a 5-HT receptor blocking agent. The results are shown in Table 9.1.

Serotonin resulted in a marked suppression of lordosis frequency. Methysergide also slightly reduced the lordosis frequency. Since these results suggest that serotonin may be inhibiting lordosis, we then tested estrogen-primed females with methysergide. If serotonin decreases lordosis by action upon 5-HT receptors, then methysergide, a 5-HT receptor blocker

Table 9.1. Influence of Intracerebral Treatment (ICT) with Serotonin or Methysergide in Hormone-Treated Female Rats

Intracerebral treatment[a]	Hormone pretreatment	*n*	Lordosis quotients (min) Pretest	30	60	120
Serotonin	E + P	7	100	7.1	24	64
Methysergide	E + P	7	100	75.7	75.7	85.7
Methysergide	E	17	1.4	18.5	15.7	31.4

[a] Bilateral implants were located in the preoptic anterior hypothalamic region. Following a pretest in which females were mounted ten times by the male, ICT was administered. Lordosis/mount ratios × 100: lordosis quotient.

should facilitate lordosis. Since Ward *et al.* (1975) had shown a facilitative effect of methysergide in five animals with implants in the preoptic anterior hypothalamus, we implanted cannulae throughout that region.

Whereas four methysergide implants appeared to facilitate lordosis, they were intermingled with thirteen other implants that had no facilitative effect. This differs from the clear suppression of lordosis by 5-HT in all implanted animals. These findings raise questions concerning the influence of 5-HT implants. Was the reduction following 5-HT implants due to activation of 5-HT receptors? If so, why does methysergide not have the opposite effect? On the other hand, are the effects of methysergide in those animals that were affected due to its action on 5-HT receptors or does it have other effects when implanted directly into the brain?

5. *Alternative to a General Inhibition Hypothesis*

While it is certainly too early to exclude entirely the possibility that inhibitory mechanisms are involved in the control of sexual receptivity in the rat, the evidence for them at either the behavioral or neurophysiological level is not compelling. In general, the only criterion used to establish a neurological concept of inhibition has been an increase or decrease in the frequency of lordosis or an increase in the percentage of animals showing one or more lordotic responses. Drug treatments that decreased lordosis were said to enhance inhibition, while treatments that increased lordosis were said to decrease inhibition. But these same kinds of behavioral changes have led many of the same investigators to noninhibitory hypotheses in other areas. For example, treatment of the neonatal female rat with testosterone will reduce the probability of lordosis; castration of the neonatal male will increase lordosis probability in the adult. These increases and decreases in lordosis are seldom interpreted as changes in inhibition but instead are explained in terms of changes in hormone sensitivity.* In any particular experiment, changes in behavior frequency or quality are explained by assumptions about what the experimental operation (lesion, drug treatment, etc.) does to the animal. Direct testing of the inhibition concept has not been achieved.

In light of these considerations and in the hope of achieving a more testable set of hypotheses, it seems fruitful to explore alternative neurological mechanisms which can explain existing experimental results. Three propositions will be summarized to suggest that the control of sexual receptivity involves activation of specific brain systems rather than suppression of specific brain activity.

5.1. Estrogen Activation of Preoptic Region

Proposition I: Estrogen activates neurons in the preoptic anterior hypothalamic region which facilitates lordosis. This proposition is based on five sets of findings:

1. Sexual receptivity requires the prior administration of estrogen. As reviewed earlier there are no studies with the rat that demonstrate the existence of receptive behavior in the absence of estrogen.

2. The preoptic anterior hypothalamus takes up and concentrates estrogen, as shown in studies with radiolabeled estrogens (Pfaff and Keiner, 1973). These studies, combined with experiments that show that implants of estradiol into these areas are most effective in the induction of sexual receptivity (Lisk, 1962), indicate that this forebrain region is a focal area for the activation of receptivity.

3. Some lesions in the POA of AH abolish lordosis (Law and Meagher, 1958; Singer, 1968; Herndon and Neill, 1973). Presumably, these lesions have their effects by destruction of a large proportion of estrogen-sensitive neurons.

4. Inhibition of protein synthesis around the time of estrogen administration blocks the appearance of sexual receptivity (Quadagno, Shryne, and

* Beach (1966) probably explores the inhibition alternative most fully in regard to sexual differentiation. Both male and female guinea pigs show lordosis soon after birth in response to pudendal stimulation. In this situation, lordosis appears related to eliminative functions and disappears once control over the bladder and colon become internal and the mother is not required to lick the genital region. It is suggested that loss of lordosis in the maturing animal reflects development of a tonic cerebral inhibition. Beach suggests that in the adult female this inhibition is reduced by the synergistic action of estrogen and progesterone. Even if we grant that loss of lordosis to maternal licking is due to development of cerebral inhibition, it does not appear obligatory that we also accept that idea that estrogen and progesterone must act to relieve that inhibition. Activation of sex-related lordosis may involve quite different mechanisms.

Gorski, 1971; Terkel, Shryne, and Gorski, 1973; Whalen, Gorzalka, and DeBold, 1974). Actinomycin-D, an inhibitor of DNA-dependent RNA synthesis, blocked lordosis when implanted directly into the POA within twelve hours of estrogen treatment, (Quadagno, Shryne, and Gorski, 1971). Actinomycin-D treatment 21 hr after estrogen injections failed to influence lordosis frequency. Thus, there appears to be a 12 hr time period when estrogen action upon cells in the preoptic area can be prevented by inhibiting DNA-dependent RNA synthesis. Similar estrogen-blocking effects were also obtained with the protein synthesis inhibitor, cyclohexamide (Quadagno and Ho, 1975). If estrogen acts in the POA to decrease an inhibitory process one would expect implants of Actinomycin-D or cyclohexamide to facilitate lordosis since such implants presumably would reduce the resident inhibitory process by virtue of reducing protein synthesis. The notion that inhibition of protein synthesis can reduce neural inhibitory processes is based upon the findings of Wallen, Goldfoot, Joslyn, and Paris (1972). Progesterone has a well-known biphasic effect upon sexual receptivity in the guinea pig, first facilitating lordosis in the estrogen-primed female then suppressing lordosis frequency (Zucker, 1966). Wallen *et al.* (1972) found that this progesterone-induced inhibitory process is blocked by intracerebral implants of cyclohexamide in the ventromedial hypothalamus. Were the preoptic region a "seat of inhibitory influences" for lordosis, then inhibition of protein synthesis in that area would be expected to mimic the influence of estrogen and increase the probability of lordosis. Instead, protein inhibitors block the facilitative effects of estrogen.

5. Facilitation of lordosis by the preoptic anterior hypothalamus appears to utilize a cholinergic transmitter system. Pharmacological studies have suggested that cholinergic systems may be involved in sexual receptivity (Lindstrom, 1973, 1975). In our laboratory, we found that bilateral intracerebral implants of the muscarinic substance bethanechol achieved a marked facilitation of lordosis in female rats that were given small amounts of estrogen systemically. Bethanechol is implanted in the awake animal with a double barreled cannula which has been surgically implanted one to two weeks before testing. By treating females with 1-μg estradiol benzoate per day for three days we obtain a preparation that does not show significant levels of lordosis. However, when bethanechol was implanted into the preoptic region these females become intensely receptive (Table 9.2). To be certain that the bethanechol was having its effect via muscarinic receptors we tested some females that had been pretreated with atropine one hour before the bethanechol treatment. Atropine blocked the effects of bethanechol. Further tests with neostigmine ruled out an influence of nicotinic receptors. That estrogen facilitates lordosis in the normal female by influencing a cholinergic transmitter system is further supported by the finding that estrogen increases choline acetyltransferase (ChAc) in the

Table 9.2. Influence of Muscarinic and Nicotinic Stimulation of the Preoptic Area in Female Rats

Intracerebral treatment + systemic treatment	*n*	Lordosis quotients (min) Pretest	30	60	120
Bethanechol + saline	6	0	6	52	80
Bethanechol + atropine	6	0	0	0	0
Neostigmine + atropine	5	0	0	0	0

medial-preoptic area; ChAc is responsible for the synthesis of acetylcholine (Luine, Khylchevskaya, and McEwen, 1975). Further testing of adrenalectomized animals with bethanechol implants indicated that the effect was not due entirely to adrenal hormone (Table 9.3) although lordosis frequencies were lower in adrenalectomized animals. However, the brain implants in these animals were 1-mm lower than in previous animals; hence the lower responding may reflect slightly different placement. This control experiment is now being repeated.

In summary, it is proposed that estrogen facilitates lordosis by action at the preoptic anterior hypothalamus and that at least a portion of this facilitation is mediated via cholinergic neurons.

5.2. *Preoptic and Septal Lesions*

Proposition II: Preoptic and septal lesions increase sensitivity of remaining estrogen-sensitive regions which escape the lesion and continue to mediate sexual behavior. While some POA-AH lesions decrease lordosis others appear to facilitate lordosis as do septal lesions. If we reject an inhibitory hypothesis, how can we explain a post lesion increase in lordosis?

Following a lesion in the CNS we see the simultaneous occurrence of several distinct processes. Degeneration occurs and results in brain regions distant to the lesion site undergoing change. There are changes in

Table 9.3. Influence of Muscarinic Stimulation of the Preoptic Area in Adrenalectomized Female Rats

Intracerebral treatment + systemic treatment	*n*	Lordosis quotients (min) Pretest	30	60	120
Bethanechol + saline	8	8	31	50	46
Bethanechol + atropine	8	8	0	6	11

neurotransmitter levels throughout regions connected to the lesion site; brain regions severed by the lesioned site can undergo changes in sensitivities to neurotransmitters quite independently of any process resembling the inhibition we discussed earlier. Some examples of these changes will provide a basis for the concept that behavior following a lesion reflects a complex reorganization of brain function, not solely the loss of function per se.

When lesions are placed in the nigrostriatal dopamine system, there is degeneration of the dopaminergic neurons (see Ungerstedt for review, 1974). These lesions can be achieved by either electrolytic lesioning or by injection of 6-hydroxydopamine, which leads to loss of the dopamine cells. In response to the eventual degeneration and loss of the dopamine neurons, there is a postsynaptic increase in sensitivity by the cells that had received input from the lesioned neurons. It has been suggested that this "supersensitivity" results from an increase in receptors at the postsynaptic site (Understedt, 1974). This would appear to be a normal process of reorganization. Internal feedback systems possibly compensate for lower levels of dopamine by an increase in receptor sensitivity. The concept of inhibition is not relevant to this type of enhanced responding subsequent to lesioning. While it is not clear what the functional significance of these sensitivity changes might be in relation to adjoining neurotransmitter systems, it is possible that destruction of one transmitter system results in multiple changes in neurotransmitter sensitivity (Costall *et al.*, 1976). Thus, chemical lesions of the medial ascending 5-HT pathway that increase the probability of lordosis and reduce 5-HT in the anterior hypothalamus (Everitt *et al.*, 1975) may thereby set the stage for compensatory development of hypersensitivity. These changes do not necessarily mean that serotonin normally exerted an inhibitory influence.

"Axonal sprouting" also occurs in some brain regions following denervation. The septal region, for example, receives afferent fibers from the hippocampus. If these fibers are sectioned, normal degenerative processes occur, but the intact innervation also shows axonal sprouting and an increase in synaptic contacts (Raisman, 1969). Although this type of reorganization does not occur in all brain areas (Moore *et al.*, 1974; Lynch and Cotman, 1975) and the functional significance of these "new" synapses is not yet clear, it is important to recognize that the lesioned brain undergoes a recuperative reorganization process. Loss of tissue is not the sole consequence of lesioning.

Given these processes of reorganization, it is suggested that facilitation of lordosis following lesioning is a consequence of altered sensitivity in neurons formerly connected to the lesion site. The increase in lordosis frequency is seen as a product of reorganization of the remaining neurons and not necessarily evidence for a loss of inhibition.

Within this context lesion studies that result in an enhancement of lordosis may differ from those resulting in decrease in lordosis frequency in the extent of damage to estrogen-responsive neurons. Current practices of reporting lesions do not allow for accurate comparisons between studies.

Septal lesions regularly produced an increased lordosis frequency to estrogen. The present formulation suggests that this results from development of hypersensitivity in the preoptic anterior hypothalamic area receiving input from the septum. Lesions in the amygdala, another estrogen-sensitive region, which sends fibers to the preoptic area, abolished the facilitative effect of septal lesions (Nance *et al.*, 1976). While under normal circumstances the amygdala may play little role in lordosis, following septal lesions the normally small influence of the amydgala is multiplied since those regions receiving its input have undergone compensatory increases in sensitivity to make up for lost septal input.

The proposed hypersensitivity in the POA following septal lesions may involve cholinergic processes. Septal lesions decrease preoptic Ach (Pepeu *et al.*, 1971, 1973). If presynaptic decreases in Ach result in an increase in postsynaptic sensitivity to Ach in the same way the lesions in the nigrostriatal system enhance the sensitivity of dopamine neurons, then septal lesions may be enhancing the sensitivity of cholinergic neurons in the preoptic area. In this regard, a sensitivity increase in estrogen–cholinergic neurons would result in enhanced lordosis. It is also intriguing to consider a possible increase in synapses following septal lesions; however, axonal sprouting has not been demonstrated in the preoptic anterior hypothalamus in response to septal lesioning.

5.3. Progesterone Activation of MRF

Proposition III: Progesterone activates a midbrain region which facilitates lordosis. The synergistic action of progesterone with estrogen may be achieved by a common cholinergic neuronal system. Implants of progesterone in the midbrain reticular formation facilitate lordosis in the estrogen-primed female (Clemens, 1972; Ross, Claybaugh, Clemens, and Gorski, 1971). Progesterone is also concentrated by the brain in this region, suggesting that the implanted progesterone may be acting upon neurons that play a role in the normal expression of sexual receptivity (Whalen and Luttge, 1971). Since this region contains cholinergic fibers from the substantia nigra, we implanted cholinergic agonists into this region, instead of progesterone, to determine whether cholinergic facilitation of lordosis could be achieved. The results are summarized in Table 9.4. Both muscarinic and nicotinic receptors were effective in facilitating lordosis; muscarinic stimulation could be blocked by atropine but nicotinic stimulation could not indicate that the action of the muscarinic compound was

Table 9.4. Influence of Muscarinic and Nicotinic Stimulation of the Mesencephalic Reticular Fformation of Female Rats

Intracerebral treatment + systemic treatment	n	Lordosis quotients (min)			
		Pretest	30	60	120
Bethanechol + saline	6	5	48	60	82
Bethanechol + atropine	6	3	5	5	6
Neostigmine + atropine	5	0	8	25	53

specific to muscarinic receptors. Nicotinic stimulation was blocked by the nicotinic blocking agent, mecamylamine. These data suggest that the synergistic effects of progesterone upon estrogen-primed females may be mediated in part by cholinergic fibers.

6. Conclusions

Although the concept of a neural system that inhibits lordosis is frequently expressed, it has been difficult to find a solid body of evidence to support it. Instead, the concept appears to arise more directly from assumptions that are made about specific experimental operations of drug treatments; specifically that lesions result in a loss of function, which is directly reflected in behavioral changes; pharmacological agents only have the effects that they have been reported to have.

It is suggested that increases and decreases in lordosis frequency are an insufficient criterion upon which to base concepts of neural inhibition. Until direct tests of this concept are achieved, the available data can be explained just as parsimoniously by other concepts that include mechanisms of neural compensatory response. Behavioral scores obtained after a lesion are seen as a reflection of both functional loss and neural reorganization.

ACKNOWLEDGMENTS

The author wishes to acknowledge the assistance of Schering Corporation and Sandoz Corporation for supplying hormones and other pharmaceuticals.

References

Ahlenius, S., Engel, J., Eriksson, H., Modigh, K., and Soderston, P. Importance of central catecholamines in the mediation of lordosis behavior in ovariectomized rats treated with estrogen and inhibitors of monoamine synthesis. *Journal of Neural Transmission*, 1972, *33*, 247–255.

Beach, F. A. Onotogeny of "coitus-related" reflexes in the female guinea pig. *Proceedings of the National Academy of Sciences USA*, 1966, *56*, 526–533.

Beach, F. A. Cerebral and hormonal control of reflexive mechanisms involved in copulatory behavior. *Physiological Reviews*, 1967, *47*, 289–316.

Clemens, L. G. Perinatal hormones and the modification of adult behavior. In C. H. Sawyer and R. A. Gorski (Eds.), *Steroid hormones and brain function*, Los Angeles: UCLA Medical Forum Service, University of California Press, 1972, pp. 203–213.

Clemens, L. G., Wallen, K., and Gorski, R. A. Mating behavior: Facilitation in the female rat following cortical application of potassium chloride. *Science*, 1967, *157*, 1208–1209.

Clemens, L. G., Shryne, J., and Gorski, R. A. Androgen and development of progesterone responsiveness in male and female rats. *Physiology and Behavior*, 1970, *5*, 673–678.

Costall, B., Naylor, R. J., and Pycock, C. Non-specific supersensitivity of striatal dopamine receptors after 6-hydroxydopamine lesion of the nigrostriatal pathway. *European Journal of Pharmacology*, 1976, *35*, 275–283.

Davidson, J. and Levine, S. Progesterone and heterotypical sexual behavior in male rats. *Journal of Endocrinology*, 1969, *44*, 129–130.

Edwards, D. A., and Warner, P. Olfactory bulb removal facilitates the hormonal inductions of sexual receptivity in the female rat. *Hormones and Behavior*, 1972, *3*, 321–332.

Eriksson, H., and Sodersten, P. A failure to facilitate lordosis behavior in adrenalectomized and gonadectomized estrogen-primed rats with monoamine-synthesis inhibitors. *Hormones and Behavior*, 1973, *4*, 89–98.

Espino, C., Sano, M., and Wade, G. N. Alpha-methyltryptamine blocks facilitation of lordosis by progesterone in spayed, estrogen-primed rats. *Pharmacology, Biochemistry and Behavior*, 1975, *3*, 557–559.

Everitt, B. J., Fuxe, K., Hokfelt, T., and Jonsson, G. Role of monoamines in the control of hormones of sexual receptivity in the female rat. *Journal of Comparative and Physiological Psychology*, 1975, *89*, 556–572.

Everitt, B. J., Fuxe, K., and Jonsson, G. The effects of 5,7-dihydroxytryptamine lesions of ascending 5-hydroxytryptamine pathways on the sexual and aggressive behaviour of female rats. *Journal of Pharmacology (Paris)*, 1975, *6*, 25–32.

Fuxe, K., and Jonsson, G. Further mapping of central 5-hydroxytryptamine neurons: Studies with the neurotoxic dehydroxytryptamine in serotonin: New Vistas. *Advances in Biochemistry and Psychopharmacology*, 1974, *10*, 1–12.

Gorzalka, B. B., and Whalen, R. E. Inhibition not facilitation of sexual behavior by PCPA. *Pharmacology, Biochemistry and Behavior*, 1975, *3*, 511–513.

Goy, R. W., and Phoenix, C. H. Hypothalamic regulation of female sexual behavior: Establishment of behavioural anestrus in spayed guinea pigs following hypothalamic lesions. *Journal of Reproduction and Fertility*, 1963, *5*, 23–24.

Gray, J. A. Effects of septal driving of the hippocampal theta rhythm on resistance to extinction. *Physiology and Behavior*, 1972, *8*, 481–490.

Hansult, C. D., Uphouse, L., Schlesinger, K., and Wilson, J. R. Induction of estrus in mice: Hypophyseal–adrenal effects. *Hormones and Behavior*, 1972, *3*, 113–121.

Hart, B. L. Gonadal hormones and sexual reflexes in the female rat. *Hormones and Behavior*, 1969, *1*, 65–72.

Hart, B. L. Facilitation by estrogen of sexual reflexes in female cats. *Physiology and Behavior*, 1971, *7*, 675–678.

Herndon, J. G. Effects of midbrain lesions on female sexual behavior in the rat. *Physiology and Behavior*, 1976, *17*, 143–148.

Herndon, J. G. and Neill, D. B. Amphetamine reversal of sexual impairment following anterior hypothalamic lesions in female rats. *Pharmacology, Biochemistry and Behavior*, 1973, *1*, 285–288.

Jackson, J. H. *Selected writings of John Hughlings Jackson*, Vol. 2. J. Taylor (Ed.), London: Staple Press, 1958.

Koe, B. K., and Weissman, A. *p*-Chlorophenylalanine: A specific depletor of brain serotonin. *Journal of Pharmacology and Experimental Therapeutics*, 1966, *154*, 499–516.

Komisaruk, B. Neural and hormonal interactions in the reproductive behavior of female rats. In W. Montagna and W. A. Sadler (Eds.), *Advances in behavioral biology: Reproductive behavior*. New York: Plenum Press, 1974, pp. 97–130.

Kow, L. M., Malsbury, C. W., and Pfaff, D. W. Effects of progesterone on female reproductive behavior in rats: Possible modes of action and role in behavioral sex differences. In W. Montagna and W. Sadler (Eds.), *Advances in behavioral biology: Reproductive behavior*. New York: Plenum Press, 1974, pp. 179–210.

Kurtz, R. G. Hippocampal and cortical activity during sexual behavior in the female rat. *Journal of Comparative and Physiological Psychology*, 1975, *89*, 158–169.

Law, Thomas, and Meagher, Walter. Hypothalamic lesions and sexual behavior in the female rat, *Science*, 1958, *218*, 1626–1627.

Lindstrom, L. H. Further studies on cholinergic mechanisms and hormone-activated copulatory behaviour in the female rat. *Journal of Endocrinology*, 1973, *56*, 275–283.

Lindstrom, L. H. Cholinergic mechanisms and sexual behavior in the female rat. In M. Sandler and G. L. Gessa (Eds.), *Sexual behavior: Pharmacology and biochemistry*. New York: Raven Press, 1975, pp. 161–167.

Luine, V. N., Khylchevskaya, R. I., and McEwen, B. S. Effects of gonadal steroids on activities of monoamine oxidase and choline acetylase in rat brain. *Brain Research*, 1975, *86*, 293–306.

Lisk, R. D. Diencephalic placement of estradiol and sexual receptivity in the female rat. *American Journal of Physiology*, 1962, *203*, 493–496.

Luttge, W. G. Intracerebral implantation of the anti-estrogen CN69, 725-27: Effects on female sexual behavior in rats. *Pharmacology, Biochemistry and Behavior*, 1976, *4*, 685–688.

Lynch, G., and Cotman, C. W. The hippocampus as a model for studying anatomical plasticity in the adult brain. In R. L. Isaacson and K. H. Pribram (Eds.), *The hippocampus, Vol. 1: Structure and development*. New York: Plenum Press, 1975, pp. 123–154.

Merari, A., Frenk, C., Hirurg, M., and Ginton, A. Female sexual behavior in the male rat: Facilitation by cortical application of potassium chloride. *Hormones and Behavior*, 1975, *6*, 159–164.

Meyerson, B. Central nervous monoamines and hormone induced estrus behavior in the spayed rat. *Acta Physiologica Scandinavia, Supplement*, 1964a, *241*, 3–32.

Meyerson, B. J. The effect of neuropharmacological agents on hormone-activated estrus behaviour in ovariectomized rats. *Archives, Internationales de Pharmacodynamic et de Therpie*, 1964b, *150*, 4–33.

Meyerson, B. J., and Lewander, T. Serotonin synthesis inhibition and estrous behaviour in female rats. *Life Sciences*, 1970, *9*, 661–671.

Modianos, D. T., Hitt, J. C., and Popolow, H. B. Habenular lesions and feminine sexual behavior of ovariectomized rats: Diminished responsiveness to the synergistic effects of estrogen and progesterone. *Journal of Comparative and Physiological Psychology*, 1975, *89*, 231–237.

Moore, R. Y., Bjorklund, A., and Stenev, V. Growth and plasticity of adrenergic neurons. In F. O. Schmitt and F. G. Worden (Eds.), *The neurosciences: Third study programs*, Cambridge, Massachusetts: MIT Press, 1974, pp. 961–978.

Moss, R. L. Modification of copulatory behavior in the female rat following olfactory bulb removal. *Journal of Comparative Physiology*, 1971, *74*, 374-382.

Moss, R. L., Paloutzian, R. F., and Law, O. T. Electrical stimulation of forebrain structures and its effect on copulatory as well as stimulus-bound behavior in ovariectomized hormone-primed rats. *Physiology and Behavior*, 1974, *12*, 997–1004.

Nance, D. M., Shryne, J., and Gorski, R. A. Septal lesions: Effects on lordosis behavior and pattern of gonadatropin release. *Hormones and Behavior*, 1974, *5*, 73–81.

Nance, D. M., Shryne, J., and Gorski, R. A. Effects of septal lesions on behavioral sensitivity of female rats to gonadal hormones. *Hormones and Behavior*, 1975a, *6*, 59–64.

Nance, D. M., Shryne, J., and Gorski, R. A. Facilitation of female sexual behavior in male rats by septal lesions: An interaction with estrogen. *Hormones and Behavior*, 1975b, *6*, 289–299.

Nance, D. M., McGinnis, M., and Gorski, R. A. Interaction of olfactory and amygdala destruction with septal lesion: Effects on lordosis behavior. In *Neuroscience abstracts, sixth annual meeting, Society for Neurosciences*, Bethesda, Md.: Society for Neuroscience, 1976, p. 653.

Paris, C. A., Resko, J. A., and Coy, R. W. A possible mechanism for the induction of lordosis by reserpine in spayed rats. *Biology of Reproduction*, 1971, *4*, 23–30.

Pepeu, G., Mulas, A., Ruffi, A., and Sotgiv, P. Brain acetylcholine levels in rats with septal lesions. *Life Sciences*, 1971, *10*, 181–184.

Pepeu, G., Mulas, A., and Mulas, M. L. Changes in acetylcholine content in the rat brain after lesions of the septum, fimbria and hippocampus. *Brain Research*, 1973, *57*, 153–164.

Pfaff, D., and Keiner, M. Atlas of estradiol concentrating cells in the central nervous system of the female rat. *Journal of Comparative Neurology*, 1973, *151*, 121–158.

Powers, J. B., and Valenstein, E. S. Sexual receptivity: Facilitation by medial preoptic lesions in female rats. *Science*, 1972, *175*, 1003–1005.

Quadagno, D. M., Shryne, J., and Gorski, R. A. The inhibition of steroid induced sexual behavior by intrahypothalamic actinomycin-D. *Hormones and Behavior*, 1971, *2*, 1–10.

Quadagno, D. M., and Ho, G. K. W. The reversible inhibition of steroid-induced sexual behavior by intracranial cycloheximide. *Hormones and Behavior*, 1975, *6*, 19–26.

Raisman, G. Neuronal plasticity in the septal nuclei of the adult rat. *Brain Research*, 1969, *14*, 25–48.

Rodgers, C. H., and Law, O. T. The effects of habenular and medial forebrain bundle lesions on sexual behavior. *Psychonomic Science*, 1967, *8*, 1–2.

Rodgers, C., and Schwartz, N. B. Differentiation between neural and hormonal control of sexual behavior and gonadatrophin secretion in the female rat. *Endocrinology*, 1976, *98*, 778–786.

Ross, J., Claybaugh, C., Clemens, L., and Gorski, R. A. Short latency induction of estrous behavior with intracerebral gonadal hormones in ovariectomized rats. *Endocrinology*, 1971, *89*, 32–38.

Segal, D. S., and Whalen, R. E. Effect of chronic administration of *p*-chlorophenylalanine on sexual receptivity of the female rat. *Psychopharmacologia*, 1970, *16*, 434–438.

Singer, J. J. Hypothalamic control of male and female sexual behavior in female rats. *Journal of Comparative and Physiological Psychology*, 1968, *66*, 738–742.

Singer, J. J. Effects of *p*-chlorophenylalanine on the male and female sexual behavior of female rats. *Psychological Reports*, 1972, *30*, 891–893.

Smith, S. W., and Lawton, I. Involvement of the amygdala in the ovarian compensatory hypertrophy response. *Neuroendocrinology*, 1972, *9*, 228–234.

Stumpf, C. Drug action on the electrical activity of the hippocampus. *International Review of Neurobiology*, 1965, *8*, 77–138.

Terkel, A. S., Shryne, J., and Gorski, R. A. Inhibition of estrogen facilitation of sexual behavior by the intracerebral infusion of Actinomycin D. *Hormones and Behavior*, 1973, *4*, 377–386.

Ungerstedt, U. Brain dopamine neurons and behavior. In F. O. Schmitt and F. G. Worden (Eds.), *The neurosciences: Third study program.* Cambridge, Massachusetts: MIT Press, 1974, pp. 695–703.

Uphouse, L. L., Wilson, J. R., and Schlesinger, K. Induction of estrus in mice a possible role of adrenal progesterone. *Hormones and Behavior*, 1970, *1*, 255–264.

Wallen, K., Goldfoot, D. A., Joslyn, W. D., and Paris, C. A. Modification of behavioral estrus in the guinea pig following intracranial cycloheximide. *Physiology and Behavior*, 1972, *8*, 221–223.

Ward, I. L., Crowley, W. R., Zemlan, R. P., and Margules, D. L. Monoaminergic mediation of female sexual behavior. *Journal of Comparative and Physiological Psychology*, 1975, *88*, 53–61.

Whalen, R. E., and Luttge, W. G. Differential localization of progesterone uptake in brain. Role of sex, estrogen pretreatment and adrenalectomy. *Brain Research*, 1971, *33*, 147–155.

Whalen, R. E., Gorzalka, B. B., DeBold, J. F., Quadagno, D. M., Ho, G. K., and Hough, J. C., Jr. Studies on the effects of intracerebral Actinomycin D implants on estrogen induced receptivity in rats. *Hormones and Behavior*, 1974, *5*, 327–344.

Whalen, R. E., Neubauer, B. L., and Gorzalka, B. B. Potassium chloride-induced lordosis behavior in rats is mediated by the adrenal gland. *Science*, 1975, *187*, 857–858.

Zasorin, N. L., Malsbury, C. W., and Pfaff, D. W. Suppression of lordosis in the hormone-primed female hamster by electrical stimulation of the septum. *Physiology and Behavior*, 1975, *14*, 595–599.

Zemlan, F. P., Ward, I. L., Crowley, W. R., and Margules, D. L. Activation of lordosis responding in female rat by suppression of the serotonergic activity. *Science*, 1973, *179*, 1010–1011.

Zucker, I. Facilitatory and inhibitory effects of progesterone on sexual responses of spayed guinea pigs. *Journal of Comparative and Physiological Psychology*, 1966, *62*, 376–381.

10

Conceptual and Neural Mechanisms of Masculine Copulatory Behavior

Benjamin D. Sachs

1. Introduction

This chapter deals with the conceptualization of masculine copulatory behavior in mammals, and with the search for the neural basis of this behavior. My thesis, in part, is that the search for the central neural control of sexual behavior has not led to the dicoveries and sense of understanding that many expected when the search began. The relatively slow rate of progress stems partly from the apparent complexity of the neural control, and from the technical difficulties in tracing that control. The difficulties in understanding the neural control might have been more easily anticipated, though perhaps not avoided, had there been greater concurrent conceptual analysis of the behavior patterns.

In this chapter I shall briefly review the development of conceptual mechanisms of masculine copulatory behavior, present some new data based on factor analyses, and indicate some implications of these analyses for our conceptualization of sexual behavior. I shall then review what seems clearest, and what foggiest, about our knowledge of the central neural control of masculine copulatory behavior. That will be followed by a report of some recent studies from my laboratory on some of these neural mechanisms, and on some implications of this research for our understand-

---

Benjamin D. Sachs • Department of Psychology, University of Connecticut, Storrs, Connecticut. The experimental work from my laboratory was supported by USPHS research grants HD-04048 and HD-08933, and by grants from The University of Connecticut Research Foundation.

ing of neural and conceptual mechanisms. Finally, although my emphasis throughout will be on the male rat, I will conclude by suggesting a way in which a set of conceptual mechanisms related to sexual behavior may be viewed in a comparative physiological perspective.

2. Conceptual Mechanisms of Masculine Copulatory Behavior

2.1. Background

Much of the current research on the sexual behavior of animals owes its origins to three 1956 publications (Beach, 1956; Beach and Jordan, 1956; Larsson, 1956). These works contained normative data on sexual behavior in male rats that later research on rats and many other species, including primates (e.g., Bielert and Goy, 1973), depended upon for comparison and inspiration.

Also of great influence were the ideas elaborated from the data, particularly Beach's concepts of the hypothetical mechanisms controlling copulatory behavior. Contrary to the prevailing tendency to view sexual motivation as a unitary construct, Beach inferred from the normative data that a *minimum* of two factors controlled the sexual behavior of male rats. The sexual arousal mechanism was believed to control the male's arousal prior to his first intromission, as well as the male's rearousal prior to the resumption of copulation after each ejaculation. The copulatory mechanism was assumed to act to maintain copulation so that repeated intromissions occurred, and to summate the excitation from the intromissions until ejaculation occurred. The basis for inferring these mechanisms has been reviewed in detail elsewhere (for an analysis, see the original sources and, e.g., Cherney and Bermant, 1970; McGill, 1965; Sachs and Barfield, 1976) and need not be repeated here. One example will suffice to give the flavor of the evidence. The sexually rested male rat is capable of several ejaculations. The number of intromissions preceding successive ejaculations declines (at least for the first few ejaculations). However, the time after each ejaculation before the resumption of copulation increases monotonically. The different slopes of these two functions suggested that two different processes were involved. The copulatory mechanism was presumed to be sensitized by ejaculation, leading to successively fewer intromissions, whereas the sexual arousal mechanism was desensitized by ejaculation. Beach (1956) emphasized that these concepts should be considered *working hypotheses* rather than formal theory. He has privately expressed the wish that they had been written in disappearing ink. This wish, I suspect, is based on a feeling that the hypotheses were appropriate at the time they were written but have, like all of us, aged considerably since their publication, and have perhaps, like many of us, become overworked in that time.

The initial hypotheses were tested and revised for several years by Beach and others. For example, Beach and Whalen (1959) concluded that the rate of initial arousal of the male rat was not necessarily related to the rate of recovery from postejaculatory refractoriness. They suggested that the latter was controlled by fatigue, whereas initial arousal was still attributed to a relatively independent sexual aroused mechanism. The fatigue hypothesis in turn was rejected (e.g., by McGill, 1965) in favor of a distinct process controlling sexual refractoriness. At about the same time Beach, Westbrook, and Clemens (1966) reviewed evidence on the control of ejaculation, and inferred that the ejaculatory mechanism must be relatively independent of the process controlling copulatory behavior *per se*. Clemens (unpublished) adduced evidence from some *Peromyscus* species and other rodents dividing the ejaculatory mechanism into an ejaculation-inhibiting system and an emission system. Recent evidence (Barfield and Geyer, 1975; Sachs and Barfield, 1974) has supported older suggestions (Beach and Holz-Tucker, 1949; Larsson, 1959) that postejaculatory refractoriness is itself divisible into absolute and relative phases that are under partially independent control.

This proliferation of mechanisms flew in the face of the prevailing taste for stringent economy in theory building, and did not always lend itself to clear predictions; hence, infirmability of some of the postulated mechanisms was not obvious. When results were examined closely for their bearing on hypothetical mechanisms the data were commonly at odds with predictions from the models (e.g., Cherney and Bermant, 1970). At the same time that mechanisms were multiplying, the variables by which copulatory behavior was measured were also proliferating. In some laboratories 12 or more measures may be derived for each ejaculatory series.

In view of the profligate multiplication of conceptual mechanisms, it seemed advisable to estimate the *minimal* number of mechanisms necessary to account for masculine copulatory behavior, and to determine how the many variables by which we measure the behavior are related to each other. The latter question can be answered in part by computation of the correlation coefficients among the variables. This approach was among those used by Beach (1956) in arriving at his working hypotheses. One approach to the question of the minimum number of factors necessary to account for the variability is to analyze the intercorrelations by the techniques of factor analysis.

2.2. *Factor Analyses*

Three sets of data were available for analysis, all based on normative tests in which there was no experimental manipulation of the animals. One set of data came from 18 Long–Evans rats tested in our laboratory. Two other sets of data were kindly supplied by Ronald Barfield. Because

Barfield's two sets were also based on few animals (12 and 11 Long–Evans males), these two groups were combined for the analysis. These sample sizes (18 and 23) are still small, an outcome of the necessity to test animals one at a time in order to obtain measures of the postejaculatory vocalization.* Still, a beginning must be made, and aspects of the factor analyses suggest that the data are reliable despite the small samples.

Ten measures of copulatory behavior were used in the analysis: mount latency (ML), intromission latency (IL), number of mounts preceding ejaculation (NM), number of intromissions preceding ejaculation (NI), hit rate [HR = NI/(NM + NI)], interintromission interval (III), ejaculation latency measured from first intromission (ELI), vocalization termination (VT, the time from ejaculation to the end of the ultrasonic vocalization), postejaculatory intromission latency (PEIL), and the time from VT to PEIL (=PEIL–VT). Only data from the first ejaculatory series and its subsequent PEIL were used. Table 10.1 gives the correlations among these ten measures in the Sachs and Barfield samples. These data served as the basis for the subsequent factor analysis.

Gordon Bermant generously made available the results of an unpublished factor analysis based on the data from the second standardization (normative) test of 74 Long–Evans male rats. His analysis was based upon two ejaculatory series and the following measures: IL, NM, NI, III, ELI, and PEIL. At the time his data were collected the ultrasonic vocalization had not yet been discovered.

Table 10.2 reports those correlations from the three samples that had a significance level of $P < .05$ in at least two of the three sets of data. The high correlation between ML and IL was expected in these experienced, well-rested rats, and is unrevealing. The very high negative correlation between NM and HR reflects the fact that NI varied a great deal less than NM, so that HR in these data was simply another measure of NM. Perhaps in part because of the relatively low variability of NI, this measure did not correlate significantly with any other. The III was significantly and positively correlated with ELI, VT, and PEIL. The correlation of III with ELI is again a reflection of the relative stability of NI, which is the other constituent of ELI (that is, $\bar{X}_{ELI} \sim \bar{X}_{III} \times \bar{X}_{NI}$). The reliable correlation of III with PEIL in all three sets of data confirms the relation reported by Beach (1956), and is one indication that the processes underlying these variables are not independent. The reliable correlation between III and VT, however, was a surprise. It has been said that the postejaculatory period may consist of absolute and relative refractory phases (Beach and Holz-

* Many male rats vocalize in the ultrasonic range, primarily about 22 kHz, during bouts of copulation. This vocalization is especially prominent during the postejaculatory interval. The function of the signal is unknown, but it is thought to reflect an inhibitory neural state in the male (Barfield and Geyer, 1970).

Table 10.1. Means, Standard Errors (SEM), and Intercorrelations among 10 Measures of the First Copulatory Series of Long–Evans Male Rats.[a]

	ML	IL	NM	NI	HR	III	ELI	VT	PEIL–VT	PEIL
Mean	+19.9	+28.1	+2.8	+7.5	+.75	+46.4	+289.8	+276.8	+238.6	+520.3
	(+17.7)	(+21.2)	(+3.6)	(+9.9)	(+.75)	(+52.0)	(+437.7)	(+184.8)	(+154.4)	(+343.5)
SEM	+5.09	+5.09	+.41	+.41	+.03	+5.86	+34.38	+25.43	+23.67	+36.11
	(+4.86)	(+5.26)	(+.70)	(+.80)	(+.04)	(+11.38)	(+78.06)	(+20.56)	(+18.80)	(+19.28)
Correlations										
ML	+1.00									
	(+1.00)									
IL	+.87[b]	+1.00								
	(+.98[b])	(+1.00)								
NM	−.15	+.15	+1.00							
	(+.06)	(+.16)	(+1.00)							
NI	−.19	−.28	+.02	+1.00						
	(−.15)	(−.20)	(−.14)	(+1.00)						
HR	−.09	−.22	−.93[b]	+.28	1.00					
	(−.15)	(−.26)	(−.91[b])	(+.43[c])	(+1.00)					
III	+.51[c]	+.24	−.26	−.37	+.10	+1.00				
	(+.26)	(+.25)	(+.26)	(−.37)	(−.34)	(+1.00)				
ELI	+.41	+.12	−.25	+.08	+.22	+.88[b]	+1.00			
	(+.23)	(+.21)	(+.21)	(−.13)	(−.21)	(+.91[b])	(+1.00)			
VT	+.37	+.16	−.28	−.34	+.20	+.78[b]	+.70[b]	+1.00		
	(−.29)	(−.24)	(+,45[c])	(−.04)	(−.45[c])	(+.52[c])	(+.61[b])	(+1.00)		
PEIL–VT	+.33	+.18	−.48[c]	+.25	+.42	+.21	+.31	+.07	+1.00	
	(+.40)	(+.39)	(−.24)	(−.35)	(+.07)	(+.13)	(+.02)	(−.46)	(+1.00)	
PEIL	+.48[c]	+.23	−.51[c]	−.09	+.40	+.71[b]	+.72[b]	+.75[b]	+.71[b]	+1.00
	(+.06)	(+.09)	(+.20)	(−.38)	(−.34)	(+.65[b])	(+.64[b])	(+.58[b])	(+.44[c])	(+1.00)

[a] Data without parentheses are based on n = 18 (Sachs); data in parentheses are based on n = 23 (Barfield). Abbreviations: ML is mount latency (sec); IL, intromission latency (sec); NM, number of mounts preceding ejaculation; NI, number of intromissions preceding ejaculation; HR, NI/(NM + NI); III, mean interintromission interval (sec); ELI, ejaculation latency (sec) from first intromission; VT, time (sec) from ejaculation to vocalization termination; PEIL, postejaculatory intromission latency (sec). The Sachs data and one subset (n = 12) of the Barfield data were derived from the males' second standardization test. The other Barfield animals (n = 11) were given several tests to sexual exhaustion (Karen and Barfield, 1975). Data for this group were selected from the rats' first or second test in counterbalanced order, with the constraint that the male ejaculated and then vocalized during the PEIL of the selected test.

[b] $p < .01$.

[c] $p < .05$.

Table 10.2. Correlations Statistically Reliable ($p < .05$) in at Least Two of the Three Data Sets

Correlated variables	Sachs (n = 18)	Barfield (n = 23)	Bermant (n = 74)
ML × IL	+.87	+.98	NA[a]
NM × HR	−.93	−.91	NA
III × ELI	+.88	+.92	+.79
III × VT	+.78	+.52	NA
III × PEI	+.71	+.65	+.64
ELI × VT	+.70	+.61	NA
ELI × PEIL	+.72	+.64	+.52
VT × PEIL	+.75	+.57	NA
(PEIL–VT) × PEIL	+.71	+.44	NA
Required r values for stated p values			
p .05, r =	.47	.41	.23
p .01, r =	.59	.53	.30

[a] NA, not available.

Tucker, 1949; Larsson, 1959). Recently, VT has been used as a measure of the absolute refractory phase and PEIL–VT has been used as a measure of the relative refractory phase (Barfield and Geyer, 1975; Parrott, 1976). Beach (1956) suggested that the III might be a minirefractory period, and the expectation had been that the III would have more in common with the relative than with the absolute refractory period. That relation was not borne out by correlations.

The PEIL was reliably correlated with ELI, VT, and PEIL–VT, as well as with III. That is, the PEIL was correlated with all the temporal measures except ML and IL.

In Bermant's analysis a number of other correlations, ranging in value from +.43 to +.48 were significant at the .01 level of confidence: IL × NM, IL × III, IL × ELI, ELI × NM, and NM × III. Only the last of these relations approached statistical significance in either of the other data sets, but in view of the much larger sample size in Bermant's data and the level of significance, these correlations cannot be dismissed.

A number of factor analytic techniques were applied to the matrices of correlation coefficients. The estimates of communality for the Sachs and the Barfield data (Table 10.3) were unusually high for such small samples, an indication of reliability in the underlying data. The data in the tables that follow are based upon "varimax" rotations, reported here to facilitate comparison with the Bermant data based on the same type of analysis. Other types of factor analysis yielded similar pictures.

For all three sets of data, extraction of four factors accounted for more than 90% of the variance, and factors accounting for the remaining 10% of

Table 10.3. Estimates of Communality for Sachs and Barfield Data

Variable	Sachs	Barfield
ML	.91	.98
IL	.96	.97
NM	.98	.91
NI	.91	.74
HR	.91	.98
III	.93	.88
ELI	.93	.95
VT	.86	.90
PEIL–VT	.76	.85
PEIL	.89	.85

the variance were ignored. The factors were assigned tentative names, primarily according to the variables that were most heavily loaded on (that is, accounted for by) that factor. (The square of a variable's loading on a factor gives the proportion of that variable's intersubject variance that is accounted for by that factor.) Factor loadings of less than .30 (i.e., accounting for less than 9% of a variable's variance) were ignored.

The strongest factor that emerged was substantially loaded by III, ELI, VT, and PEIL, and hence was termed a *copulatory rate factor* (Table 10.4). In each of the analyses this factor accounted for the greatest single portion of the variance, 37 to 43%, more than the sum of the variances accounted for by the other three factors. Clearly, copulatory rate is an important aspect of copulatory behavior. The grouping of these four variables on this factor indicates that they share some common controlling process. It does not mean, however, that the process is indivisible. For example, the rate of mounting may be unchanged even when ejaculation is delayed or prevented (Lodder and Zeilmaker, 1976; Sachs and Barfield, 1970) and by appropriate lesions PEIL may be reduced without reliably changing III or ELI (Barfield, Wilson, and McDonald, 1975; Clark, Caggiula, McConnell, and Antelman, 1975). Hence, it is likely that the

Table 10.4. Variables and Their Loading on Copulatory Rate Factor

Variable	Sachs	Barfield	Bermant
III	+.91	+.87	+.80
ELI	+.92	+.95	+.71
VT	+.84	+.73	NA[a]
PEIL	+.75	+.72	+.76
% variance	43	37	41

[a] NA, not available.

process that controls copulatory rate consists of at least three subprocesses, which are normally coordinated during copulation but which can be fractionated by appropriate experimental manipulations. The copulatory rate factor includes elements of several previous conceptualizations. Ejaculation latency was used as a measure of the copulatory mechanism by Beach and Jordan (1956), the PEIL was McGill's (1965) measure of a postejaculatory mechanism, and the III has been used as one measure of the copulatory clock hypothesized by Barfield and me (Schoelch-Krieger and Barfield, 1975; Sachs and Barfield, 1976).

An *initiation factor* accounted for 10 to 25% of the variance in the three sets of data (Table 10.5). This factor was most heavily loaded by ML and IL, but PEIL–VT was also represented, suggesting some communality between initiation latencies and what has been used as a measure of the relative refractory phase. This factor appears to be similar to the sexual arousal mechanism (Beach, 1956), as modified by McGill (1965). That is, the fact that PEIL does not load heavily on this factor supports the interpretation of McGill that postejaculatory refractoriness may be under the control of processes different from those controlling initial sexual arousal.

A *hit rate factor* accounted for 11 to 23% of the variance (Table 10.6). NM and HR were redundant measures in this analysis. Constituents of the PEIL also loaded on this factor, but not reliably across samples.

An *intromission count factor*, similar to the ejaculatory mechanism of some previous conceptualizations, accounted for 10 to 20% of the variance (Table 10.7). In addition to NI, the variables PEIL and PEIL–VT were represented on this factor in two of the three analyses, but in one case with opposite signs. The emergence of NI in a separate factor from III follows from the lack of correlation between these variables (see Table 10.1), and confirms experimental demonstrations (e.g., Dewsbury, 1968) of the disso-

Table 10.5. Variables and Their Loading on Initiation Factor

Variable	Sachs	Barfield	Bermant
ML	+.86	+.98	NA[a]
IL	+.94	+.96	+.81
NM			+.76
III			+.37
EL			+.54
VT		−.37	NA
PEIL–VT	+.37	+.33	NA
PEIL	+.30		
% variance	14	25	10

[a] NA, not available. Other missing entries represent loadings with absolute values lower than .30.

Table 10.6. Variables and Their Loading on Hit Rate Factor

Variable	Sachs	Barfield	Bermant
NM	−.97	−.93	−.49
HR	+.90	+.96	NA[a]
VT		+.42	NA
PEIL–VT	+.51		NA
PEIL	+.44		
% variance	23	16	11

[a] NA, not available. Other missing entries represent loadings with absolute values lower than .30.

ciability of these variables. This dissociation may speak to the question of whether ejaculatory potential is more a function of number of intromissions or of some function of copulatory rate (Beach, 1956).

The correlations and factor analyses establish that the many variables by which we measure the copulatory behavior of rats do not have a great deal of redundancy for normal males in a standard test situation. What redundancy there is may be further reduced in any given experimental situation; hence, despite the proliferation of measures we are not overmeasuring rat copulatory behavior—rather we are measuring a number of relatively independent processes that contribute to the integrated behavioral outcome. Furthermore, other measures not included in these analyses, such as intromission duration, vocalization latency, number of ejaculations preceding exhaustion, and all the variables associated with ejaculations subsequent to the first—all these might also contribute to our understanding of the underlying processes.

More generally, the results of these factor analyses suggest that a *minimum* of four conceptual mechanisms are likely to be needed for any theory adequately to account for the copulatory behavior of male rats. If the data are best described in terms of the interactions of several mechanisms, rather than a more parsimonious one or two or three mechanisms, then so be it. Much of nature is notable for its lack of par-

Table 10.7. Variables and Their Loading on Intromission Count Factor

Variable	Sachs	Barfield	Bermant
NI	+.98	+.65	+.68
PEIL–VT	+.46	−.83	NA[a]
PEIL		+.56	
% variance	10	12	19

[a] NA, not available. Other missing entries represent loadings with absolute values lower than .30.

simony, and it can be deceptive to make things easy by leaving out the hard parts. Among current formulations it is likely that only those of Freeman and McFarland (1974) and of Karen (1975) based upon control theory have a large enough set of constructs and a rich enough description of their interactions to constitute the beginning of an adequate representation of underlying processes.*

3. *Central Neural Mechanisms of Masculine Copulatory Behavior*

3.1. *Status*

Despite the apparent continuity of concern with conceptual mechanisms from Beach (1956) to Freeman and McFarland (1974), a declining interest in conceptual analyses of sexual behavior was evident by the mid 1960s. This trend may have reflected a broader change in attitudes toward the analysis of motivated behavior. Perhaps as a result of the perceived excesses and failure of Hull and the neo-Hullians, theory building was in disrepute. Physiological analyses of motivation were viewed as the wave of the future, as the techniques of stereotaxic brain lesions and brain stimulation were more widely used. These techniques had led to exciting developments in the analysis of eating and drinking (e.g., Teitelbaum and Epstein, 1962) and in the process of reward itself (Olds, 1958). There seemed every reason to believe that sexual behavior could soon also be understood in terms of physiological mechanisms, without recourse to hypothetical constructs or intervening variables.

In 1966, just 10 years after the key publications by Beach and Larsson, a number of the landmark papers on the central neural control of masculine copulatory behavior appeared. Heimer and Larsson (1966) reported that lesions in the medial preoptic-anterior hypothalamic continuum (MPO-AH) eliminated mating behavior in most male rats. Davidson (1966) discovered that the copulatory behavior of castrated male rats could be restored by implantation of small amounts of testosterone in the MPO-AH and, to a

* Those who may be intimidated by the prospect of confronting control theory may take heart from the critique of the Freeman–McFarland model by Carlsson and Larsson (1974). On strictly intuitive grounds Carlsson and Larsson revised a portion of the Freeman–McFarland model, and Freeman and McFarland agreed that the intuition-based revision would better fit the data. Still, it seems likely that formal theory tested at each stage against a reliable body of data will provide the soundest conceptual framework. It would also be politic and accurate to point out that the analysis in this chapter does not take account of the female's considerable contribution to the parameters of copulatory behavior.

lesser degree, by implants in the posterior hypothalamus. Caggiula and Hoebel (1966) paced copulatory behavior in male rats with electrical stimulation of the medial forebrain bundle (MFB) in the region of the posterior hypothalamus.

Other research has amply confirmed that the MPO-AH, the MFB, and areas of the posterior hypothalamus adjacent to the MFB are key structures in the central neural control of masculine copulatory behavior, not just in rats, but in birds (Åkerman, 1966; Barfield, 1969; Hutchison, 1970), dogs (Hart, 1974), cats (Hart, Haugen, and Peterson, 1973), and rhesus monkeys (J. Slimp, B. Hart, and R. Goy, 1976, personal communication). Despite the overwhelming evidence for the importance of these neural structures for copulation, we have gained little understanding of *how* these structures exert their influence upon behavior. The "How?" question is less easily answered than the "Where?" question, not only for copulatory behavior, but for all behavior.

In an attempt to deal with the "How?" question some investigators have couched their interpretation of the neural mechanisms of copulation in terms of the conceptual mechanisms of Beach (1956) and his successors. Thus, Malsbury and Pfaff (1974) ascribed to the MPO-AH the functioning of the copulatory and ejaculatory mechanisms because electrical stimulation of this structure potentiates penile responses of erection and emission, but does not lead male rats to initiate copulation unless they are sexually experienced animals. Chen and Bliss (1974) analogized the MPO-AH to the sexual arousal mechanism, as modified by McGill (1965), because lesions in this system either eliminated copulation in male rats or considerably extended the initiation latency. However, once copulation was initiated the MPO-AH lesions had no effect on other parameters of copulation. Merari and Ginton (1975) concluded that the MPO-AH funtions in the sexual arousal and the copulatory and ejaculatory mechanisms, since in their study electrical stimulation of the MPO-AH affected precopulatory, preejaculatory, and postejaculatory parameters of copulatory behavior. A similar view was expressed by Giantonio, Lund, and Gerall (1970, p. 45), who, while not referring to a particular theoretical framework, concluded that the MPO-AH ". . . serves as an integrative focus for afferent, particularly olfactory, impulses involved in the arousal and integration of sequential copulatory responses." This interpretation is consistent with some recent data which, among other things, emphasize the dissociability between sexual arousability and copulatory potential. Slimp *et al.* (1976, personal communication) found that male rhesus monkeys bearing MPO-AH lesions, like most males of other species that have been similarly treated, no longer attempted to copulate with receptive females. However, observation of these animals outside the copulation test situation revealed that these males continued to masturbate to ejaculation at their high preoperative levels.

Clearly, these males were not impaired in their inclination toward genital stimulation (sexual arousal mechanism?) or in the ability to achieve erection and ejaculation ("copulatory" and ejaculatory mechanisms?). The tentative interpretation of Slimp *et al.* is that the males were impaired either in the ability to process cues coming from the receptive females, or in their decision to act upon those cues by copulating.

Perhaps one reason that we understand so little about how the MPO-AH mediates copulatory behavior is that this area of the brain has had to carry so much of the interpretive weight. With few exceptions, primarily with respect to evidence of a midbrain inhibitory system (Barfield, Wilson, and McDonald, 1975; Clark, Caggiula, McConnell, and Antelman, 1975; Heimer and Larsson, 1964; Lisk, 1969), lesions or stimulation elsewhere in the rat's brain have not had the same magnitude of effect as in the MPO-AH or MFB, not even in those portions of the brain that project to the MPO-AH or through the MFB.

One of the major sources of this input is dorsal, via the bed nucleus of the stria terminalis (Conrad and Pfaff, 1976). The stria carries afferents to the anterior hypothalamus and ventromedial hypothalamus from the amygdala, primarily from the cortical and medial nuclei of the amygdala (Leonard and Scott, 1971; De Olmos, 1972). The cortical and medial nuclei of the amygdala in turn receive much of their input from the olfactory bulbs, and especially from the vomeronasal organ, via the accessory olfactory bulb and the accessory olfactory tract (Broadwell, 1975; Scalia and Winans, 1975). The vomeronasal organ, rather than primary olfactory epithelium, may be the critical sensory organ concerned with the olfactory control of social behavior (Powers and Winans, 1975). Indeed, Estes (1972) has gone so far as to suggest that the mammalian male's vomeronasal organ is the neural counterpart of the hypothetical sexual arousal mechanism. Figure 10.1 is a diagram of the principal connections in the rat between the olfactory bulbs and the MPO-AH.

In view of their connections with the MPO-AH, it might be expected that treatments of the olfactory bulbs and the corticomedial components of the amygdala would have effects comparable to those of similar treatments in the MPO-AH. The evidence is not clear. Removal of the olfactory bulbs eliminates copulatory behavior in the golden hamster (Murphy and Schneider, 1970), and possibly in the house mouse (Rowe and Edwards, 1972), at least in the relatively brief timed tests that are conventionally used. In rats, bulbectomy may eliminate copulatory behavior or may have minimal effects, depending upon the prior social and sexual experience of the male and the age at which the bulbs are removed (Beach, 1942; Bermant and Taylor, 1969; Larsson, 1975; Pollak and Sachs, 1975; Wilhelmsson and Larsson, 1973). However, even in sexually experienced male rats bulbectomized in adulthood, longer copulation latencies and longer ejaculation

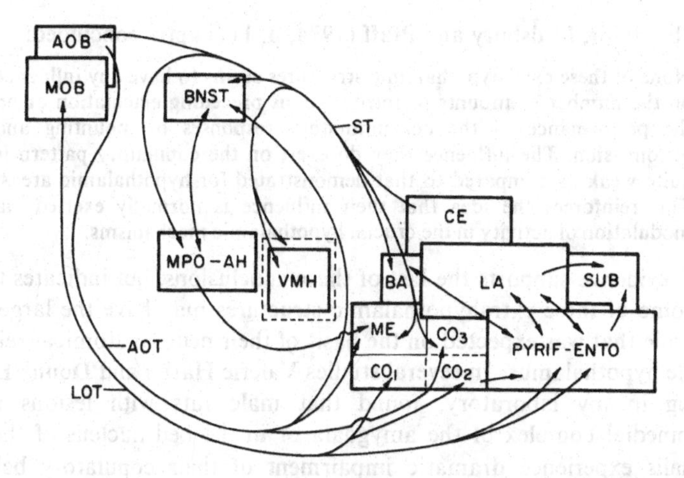

Figure 10.1. Schematic diagram of principal olfactory-bulb–amygdaloid–hypothalamic connections in the rat. Arrows indicate the direction of known projections. No attempt has been made to depict the relative number of projections. Some connections, e.g., between CO and SUB, and between SUB and VMH, have been ommitted to enhance clarity. AOB is the assessory olfactory bulb; AOT the accessory olfactory tract; BA the basal nucleus of amygdala; BNST the bed nucleus of stria terminalis; CE the central nucleus of amygdala; CO_1, CO_2, CO_3, the anterior, posterolateral, and posteromedial areas of cortical nucleus of amygdala; LA the lateral nucleus of amygdala; LOT the lateral olfactory tract; ME the medial nucleus of amygdala; MOB the main olfactory bulb; MPO-AH the medial preoptic-anterior hypothalamic continuum; PYRIF–ENTO the pyriform and entorhinal cortices; ST the stria terminalis; SUB the subiculum; and VMH the ventromedial nucleus of the hypothalamus. (Modified from Harris, 1976. Reprinted by permission.)

latencies are evident, and the proportion of males ejaculating during a test may be reduced. The number of intromissions preceding ejaculation has not been reliably affected by bulbectomy.

Few investigators have made sufficiently discrete lesions in the amygdala to be able to interpret the effects in terms of the amygdala's substructures. In one exceptional study, Giantonio, Lund, and Gerall (1970) made lesions in male rats in a number of limbic system structures. Medial preoptic lesions had their usual effect of eliminating mating behavior. Lesions in the basolateral components of the amygdala had no detectable effects. Lesions in the corticomedial amygdala or in the stria terminalis significantly, but not dramatically, reduced the number of ejaculations preceding exhaustion and increased the time from the male's first mount to ejaculation. However, the number of intromissions and the average intervals between them were not reliably affected.

In their review of the available evidence on the role of the hippocampus, amygdala, and olfactory bulbs in the control of masculine copulatory behavior, Malsbury and Pfaff (1974, p. 117) aptly concluded:

> None of these extrahypothalamic structures seems to have any influence on the number of mounts or intromissions preceding ejaculation or on the performance of the consummatory responses of mounting and intromission. The influence they do exert on the copulatory pattern is quite weak as compared to that demonstrated for hypothalamic areas. This reinforces the idea that their influence is normally exerted via modulation of activity in the crucial hypothalamic mechanisms.

Recent evidence supports the last of these conclusions, but indicates that at least some of these extrahypothalamic structures may have the large functional role that was expected on the basis of their neuroanatomical relations with the hypothalamus. In several studies Valerie Harris and Donna Emery, working in my laboratory, found that male rats with lesions in the corticomedial complex of the amygdala or in the bed nucleus of the stria terminalis experience dramatic impairment of their copulatory behavior (Emery and Sachs, 1976; Harris, 1976; Harris and Sachs, 1975).

3.2. Role for the Amygdala–Stria Terminalis System in Copulatory Behavior

Donna Emery's work dealt with the bed nucleus of the stria terminalis (BNST). Eight sexually experienced male rats had histologically verified lesions in the BNST, primarily in the dorsomedial portions, with occasional damage to other structures (Figure 10.2). Six control males were untreated

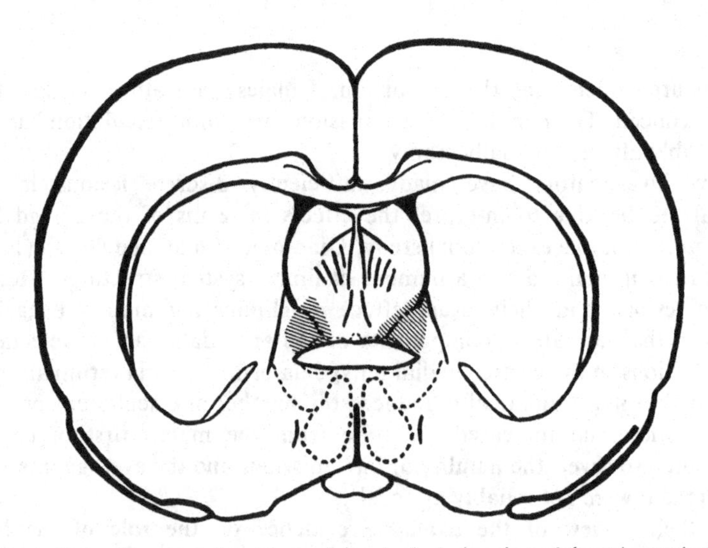

Figure 10.2. Representative effective lesion in the bed nucleus of the stria terminalis.

and six controls were given sham lesions, with the electrodes lowered but no current passed. Since the two control groups did not differ significantly on any measure, their data were combined for comparison with the BNST males.

Six weeks after surgery, in their second postoperative test, six of the eight BNST males copulated to ejaculation at least once. Some of the BNST males were slower to begin copulation, but mount and intromission latencies were not reliably longer than those of control males (Figure 10.3).

The copulatory behavior of the BNST males was most clearly marked by a 300% increase in the number of intromissions that preceded ejaculation, compared with the number required by control males (Figure 10.4). The number of mounts without intromission was also elevated (Figure 10.5). However, the proportion of copulatory attempts that led to intromission [NI/(NM + NI) = hit rate] was unchanged (Figure 10.6). The intervals between the intromissions of the BNST males were almost 100% longer than those of control males (Figure 10.7). Since long interintromission intervals are generally associated with *fewer* intromissions preceding ejaculation (the "enforced-interval effect," e.g., Bermant, 1964), the increased number of intromissions in the BNST males is all the more remarkable. Of course, ejaculation latencies were also prolonged, since their components (the number of intromissions and the intervals between them) were increased. Those BNST males that resumed copulation after ejaculation had postejaculatory intervals that were similar to control males (Figure 10.8), at least after the first two ejaculations. In part because of the extended intromission and ejaculation latencies and the restricted time of the tests (90 min after the start of copulation), 5 of 8 BNST males ejaculated only twice, and 3 of 8 ejaculated only three times.

Valerie Harris (1976) has extensively studied the behavior of males with lesions in the corticomedial complex of the amygdala (CMA, Figure 10.9). In many respects the effects of CMA lesions were similar to those of BNST lesions, and the data of those males that ejaculated during the second postoperative test are presented here. The most noteworthy characteristic of the CMA males (Harris, 1976, Experiment 1) was the increase in the number of intromissions preceding ejaculation (Figure 10.4), especially in view of a concomitant increase in the interval between intromissions (Figure 10.7). The intromission latencies of CMA males which ejaculated ($N = 5$) were significantly longer than were the latencies of controls (pooled unoperated and sham operated males; $N = 18$) (Figure 10.3), and CMA males were generally less likely to copulate, and to ejaculate if they did copulate, than control males: in one test of another study (Harris, 1976, Experiment 3), only 1 of 8 CMA males copulated without receiving additional extrinsic stimulation in the form of intermittent tail pinches.

In CMA males, as in BNST males, the increased number of

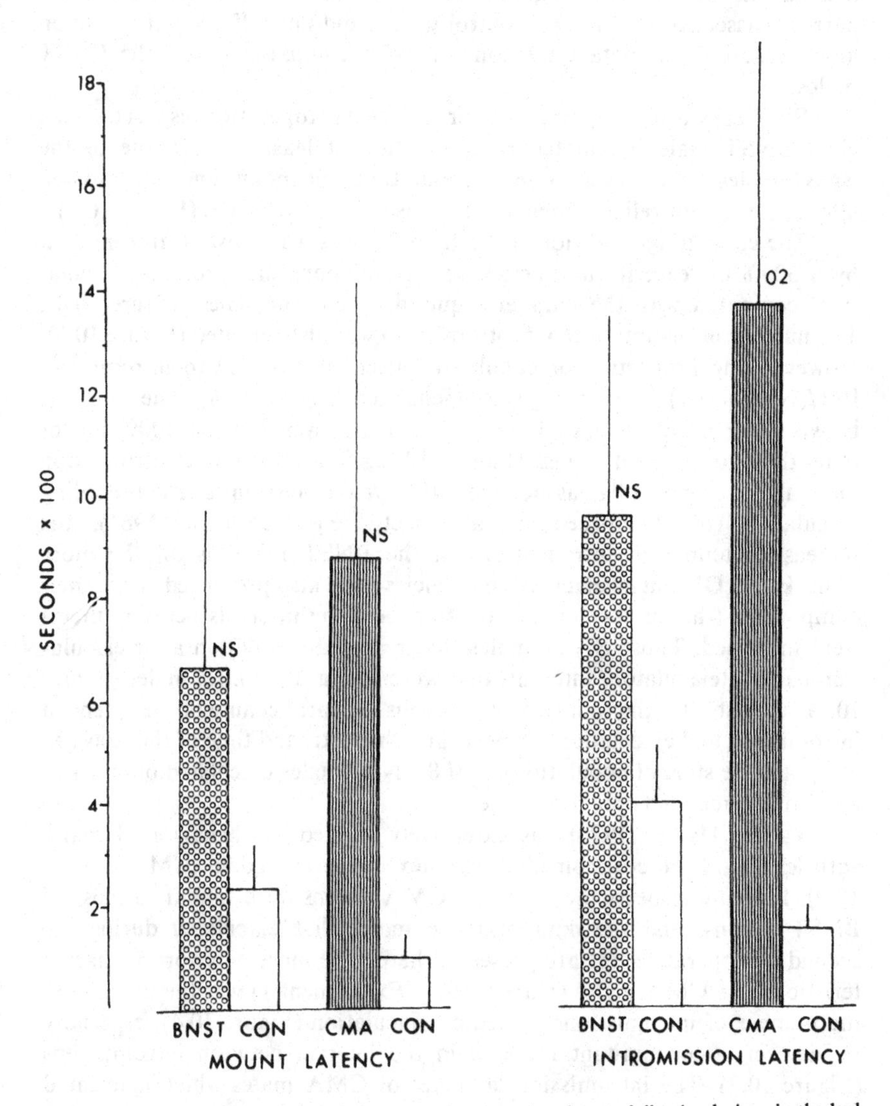

Figure 10.3 Mount latency and intromission latency of male rats following lesions in the bed nucleus of the stria terminalis (BNST) or in the corticomedial amygdala (CMA) and in control males. Lines at tops of columns indicate SEM. Significance of difference between means indicated by values between columns.

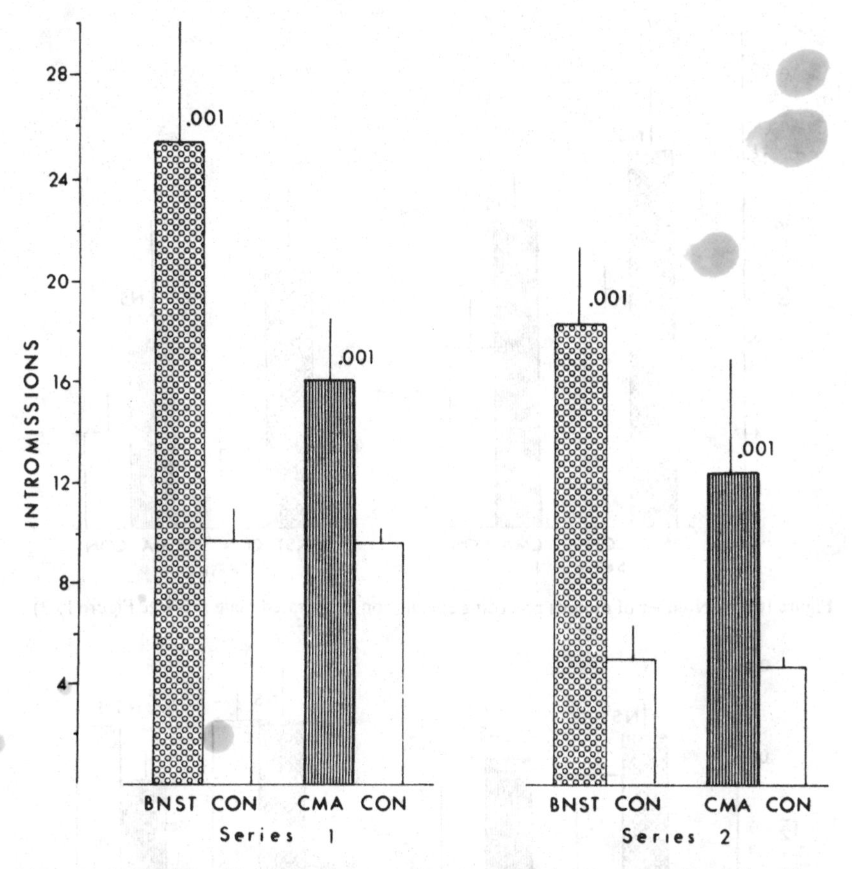

Figure 10.4 Number of intromissions preceding ejaculation in operated male rats (see Figure 10.3). Bouts of behavior associated with successive ejaculations are referred to as series.

intromissions was not attributable to a decrease in effective penile sensitivity, since the proportion of mounts in which the males gained intromission was unchanged by the surgery (Figure 10.6). Pollak and Sachs (1976) reported that under certain conditions the copulatory behavior of male rats might be misinterpreted. From the usual lateral perspective the copulatory behavior might appear to be a normal intromissive pattern. However, from a ventral perspective it could be seen that the penis did not enter the vagina. Such "extravaginal intromissive patterns" could, if they occurred frequently, "artificially" elevate the number of intromissions preceding ejaculation. This possibility was ruled out for CMA males in a separate study by Harris (1976, Experiment 2). She established that CMA

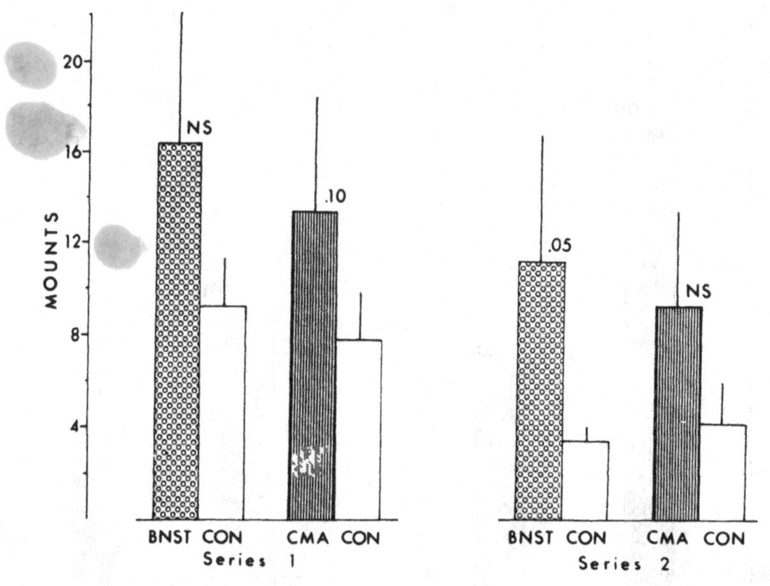

Figure 10.5. Number of mounts preceding ejaculation in operated male rats (see Figure 10.3).

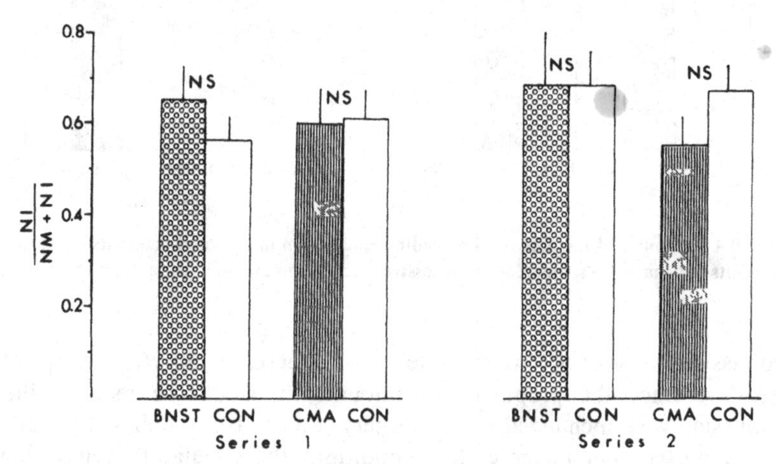

Figure 10.6. Hit rate in operated male rats (see Figure 10.3).

males were in fact gaining insertion whenever they showed the gross motor pattern of intromission.

Some BNST and CMA male rats, not included in the data summaries, had as many as 30 to 50 intromissions without ejaculation before the test was terminated after 90 min. Some of these males then had their tails pinched, and these males often ejaculated at their next intromission. In another study, Harris (1976, Experiment 3) further determined that intermittent tail pinches from the beginning of copulation could accelerate and pace the copulation of CMA males, as it could normal males. Intermittently pinched CMA males were also similar to both pinched and unpinched normal males in the number of intromissions preceding ejaculation. This was the only condition in which CMA males did not demonstrate their characteristic elevation in number of intromissions.

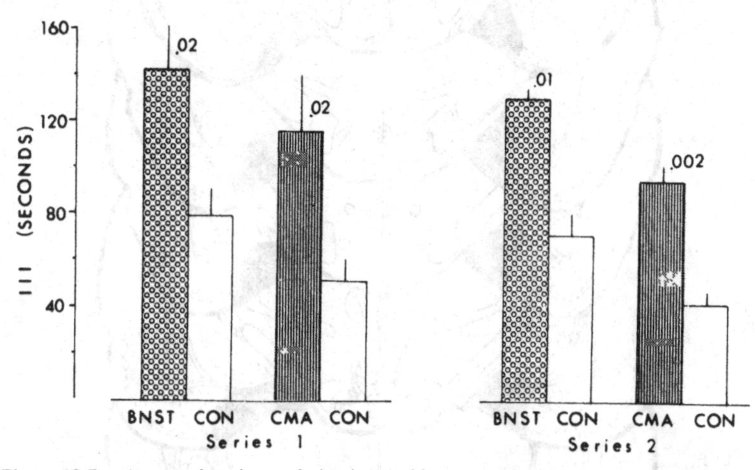

Figure 10.7. Average interintromission interval in operated male rats (see Figure 10.3)

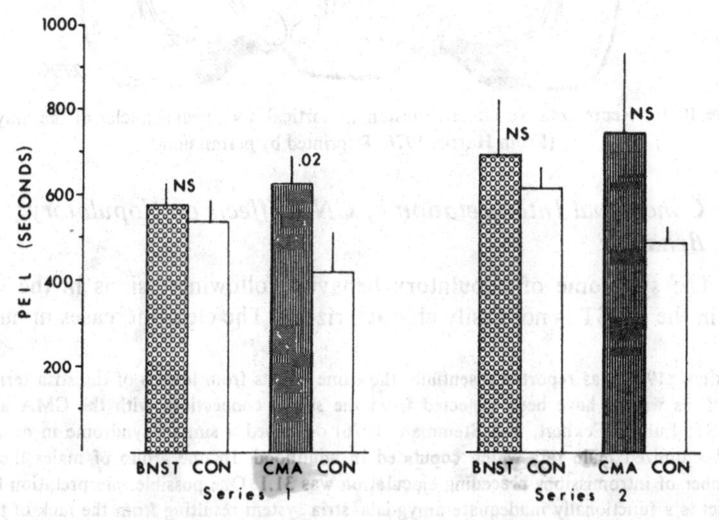

Figure 10.8. Postejaculatory intromission latency in operated male rats (see Figure 10.3).

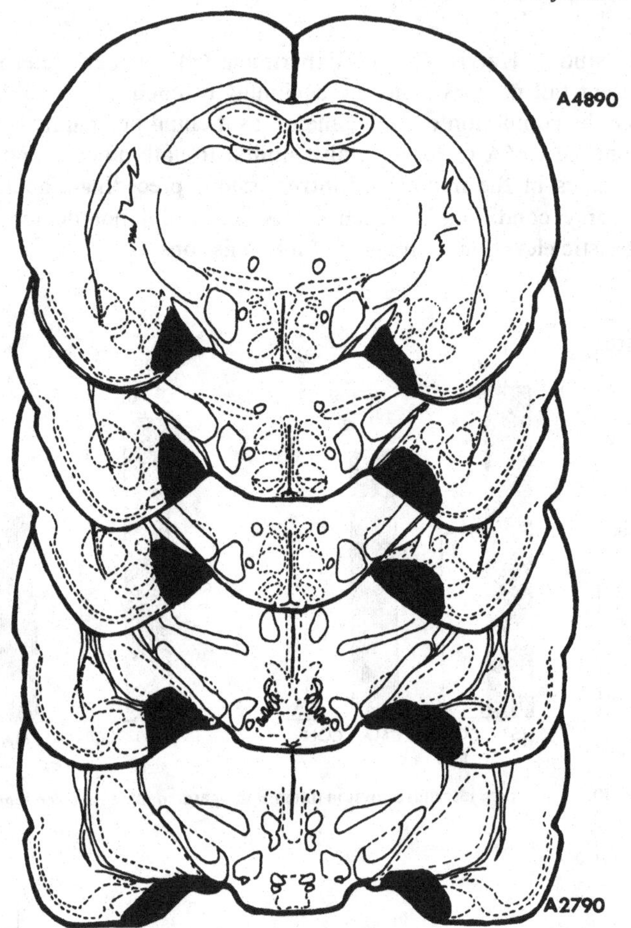

Figure 10.9. Representative effective lesion in cortical and medial nuclei of the amygdala. (From Harris, 1976. Reprinted by permission.)

3.3. *Conceptual Interpretation of CNS Effects on Copulatory Behavior*

The syndrome of copulatory behavior following lesions in the CMA and in the BNST is not easily characterized.* The clear increases in number

* Paxinos (1976) has reported essentially the same effects from lesions of the stria terminalis itself, as was to have been expected from the stria's connections with the CMA and the BNST. Lumia, Eckhert, and Steinman (1976) described a similar syndrome in neonatally bulbectomized male rats which copulated in adulthood. In one group of males the mean number of intromissions preceding ejaculation was 31.1. One possible interpretation for the effect is a functionally inadequate amygdala–stria system resulting from the lack of trophic influences from the olfactory bulbs during development.

of intromissions and interintromission intervals, and the more variable increases in intromission latencies and postejaculatory intervals, indicate a broad syndrome of reduced sexual arousability. These variables cut across the traditional sexual arousal and copulatory mechanisms, and they cut across three of the four factors revealed by our factor analyses. Only hit rate (as measured by relative number of mounts with and without intromission) was unchanged, indicating that the sensitivity of the several components responsible for triggering intromissions was unaffected by the lesions. Pottier and Baran (1973) determined that males of low sexual arousability, as defined by persistent nonmating with receptive females, were also hyporeactive in a number of nonsexual tests, including open field activity. However, CMA males do not yield concordance among different indices of arousal. Harris, (1976; Experiment 1) tested some sexually hyporesponsive CMA males in an open field, and their ambulation and defecation scores were similar to those of normal males. Furthermore, the CMA males tended to be hyperreactive to some stimuli. They were harder to handle, they tended to jump if the receptive female approached too suddenly, and unlike control males they sometimes reacted to rapidly approaching females by attacking them. These data confirm other studies, which have suggested that "general arousability" is no more unitary than "sexual arousability"; rather, animals have a complex of reactivities which may, by experimental manipulation, become dissociated from each other (Andrew, 1974). So too the several mechanisms that are normally coordinated smoothly to yield the integrated copulatory behavior of the male rat may be dissociated from each other by experimental treatments.

In view of the known feedback systems among the neural systems controlling copulation, and the hypothetical feedback systems among the conceptual mechanisms controlling copulation (Freeman and McFarland, 1974; Karen, 1975), it would be simplistic to expect a direct mapping of conceptual mechanisms onto central neural mechanisms. Still, a tentative interpretation may be assayed. The clearest effect of lesions in the amygdala–stria system, and the effect that reliably distinguished lesions there from lesions in the olfactory bulbs or in the MPO-AH, is the increased number of intromissions preceding ejaculation. Conceivably, the amygdala–stria terminalis system forms an essential portion of the hypothetical ejaculatory mechanism. Copulatory rate both prior to and after ejaculation was also affected by these lesions. To some extent this may have been a direct effect, but some of the effect on pacing could be due to a possible role of the amygdala–stria system as part of a hypothetical feedback loop between the intromission counter and the copulatory rate mechanism (Freeman and McFarland, 1974), so that reduced excitation from each intromission would have acted to slow copulation prior to ejaculation. Lesions in the midbrain structures studied by Heimer and Larsson (1964), Barfield *et al.* (1975), and

Clark *et al.* (1975), reduce the postejaculatory interval without reliably altering other measures of copulation (except interintromission interval in the Heimer and Larsson study, which employed massive lesions). Lesions in the dorsal hippocampus reduce the interintromission intervals more reliably than they affect other variables (Bermant, Glickman, and Davidson, 1968; Dewsbury, Goodman and Salis, 1968; Kim, 1960). Since there is normally a high correlation between interintromission intervals and postejaculatory intervals, the dorsal hippocampus in concert with the midbrain (and possibly other) structures may constitute essential portions of the hypothetical copulatory-rate mechanism.

Since so many of these neural structures have connections with the MPO-AH–MFB system, it is not surprising that lesions or stimulation in that system have the greatest effect on copulatory behavior, nor is it surprising that different interpretations of that system's function in terms of hypothetical constructs have been so at odds. Giantonio *et al.* (1970) concluded that the MPO-AH is the final common path of a response system that includes much of the limbic system, and that responds to the activational influences of a diffuse arousal system of which the stria terminalis and the basomedial and corticomedial amygdala are components. The data of Harris and Emery certainly substantiate the importance of the corticomedial amygdala and the bed nucleus of the stria terminalis in copulatory behavior. However, lesions in these structures have rather different effects from those in the olfactory bulbs or in the MPO-AH. This difference suggests that the amygdala–stria system is processing, not just transmitting, information between the bulbs and the MPO-AH.

It may well be that we do not yet have the physiological tools to dissect the functioning of such an integrative system. It may be that such complex interactions are more readily dissected at present by appropriate conceptual analyses, which may help inform the physiologist what he or she will find when the physiological tools become available.

4. *Cross-Species Comparisons of Conceptual and Neural Mechanisms: A Prospectus*

A final question to address is how the joint development of conceptual and neural models of the male rat's copulatory behavior might contribute to an understanding of the copulatory behavior of any other species. In 1956 (p. 18–19) Beach wrote:

> Interspecific differences in the normal [copulatory] pattern are so pronounced that it is difficult to imagine a theoretical interpretation

> which would apply with equal validity to rodents, carnivores, ungulates, primates, and other mammals, to say nothing of lower vertebrates. Nevertheless a start must be made somewhere, and the most reasonable approach is to formulate a set of hypotheses which appear applicable to a single species. If we can frame a theory that deals satisfactorily with one species, then it will prove profitable to examine the generality of that theory. To attempt a multispecies explanation from the beginning would be fruitless and ultimately frustrating.

This viewpoint strikes me as being as appropriate today as it was then, and we do not yet have a theory that deals adequately with the behavior of one species. However, in the spirit of providing a prospectus for this volume it may not be premature to suggest how a theory based on the sexual behavior of one species may be extended to the behavior of species with quite different copulatory patterns. The essence of my suggestion is that within broad taxonomic groups species do not vary qualititively in the nature of the mechanisms, conceptual or physiological, controlling copulation (or other motivated behavior patterns). Rather, species differ from each other primarily in the temporal characteristics of the mechanisms and the intensity of interactions among the mechanisms.

A convenient starting point for this analysis is Dewsbury's (1972) classification system of the copulatory behavior of rodents (this volume, Table 4.1). This system is based upon a binary choice with respect to occurrence of a penile–vaginal lock (present or absent), intravaginal thrusting (present or absent), insertions prior to ejaculation (single or multiple), and ejaculations prior to sexual exhaustion (single or multiple). Viewed in this useful schema, rodent species vary qualitatively in their masculine copulatory behavior. However, underlying the binary choices, except perhaps with respect to locking, there is at least a potential continuum. In practice the continuum is bimodal and there are large gaps in portions of the continuum. There are few multiple-thrust species that have just 2–4 thrusts, and few multiple-insertion animals have just 2–4 intromissions, at least prior to the first ejaculation. But at the physiological level of analysis such a continuum might be much more evident. Aubrey Manning (1971) and others have emphasized that small, genotype-based, quantitative variations in physiological mechanism, e.g., threshold, may give rise to apparently qualitative differences in the phenotype. The evolution of diversity in copulatory behavior may derive in part from several such multiplier effects.

There is circumstantial evidence that major differences among species in their copulatory patterns do not necessarily imply qualitative differences in the underlying mechanism. One type of such evidence is that the copulatory pattern of a male may fall into one category early in a copulatory episode and into another pattern later. For example, golden hamsters (*Mesocricetus auratus*) switch from a pattern of no intravaginal thrusting to multiple intravaginal thrusting after many ejaculations (Bunnell, Boland, and Dewsbury, 1977). Meadow voles (*Microtus pennsylvanicus*) change from a multiple insertion pattern to a single-insertion pattern after the first ejaculation (Gray and Dewsbury, 1975). Less dramatic changes characterize the changes of old-field mice (*Peromyscus polionotus*) and cotton rats (*Sigmodon hispidus*) after a few ejaculations (Dewsbury, 1975). In each of these cases it may be useful to regard the changes in copulatory behavior as resulting from quantitative changes in the underlying mechanisms and their interactions.*

What might be the nature of these variations in underlying mechanisms? Consider first a difference between the copulatory behavior of two species, Norway rats and golden hamsters, which, in Dewsbury's schema, have the same copulatory pattern at least for the first several ejaculations: no lock, no intravaginal thrust, multiple insertion, multiple ejaculation. The two species differ greatly, however, in their postejaculatory refractory periods. Hamsters have postejaculatory intervals of less than 60 seconds (Bunnell *et al.*, 1977), whereas those of rats are more than 5 minutes. One may imagine that the difference in duration of postejaculatory refractoriness is based upon quantitative differences in a hypothetical mechanism of refractoriness such as that proposed by McGill (1965). Part of the physiological counterpart of a sexual refractoriness mechanism might be the midbrain structures studied by Barfield *et al.* (1975), Clark *et al.* (1975), and Heimer and Larsson (1964). Rats with lesions in these structures had dramatically abbreviated refractory periods, which were little longer than their interintromission intervals. Comparable lesions in hamsters, which have postejaculatory intervals shorter, both absolutely and relative to the duration of their interintromission intervals, might have little or no effect. That outcome would be predicted by the assumption that this inhibitory system may be anatomically present in the male hamster but may be functionally weak with respect to copulatory behavior. Montane voles (*Microtus montanus*) might prove an interesting third species for comparison since their interintromission intervals are comparable to those of hamsters, but their postejaculatory intromission latencies are comparable to those of rats. Appropriate lesions in this species might result in reductions in the postejaculatory interval even more dramatic than have been found in male rats. Appropriate manipulation of the physiological systems controlling mating behavior may permit one to alter the behavioral parameters of one species so that it resembles the behavior of another species with a similar Dewsbury pattern.

This approach may be extended to differences between species that

* For a further consideration of this question, see Sachs and Dewsbury, this volume.

have different Dewsbury classifications. The distinction between single- and multiple-ejaculation species may be modulated by quantitative differences in a similar hypothetical (and physiological) mechanism controlling refractoriness. Rats normally have 4–8 ejaculations before sexual exhaustion under laboratory conditions. One of the midbrain-lesioned males tested by Barfield *et al.* (1975) had 17 successive copulatory series, and none of the postejaculatory intervals was longer than 390 sec (a common interval after the first ejaculation). Could males from species that normally ejaculate only once (e.g., *Mus musculus, Peromyscus eremicus,* and *Baiomys taylori*) be surgically modifed to make them reliable multiple ejaculators?

In a similar vein, several species of rodent, e.g., cotton mouse (*Peromyscus gossypinus*), plateau mouse (*Peromyscus melanophrys*), and Western harvest mouse (*Reithrodontomys megalotis*) neither lock with the female nor have multiple thrusts, and they may ejaculate during their first intromission (Dewsbury, 1975). Assuming these animals have a sensitive hypothetical ejaculatory mechanism, it would be rewarding to know whether elevation of that mechanism's threshold would result in multiple intromissions preceding ejaculation. In view of the research reviewed in this chapter on the amygdala–stria system, lesions or other manipulation of this system in single-intromission species might be an initial physiological approach to this question.

We may assume that there is also a system which processes information resulting from intravaginal thrusting, and which terminates thrusting and generates dismount from the female. [In some species, however, it is the female's movement away from the male that terminates mounts (C. Diakow and D. Dewsbury, personal communication, 1975)]. Experimental manipulations of this system might turn multiple-thrust males into single-thrust males and vice versa, and, conceivably, genetically based variations in the function of this system may be the origin of species variations in number of thrusts.

Of course nonneural systems, including penile morphology, perigenital musculature, mechanism of erection, and so on, are also involved in determining copulatory behavior patterns. Dewsbury (1972, 1975) has demonstrated that aspects of the pattern of copulatory behavior of a species can be predicted from knowledge of the species' penile anatomy. Males of species with relatively short, wide penes display locks and/or intravaginal thrusts; males with relatively long, narrow penes tend neither to lock nor to thrust. To the extent that pattern differences are based upon morphological or other nonneural factors, CNS manipulations will be ineffective in changing the pattern of copulatory behavior. However, morphological modifications often evolve after behavioral modifications; it may be that specialization of copulatory behavior initially derived from CNS-based variations in copulatory pattern, and was subsequently enhanced by changes in morphology.

Implicit in this prospectus is a prescription for coordinated development of conceptual and physiological models of copulatory behavior, enhanced by systematic exploration of homologous neural structures in species with different patterns of behavior. By extending the conceptual and physiological analyses to related species selected for specific properties of their pattern of copulatory behavior, we may arrive at a general model of the neuroethology of copulatory behavior in rodents. The value of conceptual models will fall as separate conceptual mechanisms map isomorphically on physiological systems, but the complexity of the physiological systems and our limited techniques for analyzing them leaves me confident that conceptual models will prove useful for many years.

ACKNOWLEDGMENTS

My part of the work in organizing the symposium which gave rise to this book was done during a sabbatical leave at Stanford University's Department of Physiology. There too I wrote the first draft of this chapter and did the factor analyses reported here. I am grateful to Julian Davidson and his co-workers for providing a hospitable intellectual climate. Special thanks go to David Damassa for his assistance with the computers. James R. Wilson helped with the interpretation of the factor analyses. Sarah Winans provided fine timely advice concerning neuroanatomy. Donald A. Dewsbury's careful critique helped me avoid some errors. Of course, whatever errors remain in the manuscript are my responsibility.

References

Åkerman, B. Behavioural effects of electrical stimulation in the forebrain of the pigeon. I. Reproductive behaviour. *Behaviour*, 1966, *26*, 323–338.

Andrew, R. J. Arousal and the causation of behavior. *Behaviour*, 1974, *51*, 135–165.

Barfield, R. J. Activation of copulatory behavior by androgen implanted into the preoptic area of the male fowl. *Hormones and Behavior*, 1969, *1*, 37–52.

Barfield, R. J., and Geyer, L. A. The ultrasonic postejaculatory vocalization and the postejaculatory refractory period of the male rat. *Journal of Comparative and Physiological Psychology*, 1975, *88*, 723–734.

Barfield, R. J., Wilson, C., and McDonald, P. G. Sexual behavior: Extreme reduction of the postejaculatory refractory period by posterior hypothalamic lesions in male rats. *Science*, 1975, *189*, 147–149.

Beach, F. A. Analysis of factors involved in the arousal, maintenance and manifestation of sexual excitement in male animals. *Psychosomatic Medicine*, 1942, *4*, 173–198.

Beach, F. A. Characteristics of masculine "sex drive." In M. R. Jones (Ed.), *Nebraska symposium on motivation*, Vol. 4. Lincoln: University of Nebraska Press, 1956, pp. 1–32.

Beach, F. A., and Holz-Tucker, A. M. Effects of different concentrations of androgen upon sexual behavior in castrated male rats. *Journal of Comparative and Physiological Psychology*, 1949, *42*, 433–453.

Beach, F. A., and Jordan, L. Sexual exhaustion and recovery in the male rat. *Quarterly Journal of Experimental Psychology*, 1956, *8*, 121–133.

Beach, F. A., Westbrook, W. H., and Clemens, L. G. Comparisons of the ejaculatory response in men and animals. *Psychosomatic Medicine*, 1966, *28*, 749–763.

Beach, F. A., and Whalen, R. E. Effects of ejaculation on sexual behavior in the male rat. *Journal of Comparative and Physiological Psychology*, 1959, *52*, 249–254.

Bermant, G. Effects of single and multiple enforced intercopulatory intervals on the sexual behavior of male rats. *Journal of Comparative and Physiological Psychology*, 1964, *57*, 398–403.

Bermant, G., Glickman, S. E., and Davidson, J. M. Effects of limbic lesions on copulatory behavior of male rats. *Journal of Comparative and Physiological Psychology*, 1968, *65*, 118–125.

Bermant, G., and Taylor, L. Interactive effects of experience and olfactory bulb lesions in male rat copulation. *Physiology and Behavior*, 1969, *4*, 13–17.

Bielert, C., and Goy, R. W. Sexual behavior of male rhesus: Effects of repeated ejaculation and partner's cycle stage. *Hormones and Behavior*, 1973, *4*, 109–122.

Broadwell, R. D. Olfactory relationships of the telencephalon and diencephalon in the rabbit. I. An autoradiographic study of the afferent connections of the main and accessory olfactory bulbs. *Journal of Comparative Neurology*, 1975, *163*, 329–346.

Bunnell, B. N., Boland, B. D., and Dewsbury, D. A. Copulatory behavior of golden hamsters (*Mesocricetus auratus*). *Behaviour*, 1977, *61*, 180–206.

Caggiula, A. R., and Hoebel, B. G. "Copulation-reward site" in the posterior hypothalamus. *Science*, 1966, *153*, 1284–1285.

Carlsson, S., and Larsson, K. Comments on the RATSEX model. In D. J. McFarland (Ed.), *Motivational control systems analysis*. New York: Academic Press, 1974, pp. 504–507.

Chen, J. J., and Bliss, D. K. Effects of sequential preoptic and mammillary lesions on male rat sexual behavior. *Journal of Comparative and Physiological Psychology*, 1974, *87*, 841–847.

Cherney, E. F., and Bermant, G. The role of stimulus female novelty in the rearousal of copulation in male laboratory rats (*Rattus norvegicus*). *Animal Behaviour*, 1970, *18*, 567–574.

Clark, T. K. , Caggiula, A. R., McConnell, R. A., and Antelman, S. M. Sexual inhibition is reduced by rostral midbrain lesions in the male rat. *Science*, 1975, *190*, 169–171.

Clemens, L. G. Mechanism of male sexual behavior. Unpublished.

Conrad, L. C. A., and Pfaff, D. W. Efferents from medial basal forebrain and hypothalamus in the rat. I. An autoradiographic study of the medial preoptic area. *Journal of Comparative Neurology*, 1976, *169*, 185–220.

Davidson, J. M. Activation of the male rat's sexual behavior by intracerebral implantation of androgen. *Endocrinology*, 1966, *79*, 783–794.

De Olmos, J. S. The amygdaloid projection field in the rat as studied by the cupric-silver method. In B. E. Eleftheriou (Ed.), *The neurobiology of the amygdala*. New York: Plenum Press, 1972, pp. 145–204.

Dewsbury, D. A. Copulatory behavior of rats—variations within the dark phase of the diurnal cycle. *Communications in Behavorial Biology*, 1968, Part A, *1*, 373–377.

Dewbury, D. A. Patterns of copulatory behavior in male mammals. *The Quarterly Review of Biology*, 1972, *47*, 1–33.

Dewsbury, D. A. Diversity and adaptation in rodent copulatory behavior. *Science*, 1975, *190*, 947–954.

Dewsbury, D. A., Goodman, E. D., Salis, P. J., and Bunnell, B. N. Effects of hippocampal lesions on the copulatory behavior of male rats. *Physiology and Behavior,* 1968, *3,* 651-656.

Emery, D. E., and Sachs, B. D. Copulatory behavior in male rats with lesions in the bed nucleus of the stria terminalis. *Physiology and Behavior,* 1976, *17,* 803-806.

Estes, R. D. The role of the vomeronasal organ in mammalian reproduction. *Extrait de Mammalia,* 1972, *36,* 315-341.

Freeman, S., and McFarland, D. J. RATSEX—an exercise in simulation. In D. J. McFarland (Ed.), *Motivational control systems analysis.* New York: Academic Press, 1974, pp. 479-504.

Giantonio, G. W., Lund, N. L., and Gerall, A. A. Effect of diencephalic and rhinencephalic lesions on the male rat's sexual behavior. *Journal of Comparative and Physiological Psychology,* 1970, *73,* 38-46.

Gray, G. D., and Dewsbury, D. A. A quantitative description of the copulatory behavior of meadow voles (*Microtus pennsylvanicus*). *Animal Behaviour,* 1975, *23,* 261-267.

Harris, V. S. Copulatory behavior of male rats following lesions of the corticomedial component of the amygdala. Doctoral dissertation, University of Connecticut, 1976.

Harris, V. S., and Sachs, B. D. Copulatory behavior in male rats following amygdaloid lesions. *Brain Research,* 1975, *86,* 514-518.

Hart, B. L. Medial preoptic-anterior hypothalamic area and sociosexual behavior of male dogs: A comparative neuropsychological analysis. *Journal of Comparative and Physiological Psychology,* 1974, *86,* 328-349.

Hart, B. L, Haugen, C. M., and Peterson, D. M. Effects of medial preoptic-anterior hypothalamic lesions on mating behavior of male cats. *Brain Research,* 1973, *54,* 177-191.

Heimer, L., and Larsson, K. Drastic changes in mating behaviour of male rats following lesions in the junction of diencephalon and mesencephalon. *Experientia,* 1964, *20,* 460-461.

Heimer, L., and Larsson, K. Impairment of mating behavior in male rats following lesions in the preoptic-anterior hypothalamic continuum. *Brain Research,* 1966/1967, *3,* 248-263.

Hutchison, J. B. Influence of gonadal hormones on the hypothalamic integration of courtship behaviour in the Barbary dove. *Journal of Reproduction and Fertility, Supplement.* 1970, *11,* 15-41.

Karen, L. M. Male rat sexual behavior: A theoretical interpretation of experiments on its organization. Doctoral dissertation, Rutgers University, 1975.

Karen, L. M., and Barfield, R. J. Differential rates of exhaustion and recovery of several parameters of male rat sexual behavior. *Journal of Comparative and Physiological Psychology,* 1975, *88,* 693-703.

Kim, C. Sexual activity of male rats following ablation of hippocampus. *Journal of Comparative and Physiological Psychology,* 1960, *53,* 553-557.

Larsson, K. *Conditioning and sexual behaviour in the male albino rat.* Stockholm: Almquist & Wiksell, 1956.

Larsson, K. Effects of prolonged postejaculatory intervals in the mating behaviour of the male rat. *Zeitschrift für Tierpsychologie,* 1959, *16,* 628-632.

Larsson, K. Sexual impairment of inexperienced male rats following pre- and postpuberal olfactory bulbectomy. *Physiology and Behavior,* 1975, *14,* 195-199.

Leonard, C. M. and Scott, J. W. Origin and distribution of the amygdalofugal pathways in the rat: An experimental neuroanatomical study. *Journal of Comparative Neurology,* 1971, *141,* 313-330.

Lisk, R. D. Reproductive potential of the male rat: Enhancement of copulatory levels following lesions of the mammillary body in sexually non-active and active animals. *Journal of Reproduction and Fertility,* 1969, *19,* 353-356.

Lodder, J., and Zeilmaker, G. H. Effects of pelvic nerve and pudendal nerve transection upon mating behavior in the male rat. *Physiology and Behavior,* 1976, *16,* 745–751.

Lumia, A. R., Eckhert, S., and Steinman, J. Social facilitation of sexual behavior following neonatal olfactory bulb removal in male rats. Paper presented at the meeting of the International Society for Developmental Psychobiology, Toronto, November, 1976.

Malsbury, C. W., and Pfaff, D. W. Neural and hormonal determinants of mating behavior in adult male rats. A review. In L. DiCara (Ed.), *The limbic and autonomic nervous system: Advances in research,* New York: Plenum Press, 1974, pp. 86–136.

Manning, A. Evolution of behavior. In J. L. McGaugh (Ed.), *Psychobiology, Behavior from a biological perspective.* New York: Academic Press, 1971, pp. 1–52.

McGill, T. E. Studies of the sexual behavior of male laboratory mice: Effects of genotype, recovery of sex drive, and theory. In Beach, F. A. (Ed.), *Sex and behavior.* New York: John Wiley and Sons, Inc., 1965, pp. 76–88.

Merari, A., and Ginton, A. Characteristics of exaggerated sexual behavior induced by electrical stimulation of the medial preoptic area in male rats. *Brain Research,* 1975, *86,* 97–108.

Murphy, M. R., and Schneider, G. E. Olfactory bulb removal eliminates mating behavior in male hamsters. *Science,* 1970, *167,* 302–303.

Olds, J. Self-stimulation of the brain. *Science,* 1958, *127,* 315–324.

Parrott, R. F. Effect of castration on sexual arousal in the rat, determined from records of post-ejaculatory ultrasonic vocalizations. *Physiology and Behavior,* 1976, *16,* 689–692.

Paxinos, G. Interruption of septal connections: Effects on drinking, irritability, and copulation. *Physiology and Behavior,* 1976, *17,* 81–88.

Pollak, E. I., and Sachs, B. D. Male copulatory behavior and female maternal behavior in neonatally bulbectomized rats. *Physiology and Behavior,* 1975, *14,* 337–343.

Pollak, E. I., and Sachs, B. D. Penile movements and the sensory control of copulation in the rat. *Behavioral Biology,* 1976, *17,* 177–186.

Pottier, J. J. G., and Baran, D. A general behavioral syndrome associated with persistent failure to mate in the male laboratory rat. *Journal of Comparative and Physiological Psychology,* 1973, *83,* 499–509.

Powers, J. B., and Winans, S. S. Vomeronasal organ: Critical role in mediating sexual behavior of the male hamster. *Science,* 1975, *187,* 961–963.

Rowe, F. A., and Edwards, D. A. Olfactory bulb removal: Influences on the mating behavior of male mice. *Physiology and Behavior,* 1972, *8,* 37–41.

Sachs, B. D., and Barfield, R. J. Temporal patterning of sexual behavior in the male rat. *Journal of Comparative and Physiological Psychology,* 1970, *73,* 359–364.

Sachs, B. D., and Barfield, R. J. Copulatory behavior of male rats given intermittent electrical shocks: Theoretical implications. *Journal of Comparative and Physiological Psychology,* 1974, *86,* 607–615.

Sachs, B. D., and Barfield, R. J. Functional analysis of masculine copulatory behavior in the rat. In J. S. Rosenblatt, R. A. Hinde, E. Shaw, and C. G. Beer (Eds.), *Advances in the study of behavior,* Vol. 7. New York: Academic Press, 1976, pp. 91–154.

Scalia, F., and Winans, S. S. The differential projections of the olfactory bulb and accessory olfactory bulb in mammals. *Journal of Comparative Neurology,* 1975, *161,* 31–56.

Schoelch-Krieger, M., and Barfield, R. J. Independence of temporal patterning of male mating behavior from the influence of androgen during the neonatal period. *Physiology and Behavior,* 1975, *14,* 251–254.

Teitelbaum, P., and Epstein, A. N. The lateral hypothalamic syndrome: Recovery of feeding and drinking after lateral hypothalamic lesions. *Psychological Review,* 1962, *69,* 74–90.

Wilhelmsson, M., and Larsson, K. The development of sexual behavior in anosmic male rats reared under various social conditions. *Physiology and Behavior,* 1973, *11,* 227–232.

11

The Temporal Patterning of Copulation in Laboratory Rats and Other Muroid Rodents

Benjamin D. Sachs and Donald A. Dewsbury

The value of studying a single species in the detail that has been lavished upon the laboratory rat depends in part upon the assumption that the species will prove an adequate model from which one may derive principles that will be generalizable to other species. The application of this problem to the specific case of the control of copulatory behavior in rats has been discussed (Beach, 1956; Dewsbury, 1973), and some speculative generalizations from rats to other muroid rodents were made by Sachs (this volume). A minimal basis for such generalizations includes evidence that the functional relations among different aspects of the behavior are similar among different species.

Frank Beach (1956) called attention to the reliable correlation ($r = +.50$) between the mean interval between intromissions (MIII) and the post-ejaculatory interval (PEI) of laboratory rats.* This correlation was based

* In many species of rodents the occurrence of penile insertion with sperm transfer (termed ejaculation) is preceded by several penile insertions without sperm transfer (termed intromissions). Copulatory behavior tends to occur in "series," with each series terminated by an ejaculation and separated from other such series by a postejaculatory refractory period. Included among the standard measures of copulatory behavior are the mean inter-intromission interval (MIII), which is the mean interval separating the insertions of a series, and the postejaculatory interval (PEI), the interval from an ejaculation until the next insertion. Typically, IIIs are much shorter than PEIs. MIII and PEI are measures of distinct aspects of the integrated copulatory pattern. In interpreting effects of experimental manipulations on MIII and PEI, and in the formulation of theoretical models of the control of copulatory behavior, it becomes important that the relationship between these two (and other) measures of copulatory behavior be understood.

Benjamin D. Sachs • Department of Psychology, University of Connecticut, Storrs, Connecticut. *Donald A. Dewsbury* • Department of Psychology, University of Florida, Gainesville, Florida.

on scores from a single large sample of males. Data reported by Sachs (this volume) demonstrate the reliability of this relationship. In three independent sets of data, values of +.64, +.65, and +.71 were obtained for the correlation between MIII and PEI. In the course of our correspondence about our respective chapters for this volume, it became apparent that the relationship between MIII and PEI holds not only among different individual male rats of single samples, but also among rats of different genotypes and among different species of muroid rodents.

Dewsbury (1975a) reported data from a diallel cross study of copulatory behavior in laboratory rats. Animals of four inbred strains and their twelve possible F_1 crosses were studied. The mean MIII and PEI for the first series for rats of these 16 genotypes are presented in Figure 11.1. The product moment correlation coefficient for these 16 points is +.61 (df = 14, $p < .02$). The least squares regression line fit to these data conforms to the equation: PEI = (2.58 × MIII) + 185.3. The correlation among the four parental strains taken by themselves is +.67; that among the 12 F_1 crosses is +.62 (df = 10, $p < .05$).

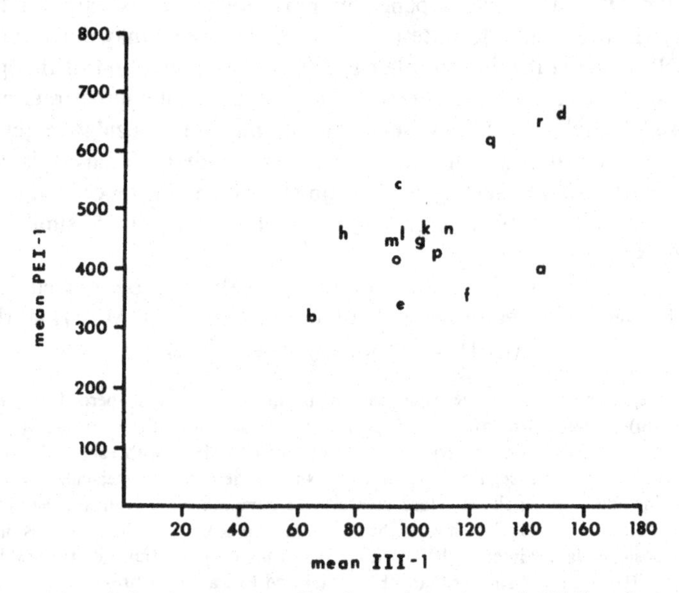

Figure 11.1. The relation between mean interintromission interval (III-1) and mean post-ejaculatory interval (PEI-1) of the first ejaculatory series of four inbred strains of laboratory rat and their reciprocal crosses. The parental strains are (a) ACI, (b) F344, (c) LEW, and (d) WF. The initial letter of the crosses reflects the maternity, and the second letter the paternity, and they are represented by letters in the figure as follows: (e) AF, (f) FA, (g) AL, (h) LA, (k) AW, (l) WA, (m) FL, (n) LF, (o) FW, (p) WF, (q) LW, (r) WL. The data are from Dewsbury (1975a).

Dewsbury (1975b) summarized data from a program of research in which the copulatory patterns of 31 species of muroid rodents were compared. It is not possible to utilize information on MIII and PEI for all species, because either (a) the males of some species virtually always ejaculate on their first mount with insertion and thus MIII is not a meaningful measure, or (b) inadequate data could be collected. We found 21 species for which both the copulatory pattern and the available data base permitted reasonable estimates of both MIII and PEI under the testing conditions of Dewsbury's laboratory. References to the sources of these data can be found in the report of Dewsbury (1975b; the data for *Peromyscus maniculatus bairdi* were obtained since that report). The mean MIII and PEI for the first series for these 21 species of muroid rodents are presented in Figure 11.2. The appearance of a strong relationship between MIII and PEI is consistent with the correlation coefficient ($r = +.65$, df $= 19$, $p < .002$). The least-squares regression equation for these points is PEI $= (7.9 \times \text{MIII}) + 113.7$.

By visual inspection, three species, *Mus musculus* (b), *Microtus ochrogaster* (l), and *Microtus pinetorum* (r), appear to depart from the regression line for the remaining species and to lie along a different regression line of their own. The MIII $\times$ PEI correlation among these three species is $+1.00$ (df $= 1$, $p < .01$), and the regression equation is PEI $= (11.2 \times \text{MIII}) + 608.2$. The correlation among the remaining 18 species is $+.82$ (df $= 16$, $p < .0001$), and the appropriate regression equation is PEI $= (5.38 \times \text{MIII}) + 140.8$. Division of these 21 species into these two groups is clearly *post hoc*. Only the addition of data from other species will permit a determination of whether there is a true dichotomy, or whether the appearance of the dichotomy is simply a function of the absence of data from extant species characterized by intermediate values.

The data in Figures 11.1 and 11.2 indicate that the correlation between MIII and PEI that has been observed among individuals in several studies of laboratory rats is also observed when comparing both different genotypes of laboratory rats and different species of muroid rodents. Two limitations are apparent. First, data from only the first ejaculatory series have been used in establishing these correlations; the relationship between MIII and PEI may be quite different in later series. This possibility is suggested by the different rates at which different individuals and different species approach sexual satiety, and by the considerable differences in the slopes of the curves for MIII and PEI as satiety is approached (Brown, 1974; Dewsbury, 1968; Karen and Barfield, 1975).

A second limitation stems from the limited range of species sampled in developing the correlation among muroid rodents. It must be remembered that the correlation between MIII and PEI can apply only to species with particular copulatory patterns. None of the 21 species reported in Figure

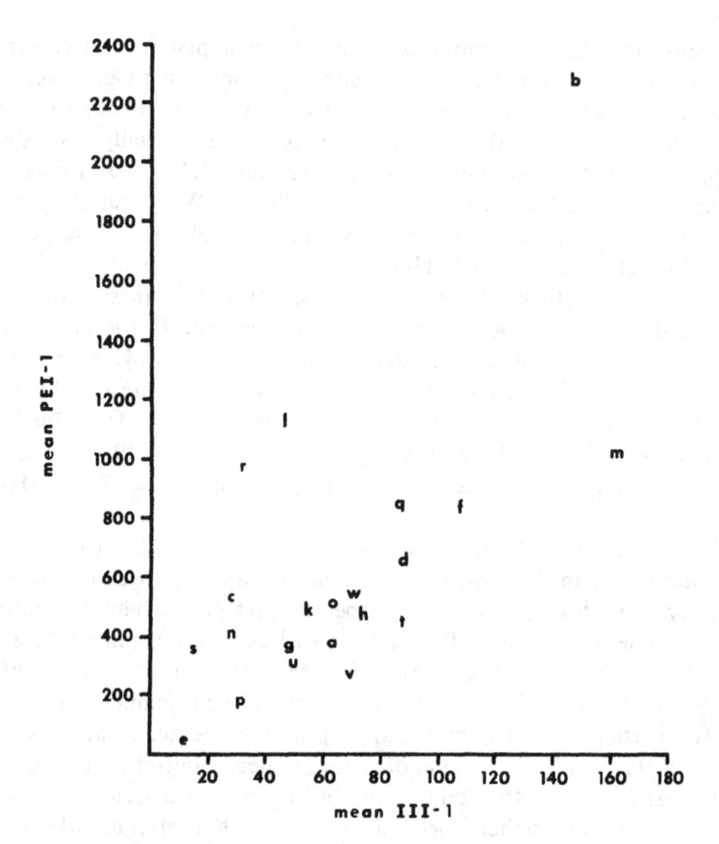

Figure 11.2 The relation between mean interintromission interval (III-1) and mean postejaculatory interval (PEI-1) of the first ejaculatory series of 21 species of muroid rodent. Each letter in the figure designates a species as follows [the number following the species indicates the species' copulatory pattern classification according to Dewsbury (1975b)]: (a) *Rattus norvegicus*, 13; (b) *Mus musculus*, 9; (c) *Microtus pennsylvanicus*, 11; (d) *Oryzomys palustris*, 13; (e) *Mesocricetus auratus*, 13; (f) *Microtus californicus*, 11; (g) *Peromyscus eremicus*, 9; (h) *Peromyscus polionotus*, 13; (k) *Sigmodon hispidus*, 13; (l) *Microtus ochrogaster*, 9; (m) *Peromyscus californicus*, 11; (n) *Meriones tristrami*, 13; (o) *Reithrodontomys megalotis*, 15; (p) *Meriones unguiculatus*, 13; (q) *Peromyscus crinitus*, 13; (r) *Microtus pinetorum*, 11; (s) *Microtus montanus*, 9; (t) *Peromyscus leucopus*, 13; (u) *Peromyscus gossypinus*, 15; (v) *Rattus rattus*, 13; (w) *Peromyscus maniculatus bairdi*, 13.

11.2 lock during copulation. In order that MIII and PEI be meaningful measures for the species, multiple intromissions must frequently precede ejaculation and copulation must be resumed after the first ejaculation of an episode. Thus, a correlation between MIII and PEI cannot meaningfully be considered for 11 of the 31 species discussed by Dewsbury (1975b). On the other hand, it should be noted that the correlation transcends several of the dichotomous classes of copulatory patterns proposed by Dewsbury (1972). Indeed, the data presented in Figure 11.2 are from animals with four different copulatory patterns: pattern #9 (intravaginal thrusting and multiple intromissions prerequisite to ejaculation), pattern #11 (thrusting, but multiple intromissions not prerequisite), pattern #13 (no thrusting, multiple intromissions prerequisite), and pattern #15 (no thrusting and multiple intromissions not prerequisite). All of these species are capable of attaining more than one ejaculation per episode.

The correlation between MIII and PEI implies at least partial sharing of underlying mechanisms. This was recognized by Beach (1956), who proposed that the III might be like a minirefractory period, a reduced version of the refractoriness that characterizes the PEI. Beach wrote that "each intromission can be thought of as temporarily reducing sexual excitation, so that the copulatory threshold has to be regained before another sexual act can occur" (p. 24). Although the potential communality of mechanisms underlying MIII and PEI has not always been considered in theoretical models of the control of copulatory behavior, such a relationship has been notable in more recent theorizing (e.g., Freeman and McFarland, 1974; Karen and Barfield, 1975; Sachs and Barfield, 1974, 1976). The present data are fully consistent with this trend. Particular attention must be given to the fact that in rats the MIII correlates more strongly with the early portion of the PEI (the absolute refractory period) than with the later portion of the PEI (the relative refractory period; Sachs, this volume). Such a distinction was not made in the data of Dewsbury (1975a; b). It would also be of interest to determine which of the individual IIIs of a series correlates most strongly with the PEI (see Dewsbury and Bolce, 1970).

The strength of the relationship between two fundamental measures of the temporal patterning of copulation in rodents of differing genotypes, species, and genera, suggests a substantial communality in the processes underlying these variables. It appears that the mechanism(s) controlling these aspects of temporal patterning have diverged less than either the copulatory patterns themselves or the absolute magnitude of the MIIIs and PEIs. Hypothetically, variation in a single parameter of a single process could account for many of the individual, strain, and species differences in MIII and PEI. To the extent that the underlying processes have been stable during evolution, neural or conceptual mechanisms of temporal patterning, inferred from research with laboratory rats, may have the generalizability to other species that helps justify focus on a single species. At the same time, comparative research is important both to the determination of the range of this generalizability, and to the discovery of relationships not present in laboratory rats.

References

Beach, F. A. Characteristics of masculine "sex drive". In M. R. Jones (Ed.), *The Nebraska symposium on motivation*. Lincoln: University of Nebraska Press, 1956, pp. 1-32.

Brown, R. Rat copulatory behavior. In D. J. McFarland (Ed.), *Motivational control systems analysis*. New York: Academic Press, 1974, pp. 461–479.

Dewsbury, D. A. Copulatory behavior in rats: Changes as satiety is approached. *Psychological Reports*, 1968, *22*, 937–943.

Dewsbury, D. A. Patterns of copulatory behavior in male mammals. *Quarterly Review of Biology*, 1972, *47*, 1–33.

Dewsbury, D. A. Comparative psychologists and their quest for uniformity. *Annals of the New York Academy of Sciences*, 1973, *223*, 147–167.

Dewsbury, D. A. A diallel cross analysis of genetic determinants of copulatory behavior in rats. *Journal of Comparative and Physiological Psychology*, 1975a, *88*, 713–722.

Dewsbury, D. A. Diversity and adaptation in rodent copulatory behavior. *Science*, 1975b, *190*, 947–954.

Dewsbury, D. A., and Bolce, S. K. Effects of prolonged postejaculatory intervals on copulatory behavior of rats. *Journal of Comparative and Physiological Psychology*, 1970, *72*, 421–425.

Freeman, S., and McFarland, D. J. RATSEX—an exercise in simulation. In D. J. McFarland (Ed.), *Motivational control systems analysis*. New York: Academic Press, 1974, pp. 479–504.

Karen, L. M., and Barfield, R. J. Differential rates of exhaustion and recovery of several parameters of male rat sexual behavior. *Journal of Comparative and Physiological Psychology*, 1975, *88*, 693–703.

Sachs, B. D., and Barfield, R. J. Copulatory behavior of male rats given intermittent electric shocks: Theoretical implications. *Journal of Comparative and Physiological Psychology*, 1974, *86*, 607–615.

Sachs, B. D., and Barfield, R. J. Functional analysis of masculine copulatory behavior in the rat. In J. S. Rosenblatt, R. A. Hinde, E. Shaw, and C. G. Beer (Eds.), *Advances in the study of behavior*, Vol. 7. New York: Academic Press, 1976, pp. 91–154.

PART III

HUMAN SEXUALITY AND SEX DIFFERENCES

12

Sex Differences in the Human Infant

Jerome Kagan

The mind's attraction to difference and automatic exaggeration of the contrast between physical stimuli is the partial consequence of reciprocal inhibition and exitatory processes in the cells of the central nervous system. But the mind also amplifies the difference between opposing categories. The English sailors who first encountered West Africans in the middle of the 16th century exaggerated both the contrast between the blackness of the Africans' skin and their own as well as the psychological qualities that were at variance with the English conception of the ideal (Jordan, 1968). When categories form antonym pairs, we are tempted to go far beyond the essential contrasts that define the concepts and assume a much larger set. The small number of physical differences between human males and females have led some to expect a large set of biologically based psychological differences to accompany them. However, existing empirical data indicate that sex differences in infants and young children are the exception rather than the rule.

1. Evidence for Sex Differences

After an exhaustive summary of the relevant literature, Maccoby and Jacklin (1974) concluded that, with the exception of the earlier onset of language in girls, there was little unambiguous evidence of sex differences in young children, although the differences in older children and adolescents

Jerome Kagan • Department of Psychology and Social Relations, Harvard University, Cambridge, Massachusetts. This research was supported by a grant from the Carnegie Corporation of New York and the Office of Child Development, and was written while the author was a Belding Scholar of the Foundation for Child Development.

were more reliable. For example, existing data did not yield consistent support for popular hypotheses concerning sex differences in sensory thresholds for visual and auditory events, rate of habituation to visual events, attentiveness, discrimination learning, disposition to imitate models, memory span, impulsivity, or sociability. Maccoby and Jacklin regarded the data on tactile sensitivity, fear, activity level, dominance, competetiveness, and nurturance as ambiguous but concluded that existing data were consistent in support of the suggestion that girls' verbal skills matured more rapidly than boys, while older boys excelled in spatial and mathematical skills. Moreover, in most cultures of the world boys and men were more aggressive and engaged in more power maintaining actions. A recent study of children in six societies is probably the best support for this conclusion. The children, who were between 3 and 11 years of age, were observed in and around their homes in the following six settings: (1) a village in northeastern Okinawa, (2) a village in north central India, (3) Oaxaca, Mexico, (4) hamlets in northern Luzon in the Philippines, (5) villages on the rolling hills in the western provinces of Kenya, and (6) a small town in New England. In all six settings, the 3–6-year-old boys displayed more aggressive behavior (assaults on others, insults, and horseplay) than did the girls; among 7–11-year olds, in five of the six settings the boys displayed more aggression than the girls, the sex differences being most marked in the Philippines, Mexico and Okinawa, and least in the Kenyan villages (Whiting and Whiting, 1975).

The apparently universal finding of earlier onset of language among girls provides a clue to other places to search for potential sex differences. Since the first appearance and rate of growth of language are monitored by maturational forces, earlier onset implies a biological precocity in cognitive functioning. Gazzaniga (1970) has suggested that the newborn may have no clear hemispheric dominance. Only gradually does the left hemisphere gain dominance over the right during the opening years of life. If growth of the central nervous system were more rapid in the infant girl, she might attain left hemisphere dominance ahead of the boy, due perhaps to the earlier myelinization of the corpus callosum and the medial surface of the temporal lobe (Gazzaniga, 1970; Lancaster, 1968). As a result, the speech function of the left hemisphere might develop at a faster rate and girls would speak earlier than boys. But since girls' physical growth is also more rapid than that of boys during the early years, the possibility is raised that other maturationally monitored psychological dimensions may be speeded up in girls during the opening years of life.

2. The Fears of Infancy

The emergence of the fears of infancy is dependent, in part, on the maturation of cognitive abilities (Kagan, 1976). Some apprehensions are the result of the child's ability to recognize a discrepancy between an event and a schema for that class of event, and the simultaneous inability to assimilate the unfamiliar experience. The infant's first cognitive structures are schemata, abstract cognitive representations of experience that allow the child to recognize past events. If females developed well-articulated schemata of their environments sooner than boys, they would be more vulnerable to fear in less familiar surroundings. Discrepant events alert infants and provoke attempts at assimilation. If the assimilation is not possible, uncertainty occurs and behavioral signs of fear may appear. There is some evidence suggesting that infant girls show fearfulness a little earlier than boys (Robson, Pedersen, and Moss, 1969); and female infants under six months are more likely than males to cry when brought to unfamiliar laboratory settings (Kagan, 1971).

A second, related class of fears that emerges toward the end of the first year seems to be the result of an enhanced ability to retrieve representations of past experience withou perceptual support in the immediate field and to generate predictions of possible future events. Evidence from many laboratories points to an improvement, from 7 to 13 months, in the infant's ability to recall past experience and to hold those recalled structures in "memory" for a period of time. Hence, the child displays object permanence, vacillation, and inhibition to events that are discrepant from past experience. But this is also the time when the infant shows behavioral signs of fear to strangers, the departure of a primary caretaker, and to unfamiliar children. It is possible that the emergence of these fears depends, in part, on the new abilities to retrieve elements of the past and to predict the future.

Upon encountering an unfamiliar person, the infant retrieves the schema for his familiar caretaker and compares it with his perception of the stranger. If the child cannot resolve the discrepancy between the two structures, he becomes uncertain and, if he has no coping response available, may cry. Similarly, following the departure of the caretaker, the child retrieves the schema of the caretaker's presence seconds earlier, compares it with the present, and may try to generate anticipations of the future. If the infant cannot resolve the uncertainty that follows inability to "understand" the caretaker's absence or the infant cannot predict when the caretaker will return, the child may display distress through inhibition of play and/or crying.

These hypothetical explanations make the "apprehension" dependent upon the maturation of these new cognitive processes. The child with a more rapid growth rate would reach the new stage of cognitive functioning a little earlier and, hence, be vulnerable to the fears earlier. Moreover, since the fears disappear during the second and third years, when the child can assimilate the discrepant event, the infant with a more rapid rate of psychological growth might be expected to conquer these apprehensions a little earlier.

As indicated above, three universal incentives for apprehension during the first two years of life include unfamiliar adults, separation from a primary caretaker, and unfamiliar peers. These fears generally emerge in that order, at 7–9 months, 9–11 months, and 11–13 months, respectively. Each lasts from 4 to 20 months before vanishing. The results of recent experiments indicate that, under some rearing conditions, girls either display the last of these fears earlier than boys or begin to shed them at an earlier age. We interpret this phenomenon as suggesting that girls may enter the new stage of cognitive functioning a little earlier than boys. We shall now consider these data.

2.1. Study 1

A sample of Chinese and Caucasian infants from the Boston area were studied from 3½ or 5½ to 29 months of age. The maximal sample size was 116 children and 33 of these infants attended an experimental day care center five days a week. The remaining infants were reared at home for most of that period. The infants were assessed at eight ages: 3½, 5½, 7½, 9½, 11½, 13, 20, and 29 months of age with a battery designed to evaluate attentiveness, reaction to discrepancy, language, memory, play with peers, and attachment to the mother. There were no important effects of care on any aspects of development, but consistent ethnic differences in signs of affectivity and inhibition.

The Chinese children were less motoric, less vocal, smiled less to interesting events, and were more fearful and inhibited than Caucasians in most unfamiliar social situations. As others have reported, there were no consistent sex differences in attentiveness, vocalization, irritability, or smiling to the interesting visual and auditory episodes that held across age and ethnic groups. Although the girls obtained higher scores on the mental items of the Bayley scale of infant development at 13 months of age (sample items include imitation of an examiner, scribbling with a crayon, imitation of words, and early speech) and the nonlanguage items on the Bayley at 20 months (sample items include imitation of the examiner, building a tower, and placing blocks in a form board), there were no sex differences at 29 months in quality of short-term memory for locations or comprehension of verbal concepts. Hence, girls did not show a general superiority to boys on all aspects of cognitive development at all ages tested.

However, the girls differed from the boys in the age of peak apprehension to an unfamiliar child, as well as the age at which they resolved that apprehension. At 13, 20, and 29 months each child was observed in a pair of temporally contiguous episodes that were similar in structure. Initially, the child and mother were brought into a room with a set of interesting age-appropriate toys, and the child was allowed to play for about 20 min while observers coded the child's play, vocalization, and overtures to the mother. The child and mother then left the room; when they returned they encountered a new set of toys and an unfamiliar child of the same sex, age, and ethnicity as the subject and, of course, the new child's mother. A second comparable period of observation ensued during which similar variables were coded. The variable of primary interest was the change in time playing from the solo to the peer play session. Inhibition of play is a sensitive index of apprehension or uncertainty in human infants as it is in primates (Dolhinow and Bishop, 1970). When a child encounters an unusual event which he cannot assimilate, he is likely to show a temporary inhibition of play (Kagan, 1976).

Table 12.1 shows the change in time playing from the solo to the peer session for the first half of the session as well as the change in tempo of play for the whole session. The latter variable is based on the following computational procedure. The duration of each continuous attentional epoch with a toy or toys is quantified and these epochs are cast into a frequency distribution. The third quartile is selected as the index of an individual child's tendency to invest long or short bouts of attention in play. When a child is uncertain, the average length of an epoch of play tends to be reduced. As Table 12.1 indicates, among girls, the relation between age and inhibition of play was an inverted U function. There was a sharp increase in inhibition of play—both total time and duration of each epoch—from 13 to 20 months, followed by a sharp decrease in inhibition from 20 to 29 months. However, many boys continued to be inhibited at 29 months, suggesting that at the oldest ages more girls than boys had overcome their uncertainty to the unfamiliar child.

2.2. *Study 2: A Replication*

The suggestion from Study 1 that girls reached peak apprehension to the peer and resolution of the apprehension earlier than boys was supported

Table 12.1. Changes in Play From Solo to Peer Session[a]

	Age	Chinese boys	Chinese girls	Caucasian boys	Caucasian girls
Duration of time playing[b]	13	15	18	17	15
(first half of session)	20	13	32	20	25
(seconds per minute)	29	20	7	18	19
Change in third quartile	13	13	−3	7	8
value[c] (whole session)	20	16	33	20	32
	29	17	3	4	1

[a] Positive score reflects inhibition to the peer.
[b] Analysis of variance: Time playing—sex by age interaction, $F = 3.93$ (2,128), $p < .05$.
[c] Analysis of variance: Third quartile value—sex by age interaction, $F = 2.93$ (2,82), $p < .05$.

by an independent study of 40 pairs of Caucasian middle class children, matched on sex and age, who were observed in an unfamiliar apartment in a large building complex in which the children lived (Shapiro, 1969). The younger children were 18–20 months of age, the older children were 28 and 30 months old. The children were observed on two different occasions one week apart. During the first session, the mother of one of the children remained while the second mother left. On the second session, the mothers alternated and the mother who had departed on the first session remained. A single attractive toy was placed in the room with the two children. The observer watched each of the two children alternately for 10-sec epochs. Each 20-min session had a total of 120 observations and each child was observed for 60 10-sec periods.

There was no important effect on the child's behavior of the presence or absence of the child's own mother because great care was taken not to initiate the experiment until both children were happy and unafraid. The major result was that the girls overcame apprehension to the unfamiliar peer earlier than boys, as revealed by a decrease in staring and an increase in joint reciprocal play with age. Among girls, the 18-month olds stared more and played less than the 28-month-old subjects, suggesting that the older girls had overcome their apprehension. But among the boys the older children played less and stared more than the younger ones (see Figure 12.1).

2.3. Study 3

A third study, conducted by Martha Zaslow, was a partial replication of the first with children living either on Israeli kibbutzim or in nuclear family households in Jerusalem. The children were observed at 14, 20, and 29 months in a cross-sectional design that used the same procedures as described in Study 1. There were eight boys and eight girls at each of the three ages in each of the two settings. The child first played in an unfamiliar room with his mother while an observer coded time playing and proximity to the mother. After this session, an unfamiliar child of the same sex and age and that child's mother were introduced into the room along with a new set of toys and time playing, proximity to mother, and staring at the unfamiliar child were coded.

As in Study 1, maximal inhibition of play following introduction of the unfamiliar peer occurred at 20 months for most of the sample. However, there was some indication that more girls than boys showed signs of inhibition at 14 months even though the sex difference was not statistically significant (see Table 12.2).

Moreover, at 14 months the girls stared at the unfamiliar child more than the boys while they were proximal to their mother (a good index of uncertainty), and among the kibbutz children this index of uncertainty was

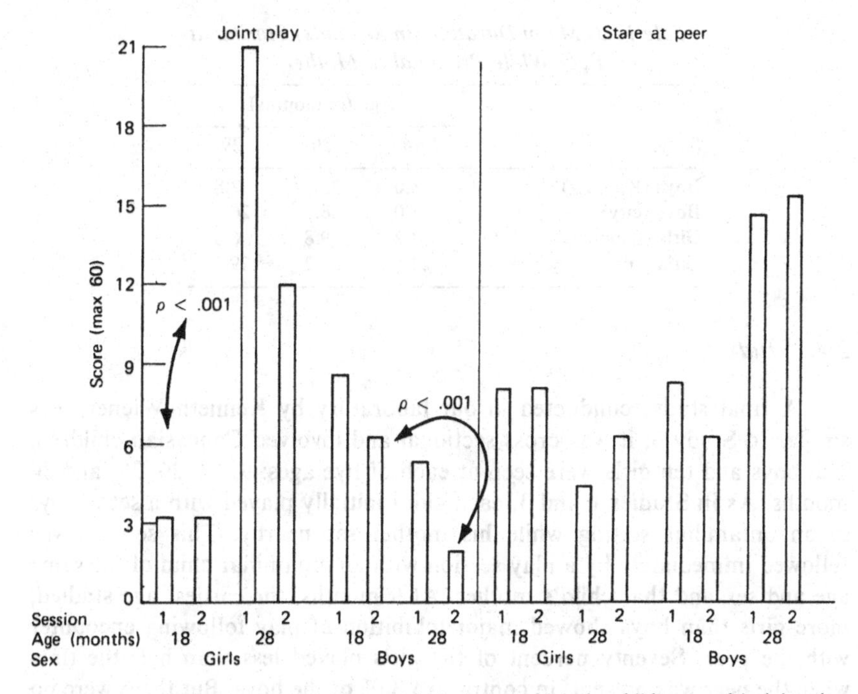

Figure 12.1. Average number of 10-sec periods during which the child engaged in joint play with the peer and stared at the peer. (From M. Lewis and L. A. Rosenblum, *Friendship and Peer Relations*. New York: John Wiley & Sons, Inc. 1975. Reproduced by permission of the publisher.)

maximal at 14 months and then declined. Among the boys, this index of uncertainty peaked at 20 months (see Table 12.3).

These data imply that uncertainty to the unfamiliar peer appears a little earlier among girls than boys even though there was no sex difference at the oldest age.

Table 12.2. Percent of Children (at Each Age) Showing a Decrease in Play From Solo to Peer of 20% or More

	Age (in months)		
Subjects	14	20	29
Boys (kibbutz)	38	88	0
Girls (Kibbutz)	63	88	12
Boys (city)	25	88	63
Girls (city)	50	75	50
Boys (both settings)	31	88	31
Girls (both settings)	56	81	31

Table 12.3. Mean Duration (in Seconds) Staring at Peer While Proximal to Mother

Subjects	Age (in months)		
	14	20	29
Boys (Kibbutz)	18.0	22.2	2.8
Boys (city)	8.0	38.6	26.5
Girls (Kibbutz)	22.2	9.8	6.3
Girls (city)	24.5	31.2	29.6

2.4. *Study 4*

A final study, conducted in our laboratory by Kenneth Wiener, was similar to Study 3; it was cross sectional and involved Caucasian children. Ten boys and ten girls were seen at each of five ages: 9, 14, 20, 29, and 36 months. As in Studies 1 and 3, each child initially played with a set of toys in an unfamiliar setting while his mother sat nearby. This session was followed immediately by a play session with an unfamiliar child of the same age and sex and that child's mother. At 9 months, the earliest age studied, more girls than boys showed major inhibition of play following encounter with the peer. Seventy percent of the girls played less than half the time when the peer was present, in contrast to 30% of the boys. But there were no sex differences at the four older ages, and inhibition of play was minimal among the 3-year olds. Although the sign of wariness occurred a little earlier in this sample than it did with the children in the first study, the results still imply earlier appearance of the apprehension among the girls.

In all four studies, the girls either displayed apprehension to the unfamiliar child a few months ahead of the boys or resolved the fear earlier, even though the specific age at which apprehension emerged, peaked, or vanished varied across the four samples. Environmental conditions obviously influence the degree of apprehension, for kibbutz boys and girls were far less inhibited at 29 months than were the children living in Jerusalem, and that environmental difference was much more robust than the sex difference. The sex difference in inhibition of play to the unfamiliar child is, from an absolute point of view, small and by three years of age many boys and girls have overcome their earlier timidity. It is not possible to state whether the earlier onset of apprehension in girls has any future consequences. Those who believe it does might reason that the earlier bouts with uncertainty might lead to a stronger predisposition to inhibition to unexpected events in the future. Others might argue that there is no reason to assume any permanent effects of the earlier appearance of fearfulness. Girls speak earlier than boys, but during later childhood there is no important sex difference in language skills, and in Western society there are more male novelists and poets than female. Hence, we are inclined to side with the more agnostic view.

3. Sex Differences in Patterns of Relations

Although sex differences in frequency or quality of behavior are not common among infants and young children, the correlational patterns for a set of variables are often different for the sexes. The most notable example is the fact the social class of the child's family usually correlates more highly with measures of cognitive development in girls than in boys, at least in American samples. Typically, the correlations between indexes of social class, like parental education and occupation on the one hand, and attentiveness during infancy or IQ and grades during school age, on the other, are higher for girls than for boys (Werner, 1969; Hindley, 1965; Kagan, 1962, 1971). This is true for both black and white children (Hess *et al.*1968, 1969), although it is not true for children living in rural Guatemalan agricultural villages (Freeman, Klein, Kagan, and Yarbrough, 1976). How might we interpret this phenomenon?

The difference in the magnitude of the correlations with class among school age children can be interpreted by assuming a stronger adoption of the family's values by girls than by boys. However, this dynamic is not operating during the first two years of life. At this early age, it is possible that there is greater variability across social class levels in maternal reactions to daughters than to sons. Most mothers in American society, whether they be high school drop-outs or college graduates, believe their sons will have to develop independence, responsibility, and a vocational expertise. When lower and middle class mothers of 4-year olds were asked to teach their child a new task, mothers of both classes were more achievement oriented toward sons than toward daughters, adopting a more business-like attitude toward boys than toward girls (Hess, Shipman, Brophy, and Bear, 1968). However, lower class mothers of daughters may project their greater sense of inadequacy onto them and, as a result, are less likely than middle class mothers to stimulate, encourage, or reward their daughters' accomplishments. If mothers from a broad range of social class backgrounds were more divergent in their concern with proper behavior and intellectual development with daughters than with sons, one would expect greater covariation between maternal social class and aspects of cognitive development among girls.

In Study 1 the sex differences in fearfulness and inhibition to the unfamiliar peer was greater for the Chinese than for the Caucasion children. We have reason to believe that this is due to the greater variability in sex role values held by Chinese families. Thus, if there is a biological basis for the sex differences we have described, and we believe this is likely, it can be muted or amplified by experience.

Parents have a conception of the ideal attributes of males and females and of the behavioral profile they wish their child to attain. These conceptions are communicated to the child and guide parental practices. By age 6 each child tries to match his or her profile to the ideal for its sex. A 6-year old knows that certain beliefs, wishes and actions are most appropriate for the category boy or girl; the child has no choice but to tailor his/her psychology to the definition of his/her sex category. Children and adults from different cultural settings agree that large size, dark hues, and pointed objects are symbolic of masculinity; small objects, light hues and rounded objects symbolic of femininity. The cross-cultural uniformity on these symbolic dimensions implies a small set of universal assumptions about the essence of maleness and femaleness. Even if we could alter the experiences of children and arrange their environments so that all adults treated boys and girls equivalently, we might not eliminate all sex differences in adolescents and adults because of the personal and symbolic interpretations of masculinity and femininity based on irrevocable differences in size, strength, anatomy, and life functions.

4. *Prospectus*

Empirical study of sex differences in infants and young children has a short history, and it is not surprising that most investigators initially examined differences on single dimensions at one point in time. That strategy has serious limitations and will probably become less popular. Maccoby and Jacklin's review did not permit any coherent set of generalizations to be drawn for the many studies that have used that paradigm. Moreover, some investigators have assumed, perhaps gratuitously, that a sex difference noted at one age was likely to be stable from that point forward. That assumption ignores a basic developmental principle; namely, that many of the behavioral phenomena of the first two years are short lived. Fearfulness to strangers or to temporary separation from a caretaker, like babbling to sounds, appears during certain delimited eras of development, holds the stage for awhile, and then disappears. Apprehension to maternal departure typically appears between 7 and 20 months and there is no intraindividual stability of the disposition to display distress to separation across the period 7–29 months in the population described in Study 1.

It is useful to view the course of development as an evolutionary tree, a history of many choices where the direction taken at one point in time alters the probability of the array of future choices by a wee bit. The observed sex differences between adult men and women in any society should be regarded as the product of a long series of such choices. It is not psychologically satisfying to explain the nurturant motives and behavior of contemporary Western women as a derivative of a small set of early childhood experiences. Such an explanation is analogous to explaining the Guatemalan earthquake in February, 1976, as being due to forces operative when the earth was cooling. We do not explain man by pointing to protozoa, even though protozoa was a remotely past event that made man a little more likely. Rather we point to the entire evolutionary sequence. Similarly, we cannot explain nurturant attitudes in 30-year-old women by listing only the experiences of early childhood. The explanation of sex differences in adults resides in a set of transitional probabilities associated with alternative itineraries over the course of development.

It is believed that future research on human sex differences is likely to examine paths in development and in the forces that determine alternative routes. For example, behavioral differences between boys and girls tend to blossom after age 2 and continue until adolescence. Study of the period between 2 and 16 years is apt to be profitable. Science is at its best when it attempts to refute hypotheses that are amenable to empirical test. The following are questions that are likely to attract attention from investigators in the future.

1. Sex differences in spatial ability. Although there is general consensus that there is a sex difference in spatial ability at adolescence, it is difficult to demonstrate that difference prior to age 13 or 14 years. This raises doubts about the popular view that this sex difference is inherited.

2. Language. The bases for the earlier onset of language in girls will be explored. Since this phenomenon is apt to be biological in origin, we suspect there will be many more studies of the central nervous system correlates of development during the first three years of life.

3. Symbolic conceptualization of sex roles. There is a mysterious agreement, across cultures, on a small set of symbolic dimensions for male and female. The association of maleness with a larger size is a natural consequence of the fact that boys and men are larger than girls or women. The association of masculinity with danger and aggression probably derives from the fact that males are more active and aggressive in most cultures, either in the service of soldiering, hunting, fighting, or as a consequence of rough and tumble play in childhood. Further, a possible link between femaleness and naturalness is reasonable, for birth and nursing are quintessential natural functions and it is easy for a child to come to the conclusion that to be female is to be closer to the experiences that nature intended. We need a better understanding of how these metaphors come to be associated with each of the sexes.

5. Caveat

It is important to emphasize that the interpretation of the significance of a particular behavior or belief in one sex is always dependent on the value for the other sex. American men are sexually more aggressive than American women but American women are more aggressive than Japanese women. When we speak of sex differences in aggression, dominance, passivity, or sexuality, we are not talking about any absolute value but about the difference in that dimension for males and females in a particular culture.

References

Dolhinow, P. J. and Bishop, N. The development of motor skills and social relationships among primates through play. In J. P. Hill (Ed.) *Minnesota symposium on child psychology,* Vol. 4. Minneapolis: University of Minnesota Press, 1970, pp. 141–198.

Freeman, H. E., Klein, R. E., Kagan, J. and Yarbrough, C. Relation between nutrition and cognition in rural Guatemala. Unpublished manuscript, 1976.

Gazzaniga, M. S. *The bisected brain.* New York: Appleton-Century-Crofts, 1970.

Hess, R. D., Shipman, V. C., Brophy, J. E. and Bear, R. M. The cognitive environments of urban preschool children. Report to the Graduate School of Education, University of Chicago, 1968–1969.

Hindley, C. B. Stability and change in abilities up to 5 years: Group trends. *Journal of Child Psychology and Psychiatry,* 1965, *6,* 85–99.

Jordon, W. D. *White over black.* Chapel Hill: University of North Carolina Press, 1968.

Kagan, J. Psychology of sex differences. In F. A. Beach (Ed.), *Human sexuality.* Baltimore: Johns Hopkins University Press, 1977.

Kagan, J. Emergent themes in human development. *American Scientist,* 1976, *64,* 186–196.

Kagan, J. *Change and continuity in infancy.* New York: Harcourt, Brace and Jovanovich, 1971.

Kagan, J. *Birth to maturity.* New York: John Wiley and Sons, Inc., 1962.

Lancaster, J. B. Primate communication systems and the emergence of human language. In P. C. Jay (Ed.), *Primates.* New York: Holt, Rinehart and Winston, 1968, pp. 439–457.

Maccoby, E. E., and Jacklin, C. N. *The psychology of sex differences.* Stanford: Stanford University Press, 1974.

Robson, K. S., Pedersen, F. and Moss, H. A. Developmental observations of diadic gazing in relation to the fear of strangers and social approach behavior. *Child Development,* 1969, *40,* 619–628.

Shapiro, L. A study of peer group interaction in 18 and 28 month old children. Unpublished doctoral dissertation, Harvard University, 1969.

Werner, E. E. Sex differences and correlations between children's IQs and measures of parental ability and environmental ratings. *Developmental Psychology,* 1969, *1,* 280–285.

Whiting, B. B., and Whiting, J. W. M. *Children of six cultures.* Cambridge, Massachusetts: Harvard University Press, 1975.

13

Sex Differences in Cognition: Evidence from the Hawaii Family Study

James R. Wilson and Steven G. Vandenberg

1. Introduction

Since 1972 we have been collecting family data in Hawaii in order to study the genetic and environmental bases of human cognition. Although data collection is now complete, it will be some time before all tests are scored and all laboratory measures are analyzed. Individual scores from the first three years of the study are available now, however, and they constitute a large and sufficient data set for many kinds of psychometric analyses. Planned analyses include the estimation of heritabilities and genetic correlations of scores on selected cognitive tests; a comparison of phenotypic and genetic cognitive factor structures; estimations of the effects of inbreeding, assortative mating, parental age, birth order, certain maternal–fetal interactions, and certain socioeconomic variables; a search for possible relationships between cognitive traits and genetic markers and between cognitive and dermatoglyphic traits; and investigations of possible heterosis, dominance, or epistasis in components of cognitive abilities. For this chapter, in honor of Frank Beach, we searched for sex differences in cognitive test performance.

Hawaii was chosen as the locale for this study because of the availability of a large, ethnically varied population using a common lan-

James R. Wilson and Steven G. Vandenberg • Institute for Behavioral Genetics, University of Colorado, Boulder, Colorado.

guage (English). A large population was needed because accurate estimation of certain of the genetic parameters of interest requires rather large samples. We wished to include ethnic variation, so that conclusions would be more broadly based than is generally the case in studies which include only one or two ethnic groups. In Hawaii, all ethnic groups are minorities. Americans of European ancestry (AEA) and Americans of Japanese ancestry (AJA) constitute about 35% and 30% of the population, respectively. The balance includes Americans of Hawaiian or part Hawaiian ancestry (about 10%), Chinese ancestry (about 6%), or a combination of these ethnic backgrounds, and persons of several other ethnic groups (Phillipino, Korean, Samoan, American Black, and others). English is used and taught in the public schools and is the standard language of government and commerce. We were aware that not all persons in Hawaii are equally facile with the English language.

Families were solicited for participation in the study in a variety of ways, including newspaper announcements, announcements in high schools and at church and club meetings, posters and bulletin boards, and personal referral by previous participants. Early attempts to recruit participants by the use of personalized letters were unsuccessful. The response rate was abysmally low, and several irritated people responded by demanding to know where we had obtained information about their families. The solicitation announcements specified the main requirements for participating in the study: both mother and father (under age 60) available for testing, along with one or more biological offspring (14 years of age or older); ability to read, write, and understand English; and willingness to spend about 3½ hours taking tests, giving a 20-ml blood sample and a small saliva sample, filling out a personal history questionnaire, and (for a subset of subjects) allowing us to take finger and palm prints. Many families called the laboratory with questions about the procedures before deciding whether or not to participate. Announcements included mention of the fact that each participating family would receive $50 (later raised by $10 for each additional offspring) for assisting in the study.

Group testing was employed, with most test sessions including about 20–30 individuals from about 5–8 different families of (usually) different ethnic backgrounds. Prior to testing, each family received by mail an information packet that included the test schedule, a map showing the route to the testing location (usually the University of Hawaii Medical School), and a copy of the text of the participation agreement. Upon reporting to a session, individuals were asked to sign the participation agreement and were given a set of gummed labels with a code number which would be used to identify all their test and laboratory results (to enhance privacy). It should be noted that essentially all subjects completed all procedures, although only results of the cognitive tests are discussed here.

The 15 cognitive tests in the Hawaii test battery are listed in Table 13.1 according to order of administration. The time limit for each test and an estimate of reliability are given in the table. These particular tests were chosen from a much larger set on the basis of results of pilot studies. Time limits are either those suggested by the authors of the tests or, in the case of modified or new tests, based upon pilot data. Further information about the tests and the methods of administration is presented in the Appendix.

DeFries *et al.* (1974) explored the dimensional aspects of the tests and compared cognitive structures in samples of the two major ethnic groups—Americans of European ancestry (AEA), and Americans of Japanese ancestry (AJA). Principal component analysis of individual scores from the AEA and AJA groups revealed four major cognitive components, or factors: verbal, spatial visualization, perceptual speed, and visual memory. Factor profiles for the AEA and AJA groups proved to be virtually identical. Using the same technique, Wilson *et al.* (1975) compared cognitive structures and factor profiles for larger samples of the same two ethnic groups and, using age stratification, for three age groups. Again, factor profiles were almost identical. It was concluded that members of these ethnic or age groups were responding to the tests in a similar and adequate

Table 13.1. The 15 Cognitive Tests in the Hawaii Test Battery[a]

Test	Test time	Reliability[b]
Primary Mental Abilities (PMA) Vocabulary	3 min	0.96 (PUBL)
Visual Memory (immediate)	1-min exposure/ 1-min recall	0.58 (KR–20)
Things (a fluency test)	2 parts/3 min each	0.74 (CRα)
Sheppard–Metzler Mental Rotations (modified for group testing by Vandenberg)	10 min	0.88 (KR–20)
Subtraction and Multiplication	2 parts/2 min each	0.96 (CRα)
Elithorn Mazes ("lines and dots"), shortened form	5 min	0.89 (PUBL)
Educational Testing Service (ETS) Word Beginnings and Endings	2 parts/3 min each	0.71 (CRα)
ETS Card Rotations	2 parts/3 min each	0.88 (CRα)
Visual Memory (delayed recall)	1 min	0.62 (KR–20)
PMA Pedigrees (a reasoning test)	4 min	0.72 (PUBL)
ETS Hidden Patterns	2 parts/2 min each	0.92 (CRα)
Paper Form Board	3 min	0.84 (KR–20)
ETS Number Comparisons	2 parts/1.5 min each	0.81 (CRα)
Whiteman Test of Special Perception	10 min	0.69 (KR–20)
Raven's Progressive Matrices, modified form	20 min	0.86 (KR–20)

[a] From Wilson *et al.* (1975).
[b] PUBL is taken from test manual; KR–20 is the Kuder–Richardson Formula 20; CRα is the Composite Reliability Coefficient.

fashion and that they had the same or nearly the same cognitive structures, as measured by this battery. Park, Johnson, and Kuse (1976), who administered a translated version of the same tests to a sample of families in the Republic of Korea (ROK), reported a similar factor structure, but with some indication that the spatial factor was not as simple in the ROK sample and perhaps should be split into two factors. The authors hypothesized that this factor might be more developed or more finely delineated in people who use an ideographic rather than an alphabetic method of written communication. This hypothesis is presently being tested further with new Hawaiian data (Johnson *et al.*, unpublished).

2. Cognitive Structure

For the present discussion, test scores are available for a total of 5,078 persons of varied ancestry. Scores for the 2,502 males in the sample were intercorrelated and subjected to principal component analysis with varimax rotations, as were the scores for the 2,576 females. Again, the same four components reported by DeFries *et al.* (1974) were indicated. The sample was then further stratified by generation to form four subsamples (fathers, sons, mothers, and daughters), and scores for each subsample were independently correlated and subjected to principal component analysis. Factor loading profiles comparing the four subsamples on each of the four main factors are shown in Figure 13.1. Confirming the visual impression, all coefficients of congruence (Tucker, 1951) are 0.99 or 1.00. We may conclude that cognitive structure is essentially identical in both sexes for the ages (14–60 years) included in this sample. It seems appropriate to proceed to the examination of sex differences, if any, on these tests.

3. Mean Differences

Table 13.2 presents test score means by sex for the total sample, including parents and offspring. Of the 15 tests, a significant sex difference is shown for all but one—the Whiteman Test of Social Perception (verbal). Females achieved a higher mean score on six of the tests, while males showed higher mean performance on eight tests.

The six tests showing female advantage are Vocabulary, Visual Memory (immediate), Word Beginnings and Endings, Visual Memory (delayed), Pedigrees, and Number Comparisons. Female advantage has often been reported for verbal tests, expecially verbal fluency, as well as for tests of perceptual speed (see, for example, Tyler, 1965). Of these six tests,

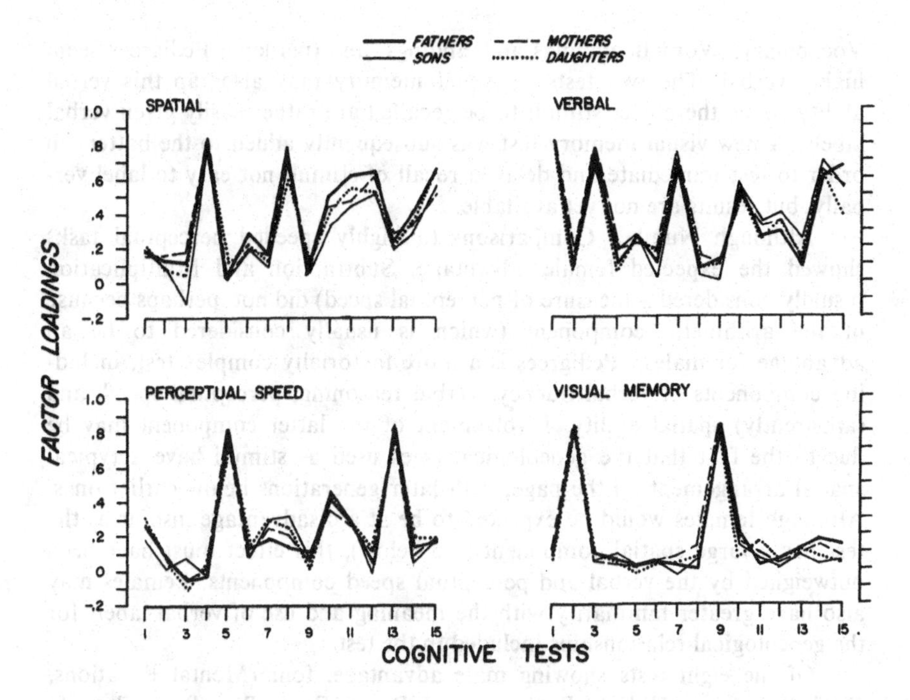

Figure 13.1. Factor loading profiles of fathers, sons, mothers, and daughters on the Verbal, Spatial Visualization, Perceptual Speed, and Visual Memory factors.

Table 13.2. Test Score Means for the Total Sample as a Function of Sex

Test	Male	Female	*F*	df	*p*<
Vocabulary	33.1	34.6	20.01	1,5055	.001
Visual Memory (immediate)	14.9	15.3	27.37	1,5044	.001
Things	34.0	31.0	95.02	1,5054	.001
Mental Rotations	19.7	10.1	943.13	1,5052	.001
Subtraction and Multiplication	59.7	58.0	7.52	1,5054	.01
Elithorn Mazes	14.1	12.1	207.00	1,5016	.001
Word Beginning and Endings	18.7	19.8	27.98	1,5053	.001
Card Rotations	105.4	87.5	285.33	1,5045	.001
Visual Memory (delayed)	11.3	11.9	37.53	1,5027	.001
Pedigrees	22.6	24.4	86.41	1,5050	.001
Hidden Patterns	70.1	65.5	42.33	1,5047	.001
Paper Form Board	12.2	11.4	44.99	1,4997	.001
Number Comparisons	38.5	42.6	224.97	1,5051	.001
Social Perception (verbal)	13.3	13.3			NS
Progressive Matrices (modified)	19.9	18.9	43.91	1,5024	.001

Vocabulary, Word Beginnings and Endings, and (perhaps) Pedigrees seem highly verbal. The two tests of visual memory may also tap this verbal ability, since the object stimuli to be recalled are rather easily given verbal labels. A new visual memory test was subsequently added to the battery in order to test immediate and delayed recall of stimuli not easy to label verbally, but results are not yet available.

Although Number Comparisons (a highly speeded perceptual task) showed the expected female advantage, Subtraction and Multiplication (usually considered a measure of perceptual speed) did not, perhaps because of the arithmetic component (which is usually considered to be an advantage for males). Pedigrees is a more factorially complex test, including components of verbal fluency, verbal reasoning, perceptual speed, and (apparently) spatial ability. Involvement of the latter component may be due to the fact that the geneological trees used as stimuli have a typical spatial arrangement on the page, with later generations below earlier ones. Although females would be expected to be at a disadvantage insofar as the test has a large spatial component (see below), this effect must have been outweighed by the verbal and perceptual speed components. Females may also have greater familiarity with the meaning and use of verbal labels for the geneological relationships included in the test.

Of the eight tests showing male advantage, four (Mental Rotations, Card Rotations, Hidden Patterns, and Paper Form Board) are heavily dependent on spatial visualization, and aspects of two others (Elithorn Mazes and Progressive Matrices) tap spatial ability. It has been an almost universal observation (Maccoby, 1974; Tyler, 1965) that males have a strong advantage on spatial tasks. Opinion differs as to why this is so, with conjectures ranging from genetic endowment to a parental tendency to allow boys but not girls to explore and build things. Because the present data were collected on a single occasion in a person's life, they have very limited utility for exploring the environmental hypothesis. From the genetic viewpoint, an interesting possibility is that a gene, or genes, relevant to the development of spatial ability may be X linked (see, for example, Bock and Kolakowski, 1973; Stafford, 1961). Although analyzed results of the Hawaii Family Study do not support this hypothesis (DeFries *et al.*, 1976), it will be further tested in analyses of the full data set. Another interesting possibility is that higher spatial ability is part of normal male differentiation begun by fetal androgen which, in turn, is responsive to a Y-linked gene (or genes). If this is the same "switch" gene, which begins male development during fetal life, this may indicate a need for examination of the effects of androgens on cerebral development related to spatial ability—perhaps by differential lateralization of cerebral function. Androgenic involvement would be consistent with the report that females whose mothers were given progesterone or androgenic steroids during pregnancy may exhibit higher

IQs (Ehrhardt and Money, 1967), since spatial ability is an important component of most IQ tests. Such girls also tend to be more active and tomboyish, indicating that the hormone may act to enhance their spatial ability by stimulating them to greater exploration and manipulation of the environment. It should be borne in mind that genetic, hormonal, and environmental hypotheses are not mutually exclusive. Also, although the sex difference in mean spatial ability may be large, the large amount of overlap in ability in the sex distributions must also be considered.

For the Elithorn Mazes, a test that has spatial, perceptual speed, and reasoning components, the balance must have yielded male advantage. A male advantage was also revealed by the modified version of Raven's Progressive Matrices used in this study. This test involves reasoning, and it seems to have perceptual speed and spatial components as well. It is the most general of the tests in the battery. Because it includes progressions of symbols, rather than explicitly verbal material, it has often been used as a "culture-fair" test of general intelligence. It has also been used as a measure of what Cattell (1971) has called "fluid" intelligence, or the ability to abstract new relationships, as opposed to "crystallized" intelligence, which is conceived to be accrual of information (such as vocabulary). The male advantage in tests of reasoning is more controversial than the male advantage in tests of spatial ability, but it is largely consistent with the literature (Tyler, 1965). Again, however, the mean difference in favor of males does not reveal the large amount of overlap in the distributions.

In addition to the six tests that involve spatial ability to some degree, the Things test and the Subtraction and Multiplication test also showed male advantage. The former, which involves generating names of things to fit in designated categories, is sometimes called a fluency test and is sometimes considered as a test of creativity. To the degree that it is a test of fluency, we would expect a female advantage—an expectation not fulfilled by our results. Tests of creativity are not well standardized (or accepted), so it is difficult to say which sex should be favored. It is possible that the two categories we used in this research (see Appendix) may have been more familiar to males. The Subtraction and Multiplication test, though usually considered a measure of perceptual speed, does contain an arithmetic component, and this must have led to the small but significant male advantage.

Because of the marked age differences in cognitive ability previously reported by Wilson *et al.* (1975), it seems useful to examine sex differences as a function of generation. These comparisons are presented in Table 13.3. Although the overall picture remains the same, a few differences are revealed. The statistically significant female advantage on the Vocabulary test is even greater in the offspring group (nearly all between 14 and 20 years of age) than in the total sample, but it is lost in the parental group

Table 13.3. Test Score Means as a Function of Generation and Sex

	Parents (n = 2,961)			Offspring (n = 2,096)		
Test	Male	Female	$p\leq$	Male	Female	$p\leq$
Vocabulary	37.5	38.1	NS	26.7	29.9	.001
Visual Memory (immediate)	14.6	15.0	.003	15.2	15.7	.001
Things	34.8	31.2	.001	32.7	30.6	.001
Mental Rotations	16.5	6.4	.001	24.3	15.1	.001
Subtraction and Multiplication	67.1	62.0	.001	48.7	52.6	.001
Elithorn Mazes	14.0	11.7	.001	14.3	12.6	.001
Word Beginning and Endings	19.4	20.1	.027	17.6	19.3	.001
Card Rotations	97.1	77.2	.001	117.5	101.5	.001
Visual Memory (delayed)	11.0	11.6	.001	11.6	12.3	.001
Pedigrees	22.4	22.7	NS	22.7	26.7	.001
Hidden Patterns	70.4	62.2	.001	69.7	70.0	NS
Paper Form Board	12.2	11.0	.001	12.3	12.0	NS
Number Comparisons	39.6	43.0	.001	36.8	42.1	.001
Social Percention (verbal)	13.6	13.1	.001	12.9	13.6	.001
Progressive Matrices (modified)	19.0	17.4	.001	21.2	21.0	NS

(mostly between 35 and 60 years of age), although the direction of the difference is still in favor of females. It should be noted that this particular test (PMA Vocabulary) proved to be too easy for our subjects (3.2% got a perfect score, and 7.7% missed only one word). A more difficult test might have yielded better differentiation among adults.

Although the male advantage on the Mental Rotations test appears in both generations, it is notable that young females scored nearly as high as did the fathers. This is the most difficult of the spatial ability tests in the battery, and it is the one showing the largest age decrement. Subtraction and Multiplication scores show a reversal, with younger females doing better than their male age mates and older males doing better than their wives. This may reflect the faster developmental rate for young females versus males, more practice by older males than females, or both. On the Pedigrees test, younger females showed a clear advantage, but adult women did not.

On the Hidden Patterns test (largely spatial, but with a perceptual speed component), younger females actually showed a slight advantage over males, although the difference is not statistically significant. The male advantage on the Progressive Matrices test does not appear in the younger group; in adults, it seems to be a reflection of a smaller age decrement in males than in females.

An overall comparison of the results of the analyses shown in Tables 13.2 and 13.3 indicates that the pattern of sex differences remains the same across generations, with females excelling on verbal fluency and perceptual speed tasks and males doing better on tests that involve spatial visualization or reasoning. In all tests, there is a vast amount of overlap in the distributions after categorization by sex.

4. *Differences (?) in Distribution*

Although the sex ratio is somewhat in favor of males at birth in *Homo sapiens* (and many other species), males die at a higher rate in infancy, childhood, and adulthood. The result is that the sexes are about equal in number throughout most of life (except in the older age groups, when females may outnumber males by two to one or more). Yet, in almost every field of human endeavor (political, scientific, literary, musical, artistic, mathematical, etc.), males outnumber females in terms of high achievement and also at the lower end of the continuum (in jail, retarded, ill, abusers of drugs, etc.). Why this is so has led to a great deal of controversy. It is doubtful if any satisfactory answer will emerge as long as females fulfill the critical biological role of incubating the young of the species plus the social role of caring for the infants. Even under these conditions, and without addressing other important issues (such as personality differences in aggression, in nurturance, in need for outward success, etc.; size and strength differences; the effects of menstrual cycles in females; etc.), we can consider one specific hypothesis: The *distribution* of abilities in males may be more dispersed, with more males appearing in the extreme tails of the curve (Hunt, 1974; Tyler, 1965). Even a moderately flat, platykurtic distribution would have a great excess of individuals in the tails (say, past ± 2 standard deviations) relative to the Gaussian curve. If the hypothesis is true, there would be a great excess of men available for jail or a presidency (or both?).

To examine this hypothesis, we can compare test variances separately for males and females. These variances are shown in Table 13.4. On 11 of the 15 tests, male variance exceeds female variance ($p < .05$, binomial test), although it should be noted that only three of the individual variance ratios are statistically significant ($p < .05$, using the F distribution with 1000, 1000 df). In any case, there is an indication that test variances tend to be greater for males. Incidentally, if we had not seen the results of this analysis, we might have expected greater variance for the females, who were virtually all postpubertal and mostly premenopausal, and thus were subject to possible cyclic disruptions of test performance.

From the variance statistic itself, it is not possible to determine if males tend to be overrepresented at the high end of a distribution, at the low end, or both. We can, however, examine the tails of each distribution and count occurrences of males and females. To do this, it is necessary to choose

Table 13.4. Test Score Variances for the Total Sample as a Function of Sex

Test	Male	Female	σ_M^2/σ_F^2	σ_F^2/σ_M^2
Vocabulary	145.58	129.95	1.12	.89
Visual Memory (immediate)	9.06	8.60	1.05	.95
Things	122.63	121.88	1.01	.99
Mental Rotations	136.13	109.83	1.24[a]	.81
Subtraction and Multiplication	491.87	428.03	1.15[a]	.87
Elithorn Mazes	24.66	22.31	1.11	.90
Word Beginning and Endings	53.71	47.46	1.13[a]	.88
Card Rotations	1391.30	1437.22	.97	1.03
Visual Memory (delayed)	13.93	13.71	1.02	.98
Pedigrees	47.06	52.36	.90	1.11
Hidden Patterns	652.93	651.33	1.00+	1.00−
Paper Form Board	17.67	17.07	1.04	.97
Number Comparisons	90.67	101.93	.89	1.12
Social Perception (verbal)	11.25	10.92	1.03	.97
Progressive Matrices (modified)	25.03	27.95	.90	1.12

[a] $p \leq .05$, by F test, using 1,000 × 1,000 df. Actual df's available were about 2,501 and 2,575 for males and females, respectively, though they varied slightly for different tests.

upper and lower cutoff points for each test. Whether the areas above and below these points should include 0.1%, 1%, 5%, or some other percentage is a matter of opinion. We decided to include approximately the upper and the lower 1% for each test. Since the test scores formed discrete distributions, it was necessary to choose actual scores which approximated the 1st and the 99th centiles. The scores used as cutoff points for each test are shown in Table 13.5, which also presents the expected and actual numbers of males and females and the χ^2 and p values computed for each tail for each generation. Data are presented separately for parents and offspring because of the large age differences on many of these tests (Wilson *et al.*, 1975).

With respect to adult performance on the Vocabulary test, there is no significant difference in the sex ratio for either tail of the distribution. There are significant differences for younger subjects, however, with females overrepresented at the high end and males overrepresented at the low end. For the Visual Memory (immediate) test, there were no differences between expected and obtained ratios. For both the Things test and Mental Rotations, mothers and daughters were overrepresented at the low end; at the high end, sons were overrepresented for the former test and males of both generations were overrepresented for the latter.

On the Hidden Patterns test, mothers were overrepresented at the low end and fathers at the high end. No differences were found between sons and daughters. There were no significant differences in sex ratios at either extreme of the distribution for the Paper Form Board test. On the Number Comparisons test, mothers and daughters were overrepresented at the high end. No differences were found for Social Perception (verbal). On the

Table 13.5. A Comparison of Males and Females in Upper and Lower Centiles (Approximate) for Each Test

Variable	Total number of scores		Number of scores in lower centile			Number of scores in upper centile		
	Lower centile	Upper centile	Expected	Actual	χ^2	Expected	Actual	χ^2
Vocabulary								
Parents (n = 2,961)	30 (9 or below)	149 (50)			0.30			0.33
Fathers (n = 1,485)			15	17		75	71	
Mothers (n = 1,476)			15	13		74	78	
Offspring (n = 2,096)	36 (5 or below)	56 (49 or above)			4.71[a]			9.46[b]
Sons (n = 1,012)			17	24		27	15	
Daughters (n = 1,084)			19	12		29	41	
Visual Memory (immediate)								
Parents (n = 2,952)	33 (5 or below)	62 (20)			0.76			0.02
Fathers (n = 1,478)			17	14		31	32	
Mothers (n = 1,474)			16	19		31	30	
Offspring (n = 2,093)	26 (7 or below)	64 (20)			0.35			0.39
Sons (n = 1,012)			13	15		31	28	
Daughters (n = 1,081)			13	11		33	36	
Things (total)								
Parents (n = 2,960)	41 (10 or below)	35 (61 or above)			10.76[b]			0.26
Fathers (n = 1,485)			21	10		20	31	
Mothers (n = 1,475)			18	20		17	15	
Offspring (n = 2,095)	20 (10 or below)	31 (56 or above)			4.05[a]			11.66[c]
Sons (n = 1,011)			10	5		15	25	
Daughters (n = 1,084)			10	15		16	6	
Mental rotations								
Parents (n = 2,957)	46 (−8 or below)	33 (37 or above)			18.28[c]			25.51[c]
Fathers (n = 1,484)			23	8		17	32	
Mothers (n = 1,473)			23	38		16	1	
Offspring (n = 2,095)	20 (−5 or below)	36 (40)			6.05[a]			17.41[c]
Sons (n = 1,012)			10	4		17	30	
Daughters (n = 1,083)			10	16		19	6	

(continued)

Table 13.5. (Continued)

Variable	Total number of scores		Number of scores in lower centile			Number of scores in upper centile		
	Lower centile	Upper centile	Expected	Actual	χ^2	Expected	Actual	χ^2
Subtraction and Multiplication								
Parents (n = 2,960)	34 (20 or below)	34 (114 or above)			6.62[a]			1.44
Fathers (n = 1,485)			17	9		17	21	
Mothers (n = 1,475)			17	25		17	13	
Offspring (n = 2,096)	24 (15 or below)	23 (97 or above)			9.37[b]			0.04
Sons (n = 1,012)			12	20		12	4	
Daughters (n = 1,084)			12	4		12	11	
Elithorn Mazes								
Parents (n = 2,935)	32 (3 or below)	27 (25 or above)			3.78			4.49[a]
Fathers (n = 1,474)			16	10		14	20	
Mothers (n = 1,461)			16	22		13	7	
Offspring (n = 2,083)	24 (3 or below)	27 (25 or above)			—			6.27[a]
Sons (n = 1,005)			12	12		13	20	
Daughters (n = 1,078)			12	12		14	7	
Word Beginnings and Endings								
Parents (n = 2,961)	34 (4 or below)	33 (38 or above)			12.97[c]			0.27
Fathers (n = 1,485)			17	28		17	19	
Mothers (n = 1,476)			17	6		16	14	
Offspring (n = 2,094)	20 (5 or below)	41 (33 or above)			0.45			1.20
Sons (n = 1,010)			10	12		20	16	
Daughters (n = 1,084)			10	8		21	25	
Card Rotations								
Parents (n = 2,954)	32 (−1 or below)	32 (176 or above)			3.78			9.03[b]
Fathers (n = 1,481)			16	10		16	25	
Mothers (n = 1,473)			16	22		16	7	
Offspring (n = 2,093)	22 (8 or below)	25 (190 or above)			5.50[a]			4.85[a]
Sons (n = 1,010)			11	5		12	18	
Daughters (n = 1,083)			11	17		13	7	

		Visual Memory (delayed)						
Parents (n = 2,945)	31 (1 or below)	25 (19 or above)			—			0.36
Fathers (n = 1,484)			16	16		13	11	
Mothers (n = 1,471)			15	15		12	14	
Offspring (n = 2,084)	25 (2 or below)	33 (19 or above)			0.36			5.13[a]
Sons (n = 1,006)			12	14		16	9	
Daughters (n = 1,078)			13	11		17	24	
		Pedigrees						
Parents (n = 2,958)	31 (3 or below)	27 (38 or above)			0.29			1.82
Fathers (n = 1,485)			16	14		14	10	
Mothers (n = 1,473)			15	17		13	17	
Offspring (n = 2,094)	25 (6 or below)	46 (39 or above)			3.25			15.88[c]
Sons (n = 1,010)			12	17		22	8	
Daughters (n = 1,084)			13	8		24	38	
		Hidden Patterns						
Parents (n = 2,957)	31 (3 or below)	31 (120 or above)			5.46[a]			17.08[c]
Fathers (n = 1,483)			16	9		16	28	
Mothers (n = 1,474)			15	22		15	3	
Offspring (n = 2,092)	20 (9 or below)	22 (120 or above)			0.05			0.41
Sons (n = 1,009)			10	9		11	13	
Daughters (n = 1,083)			10	11		11	9	
		Paper Form Board						
Parents (n = 2,915)	36 (3 or below)	35 (22 or above)			2.25			2.32
Fathers (n = 1,457)			18	13		17	22	
Mothers (n = 1,458)			18	23		18	13	
Offspring (n = 2,084)	27 (4 or below)	27 (22 or above)			0.04			0.92
Sons (n = 1,007)			13	14		13	16	
Daughters (n = 1,077)			14	13		14	11	
		Number Comparisons						
Parents (n = 2,958)	31 (16 or below)	34 (66 or above)			0.03			8.50[a]
Fathers (n = 1,482)			16	15		17	8	
Mothers (n = 1,476)			15	16		17	26	
Offspring (n = 2,095)	24 (18 or below)	22 (64 or above)			12.04[c]			3.68
Sons (n = 1,011)			12	21		11	6	
Daughters (n = 1,084)			12	3		11	16	

(continued)

Table 13.5. (Continued)

Variable	Total number of scores		Number of scores in lower centile			Number of scores in upper centile		
	Lower centile	Upper centile	Expected	Actual	χ^2	Expected	Actual	χ^2
Social percention (verbal)								
Parents (n = 2,911)	42 (3 or below)	80 (19 or above)			2.88			1.51
Fathers (n = 1,454)			21	15		40	46	
Mothers (n = 1,457)			21	27		40	34	
Offspring (n = 2,085)	28 (5 or below)	38 (19 or above)			0.32			3.19
Sons (n = 1,004)			13	15		18	12	
Daughters (n = 1,081)			15	13		20	26	
Progressive matrices (modified)								
Parents (n = 2,932)	33 (3 or below)	97 (27 or above)			6.82[b]			38.36[c]
Fathers (n = 1,479)			17	9		49	80	
Mothers (n = 1,453)			16	24		48	17	
Offspring (n = 2,094)	23 (7 or below)	27 (29 or above)			3.53			4.49[a]
Sons (n = 1,010)			11	6		13	19	
Daughters (n = 1,084)			12	17		14	8	

[a] $p < .05$.
[b] $p < .01$.
[c] $p < .001$.

Progressive Matrices test, mothers and daughters were overrepresented at the low end, while fathers and sons were overrepresented at the high end.

Since families were informed before volunteering that participation in the study would involve taking written tests, it is unlikely that our sample includes the lowest centile of the population as a whole. On the other hand, there is no apparent reason why the most able 1% would not have been sampled. Summarizing the results for the top 1%, we find that younger females are relatively more numerous than their male age mates for three of the tests (Vocabulary, Visual Memory [delayed], and Pedigrees), whereas older females are overrepresented in the top 1% for only one test (Number Comparisons). It is possible that the early advantage of the daughters on three tests is a reflection of their more rapid development, since the advantage seems to be lost with maturity. In the top 1%, younger males are overrepresented on five tests (Things, Mental Rotations, Elithorn Mazes, Card Rotations, and Progressive Matrices); their fathers are also overrepresented on four of the same tests (all but the first) and on a fifth (Hidden Patterns). Thus, the early difference in favor of males in this high ability group seems to be retained in adulthood. This could be due to less rapid loss with age in the case of males, since average performance on all these tests appears to decline with increasing years.

The data offer support for the notion that high spatial ability and high reasoning ability may be found more often in males than in females. They offer little support, however, for the specific hypothesis that this is a reflection of greater dispersion of male abilities. All tests showing male advantage in the upper centile also had higher mean scores for males, and two of these (Card Rotations and Progressive Matrices) actually showed a higher variance for females.

5. *Summary and Conclusions*

1. Principal component analysis revealed the same four major cognitive ability factors in males and females of different ages and different ethnic groups, and factor loading profiles were strikingly similar for the two sexes.

2. There was a statistically significant difference between the sexes for mean test scores on 14 of the 15 cognitive tests in the Hawaii test battery. Females had higher mean scores on tests involving verbal fluency or perceptual speed, while males were at advantage on tests involving spatial visualization or reasoning. Most of the mean differences though statistically significant, were small, and there was great overlap in the distributions.

3. Adult males were more numerous than females in the upper centile for five of the tests, and adult females were more numerous in the upper

centile for one. On nine tests, there was no significant sex difference in the high ability group of adults.

Appendix: Cognitive Tests in Hawaii Test Battery

A brief description of each of the 15 cognitive tests selected for inclusion in the Hawaii test battery is given below. For most of these tests, much more information is available in the test manual or in Buros (1975). Tests were administered in order as listed. Instructions were presented and timing was accomplished via audiotape synchronized with a 35-mm slide projector. Practice items were given prior to each test, and subjects were given the opportunity to ask questions before proceeding.

Vocabulary. This is a conventional test in which the subject is asked to pick a synonym for a stimulus work. There are 50 items to be completed in 3 min (example item: *ancient* dry long happy old).

Visual Memory (immediate). The stimulus array consists of simple line drawings of 40 common objects (e.g., chair, horse, spoon, drum) which are studied for 1 min. For the test (also 1 min), another array consisting of 20 of the same objects and 20 distractors is presented, and the subject is asked to pick out the 20 objects that were in the original array.

Things. Each subject is asked to write on the test booklet page all the things he can think of which fit a designated category. For the practice question, subjects are given the category, "things which are often red," along with "tomatoes" and "bricks" as examples. For the test, the first category is "things that are often round," and the second is "things that are often metal." Each part has a 3-min time limit.

Mental Rotations. Each of 20 items consists of a stimulus which is a line drawing depicting a three-dimensional arrangement of 10 cubical blocks, followed by four other drawings of 10-block arrangements—two of which are identical to the stimulus though in a rotated position, and two of which are different. The task is to circle the two drawings which are identical to the stimulus in each item. This test requires 10 min.

Subtraction and Multiplication. For each of two parts (2 min each), the subject is asked to complete 30 simple subtraction problems and 30 simple multiplication problems (example items: 98 − 75 = , 86 × 6 =).

Elithorn Mazes ("lines and dots"). Each maze is constructed of a grid of dotted lines in the shape of an inverted triangle, with large dots placed at some of the intersections within the grid. The subject's task is to draw a continuous series of line segments through the maze, beginning at the lower apex and finishing at the top in such a way that the solid line segments intersect as many of the dots as possible. The solid line must be superimposed on the dotted lines of the grid and must always proceed upward, with no backtracking or retracing allowed. For each maze, the maximum possible "dot" scores is indicated on the test. There are 10 mazes, and they are encountered in order of difficulty. There is a 5-min time limit.

Word Beginnings and Endings. This is a test of verbal fluency in which the task is to generate as many words as possible which begin and end with certain assigned letters. It is divided into two 3-min parts with different letters designated for each part.

Card Rotations. This is a test of spatial ability in which the stimulus for each item is a line drawing of a fairly simple, but abstract, enclosed figure. Each stimulus is followed by an array of eight figures, some of which are identical to the stimulus, though rotated within the plane, while others are mirror images of the stimulus, also rotated into a new position. The task is to discriminate the directly rotated figures from the mirror images. The test is divided into two 3-min parts, each of which includes 112 items.

Visual Memory (delayed). About an hour of testing has intervened between this test and the previously administered test of visual memory. Subjects are presented with another array of 40 line drawings, 20 from the original array and 20 distractors. On this delayed recall test, the set of 20 correct objects is different from that employed in the immediate recall test. There is a 1-min time limit.

Pedigrees. Two small geneological "trees" are presented, and the subject is asked to answer 40 multiple-choice questions about the relationships shown. A 4-min time limit is imposed.

Hidden Patterns. The same comparison stimulus is used for all items in both halves of this test. This stimulus consists of four line segments in the form of an inverted Y, with a line (or "tail") at the top. The subject's task is to discriminate whether this stimulus is or is not embedded in each of 200 somewhat more complex drawings. Each half of the test has a 2-min time limit.

Paper Form Board. Each of the 28 items in this 3-min test consists of a plane geometrical figure with "segments" which have been "cut" from this figure. The subject is asked to indicate on each figure where cuts would be made to obtain the segments shown. To be correct, all proposed segments must remain in the stimulus plane. During the practice period preceding this test, subjects are shown a mirror image solution and cautioned that this is not allowed.

Number Comparisons. This is a test of perceptual speed and accuracy; it is often called a "clerical" test. The test is divided into two 1.5-min parts, each consisting of 48 pairs of numbers. In each pair, the numbers may have from 3 to 13 digits, and the subject's task is to indicate the pairs in which the numbers are *not* identical.

Whiteman Test of Social Perception. Each item consists of either four or six drawings of people in social situations. In each item, three or five of the situations depicted are alike in some social way (for example, they may show greetings of various kinds or various sorts of competition), while the remaining picture does not share this attribute. The subject is asked to do two things: (1) cross out the mismatched picture (nonverbal score), and (2) identify (write) the social concept shared by the others (verbal score). There are 21 items to be completed in 10 min.

Raven's Progressive Matrices. This is among the better known of the tests chosen for this battery. It has been widely used as a test of general intelligence, a "culture-fair" intelligence test, or a test of "fluid" intelligence. Each item, or "matrix," consists of a 2×2 or 3×3 array of patterns, with one member of the array missing. The task is to choose (multiple-choice format) a pattern that would complete the array correctly. The early items tend to require completion of simple geometric figures. Later in the test, subjects must abstract the principles underlying the progression within the array, both left to right and top to bottom, and then employ these principles to choose the missing part. Although ours is a shortened version of the standard test (to save subject time and avoid fatigue), it has essentially the same reliability as reported for the standard version. A 20-min period was allowed for completion of the 29 items that were used; thus it is a "power" test, not a speeded test.

ACKNOWLEDGMENTS

The results reported here are made possible by collaboration of a group of investigators (G. C. Ashton, R. C. Johnson, M. P. Mi, and M. N. Rashad at the University of Hawaii, and J. C. DeFries, G. E. McClearn, S. G. Vandenberg, and J. R. Wilson at the University of Colorado) supported by NSF grant GB-34720 and grant MH-06669 from NICHD. We thank F. Ahern, A. Kuse, K. Wilson, J. Park, C. Frost, J. Ashton, J. Tamagawa, J. Nakayama, K. Kocel, C. Johnson, M. Nomi, D. Kunichika, S. Kuioka, T. Yamashita, and M. Meyer for technical assistance.

References

Bock, R. D., and Kolakowski, D. Further evidence of a sex-linked major gene influence on human spatial visualizing ability. *American Journal of Human Genetics*, 1973, *25*, 1–14.

Buros, O. K., (Ed.), *Intelligence tests and reviews*. Highland Park, New Jersey: Gryphon Press, 1975.

Cattell, R. B. The structure of intelligence in relation to the nature-nurture controversy. In R. Cancro (Ed.), *Intelligence: Genetic and environmental influences*. New York: Grune and Stratton, 1971.

DeFries, J. C., Vandenberg, S. G., McClearn, G. E., Kuse, A. R., Wilson, J. R., Ashton, G. C., and Johnson, R. C. Near identity of cognitive structure in two ethnic groups. *Science*, 1974, *183*, 338–339.

DeFries, J. C., Ashton, G. C., Johnson, R. C., Kuse, A. R., McClearn, G. E., Mi, M. P., Rashad, M. N., Vandenberg, S. G., and Wilson, J. R. Parent–offspring resemblance for specific cognitive abilities in two ethnic groups. *Nature*, 1976, *261*, 131–134.

Ehrhardt, A. A., and Money, J. *Man and woman, boy and girl*. Baltimore: Johns Hopkins University Press, 1972.

Hunt, S. M. *Sex differences: A biological and psychological analysis of the differences between men and women*. Liverpool, England: Vernon Scott Associates, 1974.

Maccoby, E. E., and Jacklin, C. N. *The psychology of sex differences*. Stanford, California: Stanford University Press, 1974.

Park, J. Y., Johnson, R. C., and Kuse, A. R. Cognitive structures in two different cultures. Unpublished manuscript, Behavioral Biology Laboratory, University of Hawaii, Honolulu, Hawaii, 1974.

Stafford, R. E. Sex differences in spatial visualization as evidence of sex-linked inheritance. *Perceptual and Motor Skills*, 1961, *13*, 428.

Tucker, L. R. A method for synthesis of factor analysis studies. *Personnel research section report no. 984*. Washington, D.C., Department of the Army, 1951.

Tyler, L. E. *The psychology of human differences*. 3rd ed. New York: Appleton-Century-Crofts, 1965.

Wilson, J. R., DeFries, J. C., McClearn, G. E., Vandenberg, S. G., Johnson, R. C., and Rashad, M. N. Cognitive abilities: Use of family data as a control to assess sex and age differences in two ethnic groups. *International Journal of Aging and Human Development*, 1975, *6*(3), 261–276.

14

Gender and Reproductive State Correlates of Taste Perception in Humans

Richard L. Doty

1. Introduction

It is well documented that numerous physiological and psychological processes of mammals are directly or indirectly related to circulating levels of reproductive hormones (e.g., Beach, 1947; Brobeck, Wheatland, and Stominger, 1947; Carr, Loeb, and Dissinger, 1965; Crocker, 1971; Doty, 1972; Komisaruk, Adler, and Hutchison, 1974; Pietras and Moulton, 1974). In humans, such relations are perhaps most clearly exemplified by studies undertaken during various phases of the menstrual cycle. A representative sample of variables which reportedly change systematically across these phases includes body temperature, blood pressure, respiratory efficiency, hemolysis, lymphocytosis, serum albumin, serum globulin, carbohydrate tolerance, urine volume, mucus consistency, immune reactivity, salt retention, circulating vitamin levels, depression, anxiety, and elation (see review by Southam and Gonzaga, 1965). Although a number of the studies reporting such fluctuations can be faulted on methodological grounds, the evidence for general metabolic and physiologic influences of gonadal hormones in humans is overwhelming. This evidence, along with anecdotal reports of changes in smell and taste perception during periods of

Richard L. Doty • Monell Chemical Senses Center and Department of Otorhinolaryngology and Human Communication, University of Pennsylvania Medical School, Philadelphia, Pennsylvania. This research was supported by USPHS Grant 1 RO1 NS 12,239-02.

heightened endocrine output, such as pregnancy, lend credence to the hypothesis that human sensory processes may be significantly affected by the level of reproductive hormones. Despite this likelihood, however, only a few studies have clearly established relations between endocrine factors and sensory function in humans, with the majority providing contradictory findings and conclusions. Indeed, only one human study has attempted a direct correlation between circulating levels of gonadal steroids and the sensitivity of any sensory system (see review by Doty, 1976).*

The lack of definitive findings on sensory/endocrine interactions in humans is not surprising. The task of establishing such relations is particularly arduous, given the complex motivational, physiological, and psychological factors that can confound psychophysical measures, as well as the limited extent to which relevant variables can be manipulated.† Thus, most experimenters using humans as subjects must content themselves with determining correlations between measures of sensory function and certain well-defined endocrine states or events (e.g., menses, pregnancy, menopause). When such effects are found, their functional significance is not always clear, as they often reflect, in epiphenomenal fashion, processes in related physiological or biochemical systems.

* Although there is strong scientific precedence from animal studies to interpret changes in sensory function mediated by hormones in causal relation to concomitant changes in social behavior, such attempts to causally associate sensory and behavioral changes in humans are problematical. This is due, in part, because (a) sensory changes noted during times of marked endocrine change, such as the menstrual cycle, are frequently not robust, accounting for only a small proportion of the total variance, (b) most human social behaviors appear to be dependent upon a number of stimuli, including learned cognitive ones, and (c) many behaviors likely influenced by reproductive hormone levels (e.g., temporal parameters of coitus) are rarely open to scientific scrutiny.

† The methodology involved in measuring human responses to taste and other types of stimuli is complex, and must be carefully evaluated when considering the results of any psychophysical study. The interested reader may wish to refer to one or more of the following general sources for additional detailed information: Engen (1971, 1972), Guilford (1954), Green and Swets (1966), Torgerson (1958), Marks (1974), Stevens (1975), Woodworth and Schlosberg (1960). A clear differentiation between measures of sensory sensitivity (e.g., absolute thresholds; measures of signal detection from basal noise), suprathreshold resolving power (e.g., differential thresholds; measures of signal detection from nonbasal signal plus noise), and perceptual change across suprathreshold stimulus gradients (e.g., values derived from scaling procedures such as magnitude estimation and fractionation) is commonly made. Although absolute sensitivity has, in the past, been estimated by threshold procedures (e.g., method of constant stimuli, method of limits), signal detection techniques, derived from statistical decision theory (e.g., forced-choice or rating evaluations of the presence or absence of signal from background noise under specific reinforcement constraints) are now widely used in most sensory modalities. Unlike the traditional threshold methods, the sensitivity measures they provide are theoretically unconfounded by response biases (e.g., motivational variables, willingness to report stimulus presence, etc.). Such procedures do not assume a fixed or even varying threshold and incorporate into their schemata models of the observer's decision processes. These techniques, however, require a considerable number of trials and are not always practical in some experimental situations.

Despite such methodological and conceptual shortcomings, however, it is possible to gain control over a number of the variables of interest and shed some light on the nature of certain sensory/endocrine interactions. In the present chapter I examine studies which present data related to potential influences (or correlates) of gender and reproductive state upon human taste perception. The knowledge of such influences is of particular interest at the present time, given (a) the all too frequent uncritical generalizations in both the popular and scientific presses about hormonal influences upon human sensory processes, (b) the surge of interest in the last few years in sexually dimorphic taste preferences in rodents (see Nance, Gorski, and Panksepp, 1976; Wade, 1975), and (c) the societal concerns over the influences of oral contraceptives and other endocrine manipulations, such as gonadectomy, upon weight gain, appetite, and various somatic factors. Questions of interest in this review include: Are men and women differentially sensitive to certain taste substances? Do they show differences in food preferences? Do hormonal changes occurring during the menstrual cycle affect food choices or preferences? What types of cravings and aversions are present during pregnancy, and do they systematically relate to nutritional needs or endocrine changes at this time? Although definitive answers to all of these questions are not presently available, reasonably congruent data do exist for some cases and will be examined in this compendium. Hopefully, the present review will stimulate other investigators to design and execute the critical studies needed on some of these points.

2. Sex Differences in Taste Perception

Differences between the sexes in their taste preferences, as well as in their abilities to detect or resolve slight differences in concentrations of specific tastants at threshold or above-threshold levels, have been noted in the literature. However, the reasons for such differences are not clear. A sex difference noted in taste perception could theoretically reflect one or more of the following factors: (a) a sampling bias resulting from a nonrandom selection of males or females for experimental participation (e.g., in some populations women not taking oral contraceptives volunteer for experiments more frequently midcycle, when estrogens and certain gonadotropins are elevated; Doty and Silverthorne, 1975); (b) differences in the frequencies of certain habits among male and female populations, which can affect the variable of interest (e.g., smoking, dieting, drug or alcohol usage; Kaplan, Glanville, and Fischer, 1965); (c) differences between the sexes in oral systems allied to taste, which affect molecular access to or interactions with the taste receptor sites (as well as the relative adaptation of the system) (e.g., salivary ions, salivary enzymes, saliva thickness, saliva flow; men typically have larger salivary glands than women; Bartoshuk, 1974; Ericson,

1968, 1971; O'Mahony, 1972, 1974); (d) differences in stimulative experiences encountered by male and female populations (e.g., women of some subgroups may encounter certain tastants more frequently than men due to cultural sex roles; see Fischer, 1968); (e) differences between the sexes in the make up of neural or endocrine systems having indirect influences upon taste systems, including modulating or arousing neural or endocrine networks (e.g., Bell, 1975); (f) differences between the sexes in the biochemistry or morphology of the taste sensory systems proper which may or may not be dependent upon reproductive hormone influences, either during early ontogeny, during later life, or both (see Nance, Gorski, and Panksepp, 1976); and (g) differences between the sexes in certain response biases, including responsivity to social characteristics of the experimental situation such as the sex of the experimenter, which can affect the measurement of the dependent variable of interest (see Doty, 1976; Orne, 1962; Stevenson and Allen, 1964). While these seven factors are not mutually exclusive and may not include all of the major influences contributing to sex differences noted in human taste studies, they do serve to point out the complexities in interpreting such differences, when present. More research is needed to establish the relative importance of each of these factors. Keeping these problems in mind, I shall now review specific reports examining sex differences in taste perception in humans.

2.1. Studies Examining Taste Thresholds in Men and Women

With the exception of only a few experiments (e.g., Minich, Mountney, and Prudent, 1966; Vaschide, 1904),* the majority of studies that have examined the taste thresholds of both men and women report lower thresholds (i.e., presumed greater sensitivity) in women. One of the earliest studies reporting this finding was that of Bailey and Nichols (1888). In this experiment, various concentrations of quinine, cane sugar, sulfuric acid, sodium chloride, and sodium bicarbonate were offered to 82 men and 46 women, whose task was to sort them into the perceptual categories of sweet, bitter, salty, acid, and alkaline. Although the differences were not marked, the females correctly sorted, on the average, lower concentrations of quinine, cane sugar, sulfuric acid, and sodium bicarbonate than did the males. However, the males performed slightly better than the females with the salt

* Both of these experiments reported lower male than female thresholds. The study by Vaschide (1904) can be discounted on the basis of statistical and sampling problems since the average thresholds were not markedly different between the sexes and the subjects were inmates of a mental institution. The study by Minich, Mountney, and Prudent (1966) employed an ascending triangle test across a dilution series of chicken broth—the only nonunitary taste stimulus examined in this literature. Whether this was a factor in producing the different finding is not known.

solutions. Since the observed differences were quite small, and statistical tests were not available at the time for analysis of the data, these results cannot be considered definitive.

Since this early work, however, other experimenters have confirmed these findings. Using similar methodology, Langwell (1948, 1949) asked 257 women and 242 men to sort filter paper strips (that had previously been dipped into low concentrations of sucrose, salt, citric acid, and quinine sulfate) into sweet, sour, bitter, and salty taste categories. Forty-seven percent of the men were able to perform this task correctly, as compared to 77% of the women, suggesting greater female sensitivity. Likewise, Pangborn (1959) reported lower female than male thresholds in four male and four female college students for both the recognition and detection of sucrose, sodium chloride, citric acid, and caffeine. Pangborn noted four general findings from her study: (1) wide individual variation in threshold measures; (2) a marked decrease in thresholds with practice; (3) wide variation between individuals in relation to the degree of improvement with practice; and (4) much lower original, as well as subsequent, thresholds in women than in men.*

The largest single set of studies consistently reporting lower female than male thresholds has used bitter-tasting stimuli, primarily phenylthiocarbamide (PTC; also called phenylthiourea).† The earliest of these studies, in which the sorting of various concentrations of PTC and several other tastants into sweet, sour, bitter and salty classes was required, reported lower PTC thresholds for women than for men (e.g., Blakeslee and Salmon, 1931; Falconer, 1947). The proportion of tasters and nontasters did not appear markedly different within each of the sexes. Subsequent work, using the somewhat more systematic Harris-Kalmus threshold procedure,‡

* The subjects were tested under both fasted (no breakfast or lunch) and nonfasted conditions at 4:30 PM on alternating days over considerable periods of time. Fasting had little effect on the threshold measures. The choice method of Richter and MacLean (1939) was used to establish thresholds. This method is essentially an ascending method of limits procedure with a forced-choice between the stimulus and a blank at each concentration level. Repeated sampling of a stimulus at each concentration level is typically allowed.

† The responsivity of humans to PTC has been of particular interest to geneticists, since it is to a large extent inherited. The mode of inheritance has been worked out by several authors (see Cavalli-Sforza and Bodmer, 1971), with "tasters" (individuals who are quite sensitive to the substance) being either homozygous or heterozygous for the single dominant gene T with incomplete penetrance and "nontasters" (individuals who are quite insensitive to it) being homozygous for the recessive allele t.

‡ In this procedure, an initial ascending dilution series is used to establish the general threshold range. In the estimated threshold range, the subject is then required to sort a set of eight tumblers into half that contain water alone and half that contain a given concentration of the tastant dissolved in water. The threshold is defined as the lowest concentration at which such a sorting can be made correctly.

generally supported these findings (e.g., Kalmus and Trotter, 1962; Harris and Kalmus, 1949; Leguebe, 1963; Mohr, 1951).

Although this sex difference appears to be statistically significant in nearly every study using PTC, several investigators have suggested that it may be an artifact of sampling procedures, and that complex relations exist between age, smoking habits, sex, and taste sensitivity to bitter compounds (see Fischer, Griffin, and Kaplan, 1963; Kaplan, Glanville, and Fischer, 1964; Kaplan and Fischer, 1965; Krut, Perrin, and Bronte-Stewart, 1961). Thus, Glanville, Kaplan, and Fischer (1964) initially reported that taste thresholds for PROP (6-*n*-propylthiouracil, a phenylthiourea-type compound), quinine, and hydrochloric acid increased with age, with a greater increase in males than in females. However, in a later study in which only nonsmokers were tested, these authors found no influence of either age or sex on taste thresholds ($n = 268$, ages ranging from 16 to 55 yrs; Kaplan, Glanville, and Fischer, 1965). They concluded (p. 337) that "... aging within the 16 to 55-year age span did not seem to be associated with a significant deterioration in taste sensitivity unless combined with smoking. The sex difference sometimes observed in population surveys seemed to be an artifact associated with differential smoking habits of men and women."

In addition to the reports of lower female than male thresholds within the four major taste modalities, sex differences to electrical stimulation of the tongue have also been noted. Coats (1974) examined, by an ascending method of limits procedure, electrical taste thresholds of 84 males (51 nonsmokers, 33 smokers) and 165 females (115 nonsmokers, 50 smokers). The thresholds of the females were, on the average, lower than those of the males, regardless of whether smoking or nonsmoking groups were examined. Smoking was associated with a statistically significant progressive increase in taste thresholds with advancing age in males, but not in females, in apparent support of one of the results of Glanville, Kaplan, and Fischer (1964).

2.2. *Studies Examining Taste Sensitivity Measures Derived from Signal Detection Theory in Men and Women*

It is now well documented, in a host of perceptual fields, that differences noted between various treatment groups by studies using traditional threshold procedures often do not reflect differences in sensory sensitivity per se, but differences in the choice of response criteria (e.g., Clark, 1966; DeMarchi and Tong, 1972; Price, 1966). In other cases such threshold measures reflect differences in both response criterion placement and sensory sensitivity (e.g., Chapman, Murphy, and Butler, 1973). Thus, studies utilizing methods that do not confound response factors with sensitivity ones (e.g., signal detection) are sorely needed to establish whether the taste threshold differences noted between men and women reflect actual differences in sensitivity.*

While a criterion-free signal detection approach has been applied to taste perception in several instances (e.g., Linker, Moore, and Galanter, 1964; Moore, Linker, and Purcell, 1965; O'Mahony, 1972; Pursell, Sanders, and Haude, 1973), only one study presents data pertinent to the question of sex differences. Fergenson, Moss, Dzendolet, Sawyer, and Moore (1975) examined the sensitivity of 24 male and 24 female college students to sucrose solutions by a signal detection procedure. The primary purpose of their experiment was to determine whether sensitivity was influenced by smoking, changes in signal probability, or food intake (confounded by time of day). No influences of sex, smoking, signal probability, or food intake were found for the sensitivity measure (d') or for either of the measures of response criteria employed (false alarm rate, percentage of misses), although a sex by signal probability interaction was found for the percentage of misses measure. Thus, females showed a marked decrease in misses as signal probability decreased, whereas males showed a slight increase.

While a single signal detection study reporting evidence different from the majority of the threshold literature requires additional verification, it does suggest that more such studies, using a variety of taste stimuli, are needed before it can be definitely stated that females are more sensitive than males to taste stimulation.†

2.3. *Sex Differences in Taste Preferences*

Data from several studies suggest that taste sensitivity for some chemicals can influence taste preferences. A case in point is the excellent study by Greene (1974), in which taste thresholds for PTC were examined in Indians from two Ecuadorian Andean communities in which goiter and cretinism are endemic. Apparently, individuals of the communities who can detect low levels of certain bitter goitrogens (typically PTC tasters) reject or minimize the intake of foodstuffs that contain them, whereas those who cannot detect them (typically PTC nontasters) do not minimize the intake of such foods and, thus, develop goiters. In apparent support of the relationship between taste sensitivity and food preferences, Fischer, Griffin, England, and Garn (1961) noted a weak, statistically significant, negative

* Several recent reviews of the applicability of signal detection methodology within the chemical senses have recently appeared: Doty (1976), Engen (1971), Köster (1975), Pfaffman, Bartoshuk, and McBurney (1971).

† As indicated by the critical and enlightening study by Bartoshuk (1974), workers should strive to standardize a number of their test procedures (e.g., intertrial intervals, stimulus, and rinse presentation durations) to allow for valid comparisons of data across studies.

relationship between the percentage of foods checked as disliked on a 118-item food questionnaire and taste thresholds for both quinine and 6-*n*-propylthiouracil. Taste thresholds for *d*-sucrose, sodium chloride, and hydrochloric acid were not related to the reported dislikes.

If, as the majority of the threshold studies suggest, women are more sensitive than men in the recognition of bitter and sweet tasting stimuli, several hypotheses would appear plausible concerning their probable taste preferences. First, as suggested by the studies of Greene (1974) and Fischer, Griffin, England, and Garn (1961), women would be expected to have more frequent dislikes than men for foods containing appreciable amounts of bitter components. Second, they would be expected to exhibit stronger preferences than men for low concentrations of sweet stimuli such as sucrose (which are not detected by the men or are detected in lesser magnitude). Third, they would be expected to exhibit weaker preferences than men for higher concentrations of such sweet-tasting stimuli.*

The literature on taste preferences of men and women is potentially in support of some of these hypotheses. Thus, numerous food preference surveys indicate that women report a larger number of food aversions than men, although the foods included in these surveys have not been chosen to fall primarily into sweet, sour, bitter, and salty categories (see Bryne, Golightly, and Capaldi, 1963; Hawel and Wirths, 1969; Smith, Powell, and Ross, 1955a,b; Wallen, 1943, 1945). Unfortunately, the results of such surveys probably reflect, to a large extent, cultural sex-role conditioning related to what the food looks like or comes from, in addition to taste preferences *per se*. Likewise, other sex-role influenced factors affect the outcome of such surveys, including familiarity with the food items, concerns with caloric intake and weight gain, and conscientiousness in accurately completing the items. Data from a 28-item food aversion checklist administered by Smith, Powell, and Ross (1955b) to 318 high school and college students (Tables 14.1 and 14.2) illustrate the difficulties in interpreting responses to such question-

* This latter, less tenable, hypothesis follows directly from the assumption that the preference/concentration curves of men and women run parallel from different threshold starting points, exhibiting reversals somewhere in the neighborhood of 0.3–1.0*M* concentration. However, this assumption may not be correct, since recruitment-like processes may bring the above-threshold preference estimates of observers with high thresholds into equivalence with those of observers with low thresholds (L. Bartoshuk and D. McBurney, personal communications). Thus, the preference/concentration curves of the two sexes may overlap across most of the above-threshold concentrations. Since complex relations exist between above-threshold measures of hedonicity and intraorganismal states, such as satiety or body-fat content (e.g., Cabanac, 1971; Grinker, Price and Greenwood, 1976; Moskowitz, Kumraiah, Sharma, Jacobs, and Sharma, 1976; Thompson, Moskowitz, and Campbell, 1976), a preference difference between the sexes in the hypothesized direction could well be explained on other bases. For example, a greater metabolic rate on the part of males might be exhibited as a hedonic correlate to a greater caloric need. More data are needed on these points.

Table 14.1. Percentage of Males and Females Reporting Dislike for Each of 28 Food Items Presented on a Food Aversion Questionnaire[a]

Food item	% males	% females		p^b
Brains	49.31	67.61		.001
Kidneys	35.21	57.95		.001
Buttermilk	42.25	56.82		.01
Mush	16.90	29.55		.01
Beer	14.79	24.43		.01
Liver	22.53	24.43		
Mushrooms	19.01	22.73		
Cottage cheese	21.13	16.48	+[c]	.001
Potato soup	8.45	15.34		.05
Bean soup	9.85	9.09		
Swiss cheese	8.45	9.09		
Radishes	9.85	8.52		
Cabbage	9.15	8.52		
Salt pork	5.63	8.52		
Onions	6.34	7.95		
Chili	14.79	6.25	+	.05
Fried eggs	2.11	5.68		
Salmon	6.33	5.68		
Lima beans	4.92	5.68		
Cantaloupe	2.11	3.98		
Tea	4.23	3.41		
Gravy	0.00	2.27		
Grapefruit juice	0.70	1.40		
Veal chops	2.11	0.57		
Tomatoes	0.70	0.57		
Pears	0.00	0.57		
Chocolate	0.00	0.57		
Cherries	1.40	0.00		

[a] Total sample size was 318 (sample size for each sex not reported). Modified from Smith, Powell, and Ross (1955).
[b] Based upon a χ^2 procedure.
[c] Indicates food items which received significantly more male than female dislike responses.

naires. Significantly more females than males reported aversions to brains and kidneys (I wonder how many of the respondents had ever tasted brains and kidneys), whereas the reverse was true for cottage cheese (which is presumably cognitively related to low caloric intake).

There seems to be a general consensus that men prefer, in brief preference tests, relatively stronger concentrations of sweet-tasting stimuli than do women. Thus, Desor, Greene, and Maller (1975) found, in a group of 311 Caucasian and Black American twin pairs aged 9–15 years, significantly greater preferences in males than in females for the more

Table 14.2. Percentage of Males and Females Reporting Dislike for a Particular Number of Foods on the Food Item Inventory of Smith, Powell, and Ross (1955)

Number of foods reported disliked	% males	% females
0	15.11	3.35
1	17.27	11.73
2	18.71	11.73
3	16.55	20.67
4	10.07	18.99
5	5.04	7.26
6	7.19	9.50
7	5.04	12.29

concentrated sucrose solutions (0.6*M* was the strongest concentration presented). Caucasian males, but not black ones, showed a greater preference for more concentrated lactose solutions than did their female counterparts. No sex differences were noted for sodium chloride preferences. A later study by these same authors (Desor, Greene, and Maller, 1975) found similar, although not statistically significant, trends in adults.

Results similar to those noted in the preceding studies have been found in numerous consumer surveys. For example, Pangborn and Simone (1958) evaluated the hedonic responses of a total of 2,505 people in four surveys held at annual California State fairs to apricots, pears, peaches, and vanilla ice creams differing in sugar content. In all four surveys, the sweeter samples were given higher ratings by men than by women. Likewise, Simone, Leonard, Hinreiner, and Valdés (1956) reported a distinct tendency for males to rate sweeter samples of canned cling peaches as more preferable than did females. Laird and Breen (1959) noted a similar sex difference for the preference of differentially sweetened pineapple juice samples.

While the greater preference on the part of adolescent and adult males for more strongly sweetened stimuli seems to be relatively well established, its physiological basis is unknown. If its emergence is dependent upon adult levels of male or female gonadal steroids, then it should appear at puberty. Unfortunately, few data are available concerning its ontogeny. Although this sex difference was noted by Greene, Desor, and Maller (1975) in children ranging from 9 to 15 years of age, the data from the prepubescent and postpubescent subjects were not presented separately. The sole study reporting sucrose preferences of prepubescent children (8–10 years of age) reports no such sex difference (Grinker, Price, and Greenwood, 1976).

The only other studies available which shed light on the ontogeny of this sex difference are those using newborn infants, although the results of these studies are conflicting. Thus, in a study of 22 male and 20 female newborns, Nisbett and Gurwitz (1970) reported that females consumed 24% more of a sweetened (4.7% sucrose) than unsweetened formula, whereas males consumed only 6% more. On the other hand, Dubignon, Campbell, Curtis, and Parrington (1969) found no sex differences in sucking time, number of sucks, or sucking rate in 210 newborns to nonnutritive reflexive sucking (2-min tests) or in 94 newborns sucking 0.5 ml of 5% dextrose ($\approx 0.28M$) delivered on every 10th suck. Likewise, Desor, Maller, and Turner (1973) reported no differences between the sexes in their relative intake of glucose, fructose, lactose and sucrose (0.05, 0.10, 0.20, and $0.30M$ concentrations) in 96 male and 96 female newborn infants.* In a subsequent study (Desor, Maller, and Andrews, 1975), no sex differences were observed in the ingestive responses of 48 newborns to low concentrations of urea, citric acid, sodium chloride, or sterile water alone, although total fluid intake did correlate slightly with birth weight. The lack of a sex difference persisted following the addition of sucrose to the stimuli to raise their baseline ingestion levels above that of water alone.

The discrepancy between the results of Nisbett and Gurwitz (1970) and those of the other authors may be related to textural differences of the ingestants, the choices of stimulus concentrations, or other less obvious aspects of the test paradigms (e.g., testing duration, time of day, the race, and birth weights of the infants, etc.). If a sex difference does exist for taste preferences of newborns for sweet stimuli, it would appear to be present only under very specific test conditions.

In summary, the available evidence suggests that men and women differ in their reported taste preferences to strong concentrations of certain sweet-tasting stimuli. Whether their preferences differ for other tastants is not known. Although the data available do not conflict with the hypothesis that this sex difference emerges at puberty, more work is needed on this point. Likewise, additional studies are needed to establish to what extent taste preferences and taste sensitivity are linked to one another.†

* Although an unexpected sex difference was found in response to the presentation order of the water and sugar solutions (relative to the males, the females consumed more of the solution presented first), this sex difference was not observed in later studies (Desor, personal communication).

† Caution would appear warranted in the interpretation of relationships between taste sensitivity and food preferences, since some taste sensitivity measures may be influenced by repeated testing or exposure (e.g., Fischer, 1967; Glanville and Kaplan, 1965; Kahn, 1968; Pangborn, 1959). A recent study of twins' preferences for nonbitter compounds (sucrose, lactose, sodium chloride) reported nearly null heritability coefficients, suggesting that many preference differences in humans to such substances may largely depend upon experiential factors (Greene, Desor, and Maller, 1975).

2.4. Comparison of Human Preference Data with Rodent Preference Data

Although studies comparing the absolute sensitivity of both male and female rodents across basic taste stimuli are apparently lacking, a number of rodent studies suggest that sex differences are present for the ingestion preferences for sucrose, saccharin, quinine sulfate, sodium chloride and 6-*n*-propylthiouracil. Thus, adult female rats typically ingest more (and show greater relative preferences for) glucose, saccharin, and sodium chloride, and ingest less (and show greater relative aversions to) quinine sulfate than do adult male rats in two-bottle preference tests (Krecek, Novakova, and Stibral, 1972; Nance, Gorski, and Panksepp, 1976; Valenstein, Cox, and Kakolewski, 1967a,b; Wade and Zucker, 1969a, 1970b). The sex difference in saccharin preference occurs primarily for the more highly concentrated, less generally preferred, solutions (Nance, Gorski, and Panksepp, 1976). Castration of adult male rats produces only a small increase in saccharin preference, whereas ovariectomy of adult females markedly diminishes the preference differential, indicating that the sex difference is most influenced by ovarian hormones. The saccharin preference is restored in ovariectomized females by daily injections of estradiol benzoate in combination with progesterone, but neither of these hormones alone alters the preference significantly (Zucker, 1969). Interestingly, the sex difference for the intake of sodium chloride is apparently not eliminated by gonadectomy of the female at 10 days of age, although such a procedure appears to delay the expression of the sex difference by approximately one month (Krecek, Novakova, and Stribral, 1972). The sex differences for both the saccharin and sodium chloride preferences are due, in large measure, to early organizing effects of neonatal hormones, since a single testosterone injection (1 mg) on day 2 (for sodium chloride) or day 5 (for saccharin) significantly reduces the preferences of the females when tested in adulthood (Krecek, 1973; Krecek, Novakova, and Stribral, 1972; Wade and Zucker, 1969b).

Sex differences noted in the preferences for several bitter substances (quinine sulfate, 6-*n*-propylthiouracil) are also influenced by gonadal steroid manipulations. Thus, gonadectomy of female rats in adulthood results in an enhanced intake of quinine sulfate (Wade and Zucker, 1970b), whereas such a procedure in house mice enhances the intake of 6-*n*-propylthiouracil (Hoshishima, 1967). Administration of estrogen and progesterone restores the aversion to quinine in rats although, as in the case with saccharin, neither hormone alone is effective in producing the restoration (Wade and Zucker, 1970b). However, estrogen administration alone reportedly restores the normal responsiveness of house mice to 6-*n*-propylthiouracil (Hoshishima, 1967).

It is apparent that the results of a number of the human preference studies, i.e., greater male than female preferences for more concentrated sweet stimuli, are not mirrored by the rodent data, which indicate greater female than male preferences for sweet tastants across a relatively broad range of concentrations. Several explanations of this discrepancy can be hypothesized. First, it may simply reflect a species difference and should be noted as such. Second, it is conceivable that preference results from two-bottle preference tests lasting for 16 hr or longer in ad lib test situations are not comparable to results derived from brief human preference tests in which several samples are directly compared. Perhaps longer-term ingestion tests with humans would show different results than noted in the preference studies reviewed in this chapter. Third, it is possible that cultural or experiential factors outweigh hormonal factors in determining human taste preferences, and that without such factors a sex difference that is consistent with the rodent data would appear in humans. More studies under a variety of test situations, hopefully with different cultures, appear needed to determine which, if any, of these alternatives is most plausible. Clearly, this discrepancy should lead investigators to be cautious in suggesting human implications from work on rodent taste preferences to sweet stimuli.

If women exhibit more food aversions than men to bitter-tasting substances, then the sex differences noted in rodents for quinine preferences would appear to be in agreement with the sex difference noted in the human literature. More work is needed on this point, however, possibly utilizing women with gonadal disorders or following menopause, although experience is conceivably so important that many food preferences may not change markedly during adulthood regardless of the levels of gonadal hormones present (see Doty, 1977).

3. *Menstrual Cycle Correlates of Taste Perception*

3.1. *Taste Thresholds Established During Different Menstrual Cycle Phases*

The majority of studies examining taste thresholds during the menstrual cycle report finding no marked changes across this time. Of the few studies reporting significant menstrual cycle influences, the most widely cited is the report by Glanville and Kaplan (1965). These investigators used the Harris–Kalmus procedure to establish recognition thresholds for both 6-*n*-propylthiouracil and quinine sulfate in 19 women. The subjects were tested at the same time each day, three days per week, through "one or more menstrual cycles." The threshold values were grouped into three phases: premenstrual (9–5 days before the onset of menses), menstrual (day before menstrual bleeding and the subsequent 4 days), and postmenstrual (6–10 days after the onset of menses). The average thresholds for both compounds were significantly lower during menses than during the other phases. Both 6-*n*-propylthiouracil "tasters" and "nontasters" (respectively defined as having thresholds below or above the antimode at the Harris–Kalmus dilution steps 9 and 10) showed the decrease in threshold during menstruation. These authors acknowledged the presence of a test order effect in some subjects (repeated testing decreased the threshold measures) and suggested that the comparison between the threshold values established during menses and those established during the postmenstrual period was the most valid one to make. Thus, counterbalancing of the testing order across the three test phases was apparently not performed. Individuals with the most irregular cycles reportedly exhibited the lowest thresholds during the menses.

Henkin (1974) also reported finding significant taste threshold changes across the menstrual cycle phases, although sodium chloride was the tastant utilized. Using his three-drop ascending method of limits procedure, he found lowest thresholds during the follicular phase ($n = 5$ women), and highest thresholds during the luteal phase, with thresholds during the menses falling more or less in between.

Aside from these two studies, most experiments have been unable to establish any significant changes in taste thresholds during the menstrual cycle. Beiguelman (1964) used the Harris–Kalmus procedure to obtain PTC thresholds on 100 women twice: once during menses and once 15 days before or after this period. The testing order was counterbalanced, since 50 of the women received the first test during menses and 50 the first test during the other time periods. Although a test order effect was noted, no significant difference between the menstrual and nonmenstrual threshold measures was apparent. "The results cannot support any influence of menstruation on taste thresholds for phenylthiourea" (p. 199).

Similarly, Pangborn (1967), using the threshold procedure of Richter and MacLean (1939), found no systematic differences in the taste thresholds of eight females, over a seven-month period, for sodium chloride, sucrose, glucose, and citric acid. Two male controls were also tested over this time period. "We noted as much variability among the eight females as between the two male controls, with no consistent pattern of response that could be related to the monthly cycle." (p. 149). The threshold values did decrease, however, as a result of repeated testing.

Kahn (1965), using an ascending method of limits with a forced-choice four-alternative procedure at each concentration level, also reported no significant influences of menstrual cycle phase on taste thresholds in 16 college students tested over the phases of two complete menstrual cycles. The tastants utilized were sucrose, sodium chloride, and hydrochloric acid. However, lower thresholds appeared during the second than during the first cycle, suggesting the probable influence of repeated testing.

Although Fornasari (1951) reported an increase in taste thresholds during the menses for 20 women using sucrose, tartaric acid, hydrochloric acid, and sodium chloride as stimuli, statistical analysis of his data revealed no significant differences (Binomial tests, $p < .20$; t tests, $p < .20$). Thus, his findings are in agreement with those of the aforementioned studies.

In summary, there is no convincing consensus regarding the influences of the menstrual cycle upon taste thresholds, although the majority of experiments suggest no significant effects. The two studies that have reported threshold changes across the menstrual cycle apparently did not counterbalance the testing orders, producing a potential confounding of test order with menstrual cycle phase. The studies that report no significant influences of the menstrual cycle upon taste thresholds appear to have minimized the potential influences of testing order by either counterbalancing (Beiguelman, 1964), using two consecutive cycles (Kahn, 1965), or performing a large number of tests on consecutive cycles until a reasonable stabilization of threshold values was obtained (Pangborn, 1959).

It should be noted that all of the aforementioned studies, with the exception of the one by Henkin (1974), provided no basic biological information, such as basal body temperature or levels of circulating hormones, to establish the test phases or to determine whether ovulatory cycles were actually being examined. Since none of these studies utilized psychophysical measures free of response biases, it is not known to what extent their reported thresholds are indicative of sensory sensitivity, response biases, or some combination of both. Clearly, additional studies are needed to answer the question of whether or not taste sensitivity for any tastant fluctuates in a consistent fashion across the phases of the human menstrual cycle.

3.2. *Taste Preferences Established During Different Menstrual Cycle Phases*

Several authors have reported that some women (particularly those who commonly experience symptoms of premenstrual tension) crave sweets during the premenstrual phases of their cycles. Thus, Morton, Additon, Addison, Hunt, and Sullivan (1953) noted that of 249 patients examined, 37% reported a craving for sweets premenstrually. These authors stated, "The hypoglycemia is a recent and striking finding in premenstrual tension. It is clinically manifested by increased appetite or a craving for sweets." (p. 1182). Similarly, Fortin, Wittkower, and Kalz (1958) noted that a number of their subjects craved chocolate and other sweets premenstrually. The possible influence of hypoglycemia upon human preferences for sweet stimuli was suggested by the early study of Mayer-Gross and Walker (1946), which found a close relation between blood glucose levels and preferences for certain sucrose solutions (but not saccharin solutions matched for perceived sweetness to the sucrose).

Despite these interesting and suggestive observations, only two studies have examined the hedonic ratings of humans to sweet tastants as a function of menstrual cycle phase (Aaron, 1975; Wright and Crow, 1975). However, neither of these studies reported a premenstrual peak in sugar pleasantness in their subjects. Wright and Crow (1975) obtained pleasantness ratings for five sucrose and five sodium chloride solutions from 94 women (16–36 years of age). The various solutions were presented randomly, and the tests were run in the morning. The solutions were rated three times: before, 10 min after, and one hour after the ingestion of 200 ml of a 25% glucose solution. For data analysis, each cycle was arbitrarily divided into five equal phases. In general, the solutions were rated less pleasant during the presumed early luteal phase. In all phases except the early luteal one, a downward shift in preference occurred 10 min after the glucose load. Aaron (1975), in a quite similar study, presents data that confirm the early luteal depression in pleasantness ratings. However, about half of the 74 women tested in this study were taking oral contraceptives.

It is apparent that more studies are needed to determine if taste preferences change during the menstrual cycle. The few experiments available on this topic, along with the recent studies suggesting a role of internal factors in affecting hedonic ratings to taste stimuli (e.g., Cabanac, Minaire, and Adair, 1968; Cabanac and Duclaux, 1970; Mower, Mair, and Engen, 1976; Moskowitz, Kumraiah, Sharma, Jacobs, and Sharma, 1976; Thompson, Moskowitz, and Campbell, 1976), suggest that carefully controlled studies of taste perception during the menstrual cycle may prove fruitful. Biological correlates of endocrine function (e.g., basal body temperature, circulating steroid levels) should be used to establish menstrual cycle phase, since marked variability in cycle length, ovulation time, and the proportion of ovulatory cycles is present in many populations (e.g., Malek, Gleich, and Maly, 1972).

4. *Taste Perception Changes During Pregnancy*

It is widely known that during pregnancy many women experience increases in appetite and thirst, as well as specific food aversions or cravings. Although the data are limited, a number of studies estimate that between 30% and 70% of all pregnant women exhibit food cravings or aversions (Dickens and Trethowan, 1971; Giles, 1893; Marcus, 1965; Posner,

McCottrey and Posner, 1957; Taggart, 1961). Fruit and fruit drinks are the most common cravings, with sweets and vegetables typically taking the next highest place. Table 14.3 illustrates the relative frequencies of various food cravings and aversions in 100 primigravida British women sampled by Dickens and Trethowan (1971).

Despite the apparent widespread occurrence of aversions and cravings in women during early pregnancy, little is known about the factors producing them and whether they relate to changes in taste or smell perception. Although a number of authors have hypothesized that such phenomena may be related to nutritional needs, hard data on this point are lacking. Interestingly, several lines of evidence suggest that psychological factors are intimately related to cravings and aversions. Thus, social class apparently influences the specific items that are craved (Taggart, 1951). A history of appetite change under stress before pregnancy (e.g., anorexia or overeating) appears to be related to the appearance of cravings and aversions during pregnancy (Trethowan and Dickens, 1972). Trethowan and Dickens (1972) suggest that dietary deficiencies are not strongly related to either cravings or aversions, and pose the hypothesis that cravings are most often exhibited for strong-tasting, sharp, sour, salty, or spicy foods, as well as foods that are crisp and have to be chewed or bitten. Thus, they suggest it is not fruit, per se, which is craved, but foods which have the common property of stimulating taste. The sparse data available suggest that cravings become less pronounced or less frequent as a function of successive pregnancies (e.g., Giles, 1898).

Given the widespread reports of cravings and aversions during pregnancy, it is surprising that so few studies have examined taste sensitivity or

Table 14.3. Cravings and Aversions by Food Type during Pregnancy in 100 Primigravidae British Women[a]

Food item class	Number of women reporting	
	Aversions	Cravings
Tea, coffee, cocoa	32	3
Vegetables	18	11
Meat, fish, eggs	16	7
Alcoholic drinks	9	3
Bread, cake, biscuits, crisps	6	2
Milk, dairy products	6	11
Sweets, chocolate, ice cream	5	21
Fruit, fruit drinks	2	29
Miscellaneous		2[b]

[a] Adapted from Dickens and Trethowan (1971).
[b] Includes one instance of craving for "salty things" and another for "food advertised on TV."

preferences during that time. Indeed, to the best of my knowledge, only one study is available in the English literature on this topic (Jørgenson and Buch, 1961). Although this study reported normal sensitivity in pregnancy, the subjects were also diabetic, making its findings difficult to interpret. A number of studies have appeared in other languages, however, and a few of these are briefly examined in this section.

An early and well-controlled study by Hansen and Langer (1935) examined recognition thresholds for sodium chloride, tartaric acid, cane sugar, and quinine hydrochloride in 12 nonpregnant and 28 pregnant women (late pregnancy). Heightened thresholds were noted during pregnancy for all four stimuli. Mann–Whitney U tests performed by me on their data proved significant for the salty, sour, and bitter stimuli ($p < .005$), but not for the sugar stimulus ($p > .20$).

Using natural lemon juice as the test stimulus, Giudizi and Noferi (1946) established recognition thresholds for lemon taste in 49 women during late pregnancy, 15 nonpregnant women, and 15 women in the immediate postpartum period. Although the Richter and MacLean (1939) median thresholds were slightly lower for the pregnant group than for the other two groups, these data were not statistically significant (Mann–Whitney U tests, $p > .10$).

Although the data were not presented in a form that would allow statistical analysis, Panebianco (1950) reported that the recognition and detection thresholds of 23 pregnant women and 11 nonpregnant controls supported the findings of Hansen and Langer (1935)—i.e., that during late pregnancy taste sensitivity was diminished. However, thresholds established in early pregnany were either normal or slightly lower than those observed in controls. In late pregnancy, the thresholds were 50% higher for citric acid, 70% higher for sodium chloride, 40% higher for glucose, and 35% higher for quinine sulfate.

In agreement with the aforementioned study, Baravelli (1961) reported significantly higher recognition thresholds for sucrose in 115 pregnant women than in 100 nonpregnant ones. Although slightly lower median thresholds appeared for hydrochloric acid and quinine in the pregnant women, these differences were not statistically significant. No differences were noted for the sodium chloride thresholds. The trimester of pregnancy did not appear to be significantly related to the threshold measures.

While it is clear that considerably more data are needed, the bulk of the available evidence suggests that either normal or heightened taste thresholds are present during the time of pregnancy. Interestingly, this apparent lessened sensitivity to taste stimultion parallels the reports of lessened olfactory sensitivity noted by a number of authors. For example, Hansen and Glass (1936), using a Zwaardemarker stimulus presentation apparatus, reported heightened recognition thresholds for the vapors of rose oil, nitrobenzene, and rubber in 22 women during the last months of pregnancy. Likewise, Noferi and Giudizi (1946), using a blast-injection procedure, found recognition thresholds for lemon odor to be higher in 15 women tested during the last two months of pregnancy than in 15 nonpregnant women and 15 women in the two week period following birth. Hyposmia during the late stages of pregnancy has also been reported by Luvara and Maurizi (1961) and Le Magnen (1952). Future work, using signal detection methodology and more frequent tests during the time of pregnancy, should help to determine whether systematic changes in taste sensitivity or quality occur at this time.

5. Summary and Conclusions

As evidenced by this review, a large number of studies have addressed the question of whether or not sex or hormonal factors influence human taste perception. Unfortunately, few sound generalizations concerning the possible effects of such factors on taste perception can be made from this literature. A few tentative conclusions do appear warranted, however, and are presented below with the realization that additional empirical verification is needed before they can be generally accepted as true.

First, the majority of taste threshold studies report lower thresholds for women than for men, although the reason(s) for this sex difference are not known. Such a difference could be due to a number of factors, including (a) a sampling bias resulting from nonrandom selection of males or females for experimental participation; (b) differences in the frequencies of certain habits among male and female populations, which can affect the sensitivity measure (e.g., smoking, dieting); (c) differences between the sexes in oral systems allied to taste, which affect adaptation of or molecular access to the taste system under investigation (e.g., salivary ions, enzymes, flow rates); (d) differences between the sexes in their experiences with specific tastants; (e) differences between the sexes in neural or endocrine systems indirectly influencing taste perception; (f) differences between the sexes in the biochemistry or morphology of the taste systems proper, which may or may not be dependent upon reproductive hormone influences either during early ontogeny, later life, or both; and (g) differences between the sexes in their response biases, which might affect the threshold measure (e.g., needs to please an experimenter of the opposite sex). It is clear that sophisticated criterion-free psychophysical procedures are needed to definitively establish whether men or women are differentially sensitive to specific tastants.

Second, the majority of taste preference studies report sex differences either in the number of foods checked as disliked on taste preference inventories or as to the relative concentrations of taste stimuli preferred in direct taste tests. In the former case, additional work is needed to establish to what extent the tendency for females to report more food dislikes than males reflects an actual difference in taste sensitivity or other factors, including those related to learned sex-role expectations regarding what the food looks like or comes from. Additional studies are needed to determine whether the tendency of males to report greater preferences for more highly concentrated sweet stimuli reflects influences from internal metabolic factors (e.g., higher metabolic rate) and/or a sex difference in the perception of the intensity of such stimuli.

Third, no consensus appears to be present in the literature concerning the influences of phases of the menstrual cycle upon taste thresholds, although the majority of studies report finding no systematic influences. Only two experiments have examined the pleasantness of taste stimuli as a function of menstrual cycle stage and both indicate a slight decrement in the pleasantness of sweet stimuli during the early luteal phase. Clearly, more work utilizing a variety of taste stimuli is needed on these points.

Finally, the scant literature on the influences of pregnancy upon taste sensitivity suggests a decrement in sensitivity during this time, at least during the latter stages. No convincing evidence of a close relationship between food cravings and aversions and changes in taste and smell function has yet been presented.

It is apparent that modern hormonal (e.g., plasma radioimmunoassay) and psychophysical (e.g., signal detection) measurements are needed to shed light on these very interesting possible relationships. Hopefully, the present review will stimulate more critical and decisive research on taste/endocrine interactions.

Acknowledgments

The author thanks P. Balnave, R. Bernard, M. Kare, J. Kenny, D. McBurney, G. Preti, G. Settle, and P. Rozin for their comments on a previous version of the manuscript.

References

Aaron, M. Effect of menstrual cycle on subjective ratings of sweetness. *Perceptual and Motor Skills*, 1975, *40*, 974.

Bailey, E. H. S., and Nichols, E. L. On the sense of taste. *Science*, 1888, *11*, 145–146.

Bailey, E. H. S., and Nichols, E. L. The delicacy of the sense of taste. *Nature*, 1888, *37*, 557–558.

Baravelli, P. Le alterazioni della sensibilita gustativa nella gravidanza. *Rivista Italiana di Ginecologia*, 1961, *45*, 171–188.

Bartoshuk, L. M. NaCl thresholds in man: Thresholds for water taste or NaCl taste? *Journal of Comparative and Physiological Psychology*, 1974, *87*, 310–325.
Beach, F. A. A review of physiological and psychological studies of sexual behavior in mammals. *Physiological Reviews*, 1947, *27*, 240–307.
Beiguelman, B. Taste sensitivity to phenylthiourea and menstruation. *Acta Geneticae Medicae et Gemellologiae (Rome)*, 1964, *13*, 197–199.
Bell, C. Sexual modification of cardiovascular regulation. In: M. Sandler and G. L. Gessa (Eds.), *Sexual behavior: Pharmacology and biochemistry*. New York: Raven Press, 1975, pp. 241–246.
Blakeslee, A. F., and Salmon, M. R. Odor and taste blindness. *Eugenics*, 1931, *16*, 105–109.
Brobeck, J. R., Wheatland, M., and Stominger, J. L. Variations in regulation of energy exchange associated with estrus, diestrus, and pseudopregnancy in rats. *Endocrinology*, 1947, *40*, 65–72.
Byrne, D., Golightly, C., and Capaldi, E. J. Construction and validation of the food attitude scale. *Journal of Consulting Psychology*, 1963, *27*, 215–222.
Cabanac, M. Physiological role of pleasure. *Science*, 1971, *173*, 1103–1107.
Cabanac, M., and Duclaux, R. Obesity in absence of satiety aversion to sucrose. *Science*, 1970, *168*, 486–487.
Cabanac, M., Minaire, Y., and Adair, E. R. Influence of internal factors on the pleasantness of a gustative sweet sensation. *Communications in Behavioral Biology*, 1968, Part A, *1*, 77–82.
Carr, W. J., Loeb, L. S., and Dissinger, M. L. Responses of rats to sex odors. *Journal of Comparative and Physiological Psychology*, 1965, *59*, 370–377.
Cavalli-Sforza, L. L., and Bodmer, W. F. *The genetics of human populations*. San Francisco: Freeman, 1971, p. 122.
Chapman, C. R., Murphy, T. M., and Butler, S. H. Analgesic strength of 33 percent nitrous oxide: A signal detection theory evaluation. *Science*, 1973, *179*, 1246–1248.
Clark, W. C. The *psyche* in psychophysics: A sensory-decision theory analysis of the effect of instructions on flicker sensitivity and response bias. *Psychological Bulletin*, 1966, *65*, 358–366.
Coats, A. C. Effects of age, sex, and smoking on electrical taste threshold. *Annals of Otology, Rhinology and Laryngology*, 1974, *83*, 365–369.
Crocker, A. D. Variations in mucosal water and sodium transfer associated with the rat estrous cycle. *Journal of Physiology*, 1971, *214*, 257–264.
DeMarchi, G. W., and Tong, J. E. Menstrual, diurnal, and activation effects on the resolution of temporally paired flashes. *Psychophysiology*, 1972, *9*, 362–367.
Desor, J. A., Greene, L. S., and Maller, O. Preferences for sweet and salty in 9 to 15 year-old and adult humans. *Science*, 1975, *190*, 686–687.
Desor, J. A., Maller, O., and Andrews, K. Ingestive responses of human newborns to salty, sour and bitter stimuli. *Journal of Comparative and Physiological Psychology*, 1975, *89*, 966–970.
Dickens, G., and Trethowan, W. H. Cravings and aversions during pregnancy. *Journal of Psychosomatic Research*, 1971, *15*, 259–268.
Doty, R. L. Odor preferences of female *Peromyscus maniculatus bairdi* for male mouse odors of *P. m. bairdi* and *P. leucopus noveboracensis* as a function of estrous state. *Journal of Comparative and Physiological Psychology*, 1972, *81*, 191–197.
Doty, R. L. Reproductive endocrine influences upon human nasal chemoreception: A review. In R. L. Doty (Ed.), *Mammalian olfaction, reproductive processes and behavior*. New York: Academic Press, 1976, pp. 295–321.
Doty, R. L. Food preference ratings of congenitally anosmic humans. In M. R. Kare and O. Maller (Eds.), *The chemical senses and nutrition II*. New York: Academic Press, 1977, pp. 315–325.

Doty, R. L., and Silverthorne, C. Influence of menstrual cycle on volunteering behaviour. *Nature (London)*, 1975, *254*, 139–140.

Dubignon, J., Cambell, D., Curtis, M., and Parrington, M. The relation between laboratory measures of sucking, food intake, and perinatal factors during the newborn period. *Child Development*, 1969, *40*, 1107–1120.

Engen, T. Olfactory psychophysics. In L. M. Beidler (Ed), *Handbook of sensory physiology, Vol. 4, Chemical senses. I. Olfaction.* Berlin: Springer–Verlag, 1971, pp. 216–244.

Engen, T. Psychophysics I. Discrimination and detection. In J. W. Kling and L. A. Riggs (Eds.), *Woodworth & Schlosberg's experimental psychology.* New York: Holt, Rinehart, and Winston, 1972, pp. 11–46.

Ericson, S. The parotid gland in subjects with and without rheumatoid arthritis. *Acta Radiologica*, 1968, Suppl. 275, 1–167.

Ericson, S. The variability of the human parotid flow rate on stimulation with citric acid, with special reference to taste. *Archives of Oral Biology*, 1971, *16*, 9–19.

Falconer, D. S. Sensory thresholds for solutions of phenyl thiocarbamide: Results of tests on a large sample made by R. A. Fisher. *Annals of Eugenics (Cambridge)*, 1947, *13*, 211–222.

Fergenson, P. E., Moss, S., Dzendolet, E., Sawyer, F. M., and Moore, J. W. The effect of signal probability, food intake, sex, and smoking on gustation, as measured by the theory of signal detection. *Journal of General Psychology*, 1975, *92*, 109–127.

Fischer, R. In discussion following R. I. Henkin, abnormalities of taste and olfaction in various disease states. In M. R. Kare and O. Maller (Eds.), *The chemical senses and nutrition.* Baltimore: Johns Hopkins University Press, 1967, p. 111.

Fischer, R., Griffin, F., England, S., and Garn, S. Taste thresholds and food dislikes. *Nature (London)*, 1961, *191*, 1328.

Fischer, R., Griffin, F., and Kaplan, A. R. Taste thresholds, cigarette smoking and food dislikes. *Medicina Experimentalis*, 1963, *9*, 151–167.

Fornasari, G. Variazioni della soglia gustativa in donne con ciclo menstruale normale. *Bolletino della Malattie dell'Orecchio, della Gola, del Naso*, 1951, *69*, 139–147.

Fortin, J. N., Wittkower, E. D., and Kalz, F. A psychosomatic approach to the pre-menstrual tension syndrome: A preliminary report. *Canadian Medical Association Journal*, 1958, *79*, 978–981.

Giles, A. The longings of pregnant women. *Transactions of the Obstetrics Society*, 1893, *35*, 242–249.

Giudizi, S., and Foferi, G. Le variazioni della sensibilita gustativa in particolari situazioni fisiologiche ed in alcuni stati morbosi. *Rivista de Clinica Medica*, 1946, *46*, 78–88.

Glanville, E. V., and Kaplan, A. R. The menstrual cycle and sensitivity of taste perception. *American Journal of Obstetrics and Gynecology*, 1965, *92*, 189–194.

Glanville, E. V., Kaplan, A. R., and Fischer, R. Age, sex and taste sensitivity. *Journal of Gerontology*, 1964, *19*, 474–478.

Green, D. M., and Swets, J. A. *Signal detection theory and psychophysics.* New York: Wiley, 1966.

Greene, L. S. Physical growth and development, neurological maturation, and behavioral functioning in two Ecuadorian Andean communities in which goiter is endemic. *American Journal of Physical Anthropology*, 1974, *41*, 139–152.

Greene, L. S., Desor, J. A., and Maller, O. Heredity and experience: Their relative importance in the development of taste preference in man. *Journal of Comparative and Physiological Psychology*, 1975, *89*, 279–284.

Grinker, J. A., Price, J. M., and Greenwood, M. R. C. Studies of taste in childhood obesity. In D. Novin, W. Wyrwicka, and G. Bray (Eds.), *Hunger: Basic mechanisms and clinical implications.* New York: Ravin Press, 1976, pp. 441–457.

Guilford, J. P. *Psychometric methods.* New York: McGraw-Hill, 1954.

Hansen, R., and Glass, L. Über den Geruchssin in der Schwangerschaft. *Klinische Wochenschrift*, 1936, *15*, 891–894.

Hansen, R., and Langer, W. Über Geschmacksveranderungen in der Schwangerschaft. *Klinische Wochenschrift*, 1935, *14*, 1173–1176.

Harris, H., and Kalmus, H. The measurement of taste sensitivity to phenylthiourea (P.T.C.). *Annals of Eugenics (Cambridge)*, 1949, *15*, 24–31.

Hawel, W., and Wirths, W. Untersuchen psychologischer Variablen bei Männern und Frauen während vierzehn Tage dauern dem Aufenthalt in Schutzraumen. *Psychologie und Praxis*, 1969, *13*, 19–32.

Henkin, R. I. Sensory changes during the menstrual cycle. In M. Ferin, F. Halberg, R. M. Richart, and R. L. Vande Wiele (Eds.), *Biorhythms and human reproduction*. New York: Wiley, 1974, pp. 277–285.

Hoshishima, K. Endocrines and taste. In M. R. Kare and O. Maller (Eds.), *The chemical senses and nutrition*. Baltimore: Johns Hopkins University Press, 1967, pp. 139–153.

Jørgensen, M. B., and Buch, N. H. Sense of smell and taste in pregnant diabetics: Clinical studies. *Practica-Oto-Rhino-Laryngologica*, 1961, *23*, 390–396.

Kahn, R. J. Sensory threshold fluctuations as a function of the menstrual cycle. Doctoral dissertation, Yeshiva University. Ann Arbor: University Microfilms, 1965, No. 65–11, 128 pp.

Kalmus, H., and Trotter, W. R. Direct assessment of the effect of age on P.T.C. sensitivity. *Annals of Human Genetics*, 1962, *26*, 145–149.

Kaplan, A. R., and Fischer, R. Taste sensitivity for bitterness: Some biological and clinical implications. In W. Wortis (Ed.), *Recent advances in biological psychiatry*, Vol. VII. New York: Plenum Press, 1965, pp. 183–196.

Kaplan, A. R., Glanville, E. V., and Fischer, R. Taste thresholds for bitterness and cigarette smoking. *Nature (London)*, 1964, *202*, 1366.

Kaplan, A. R., Glanville, E. V., and Fischer, R. Cumulative effect of age and smoking on taste sensitivity in males and females. *Journal of Gerontology*, 1965, *20*, 334–337.

Komisaruk, B. R., Adler, N. T., and Hutchison, J. Genital sensory field: Enlargement by estrogen treatment in female rats. *Science*, 1972, *178*, 1295–1298.

Köster, E. P. Human psychophysics in olfaction. In D. G. Moulton, A. Turk, and J. W. Johnston, Jr. (Eds.), *Methods in olfactory research*. London: Academic Press, 1975, 345–374.

Krecek, J. Sex differences in salt taste: The effect of testosterone. *Physiology and Behavior*, 1973, *10*, 683–688.

Krecek, J., Nováková, V., and Stibral, K. Sex differences in the taste preference for a salt solution in the rat. *Physiology and Behavior*, 1972, *8*, 183–188.

Krut, L. H., Perrin, M. J., and Bronte-Stewart, B. Taste perception in smokers and non-smokers. *British Medical Journal*, 1961, *1*, 384–387.

Laird, D. A., and Breen, W. J. Sex and age alterations in taste preference. *Journal of the American Dietetic Association*, 1959, *15*, 549–550.

Langwell, K. E. Women more sensitive to taste than the men. *Frosted Foods Field*, 1948, *7*, 2–4.

Langwell, K. E. Taste perception and taste preferences of the consumer. *Food Technology*, 1959, *3*, 136–139.

Leguebe, A. Sex differences in tasting P.T.C. *Life Sciences*, 1963, *2*, 337–342.

Le Magnen, J. Les phénomènes olfacto-sexuels chez l'homme. *Archives des Sciences Physiologiques*, 1952, *6*, 125–160.

Linker, E., Moore, M. E., and Galanter, E. Taste thresholds, detection models, and disparate results. *Journal of Experimental Psychology*, 1964, *67*, 59–66.

Luvara, A., and Maurizi, M. Ricerche di olfattometria in gravidanza. *Bollettino delle Malattie dell'Orecchio, della Gola, del Naso,* 1961, *79,* 367–375.

Malek, J., Gleich, J., and Maly, V. Characteristics of the daily rhythm of menstruation and labor. *Annals of the New York Academy of Sciences,* 1962, *98,* 1042–1055.

Marcus, R. L. Cravings for food in pregnancy. *Manchester Medical Gazette,* 1965, *44,* 16–18.

Marks, L. E. *Sensory processes: The new psychophysics.* New York: Academic Press, 1974.

Mayer-Gross, W., and Walker, J. W. Taste and selection of food in hypoglycaemia. *British Journal of Experimental Pathology,* 1946, *27,* 297–305.

Minich, S. K., Mountney, G. J., and Prudent, I. The effect of age and sex of panel members on taste thresholds for standard chicken broth. *Food Technology,* 1966, *20,* 970–973.

Mohr, J. Taste sensitivity to phenylthiourea in Denmark. *Annals of Eugenics (Cambridge),* 1951, *16,* 282–286.

Moore, M. E., Linker, R., and Purcell, M. Taste sensitivity after eating: A signal detection approach. *American Journal of Psychology,* 1965, *78,* 107–111.

Morton, J. H., Additon, H., Addison, R. G., Hunt, L., and Sullivan, J. J. A clinical study of premenstrual tension. *American Journal of Obstetrics and Gynecology,* 1953, *65,* 1182–1185.

Moskowitz, H. R. The sweetness and pleasantness of sugars. *American Journal of Psychology,* 1971, *84,* 387–405.

Moskowitz, H. R., Kumraiah, V., Sharma, K. N., Jacobs, H. L., and Sharma, S. D. Effects of hunger, satiety and glucose load upon taste intensity and taste hedonics. *Physiology and Behavior,* 1976, *16,* 471–475.

Mower, G. D., Mair, R. G., and Engen, T. Influence of internal factors on the perceived intensity and pleasantness of gustatory and olfactory stimuli. In M. R. Kare and O. Maller (Eds.), *Chemical senses and nutrition II.* New York: Academic Press, 1977, pp. 103–121.

Nance, D. M., Gorski, R. A., and Panksepp, J. Neural and hormonal determinants of sex differences in food intake and body weight. In D. Novin, W. Wyrwicka, and G. Bray (Eds.), *Hunger: Basic mechanisms and clinical implications.* New York: Raven Press, 1976, pp. 257–271.

Nisbett, R. E., and Gurwitz, S. B. Weight, sex, and the eating behavior of human newborns. *Journal of Comparative and Physiological Psychology,* 1970, *73,* 245–253.

Noferi, G., and Giudizi, S. Le variazioni della sensibilita gustativa in particolari situazioni fisiologiche ed in alcuni stati morbosi. Nota IV. Le variazioni della soglia gustativa per l'acido e della soglia olfattiva per l'odore limone durate la gravidanza. *Rivista di Clinica Medica* (Supp. 1, Marginalia Otolaryngologica), 1946, *5,* 89–100.

O'Mahony, M. Salt taste sensitivity: A signal detection approach. *Perception,* 1972a, *1,* 459–464.

O'Mahony, M. The interstimulus interval for taste: 1. The efficiency of expectoration and mouthrinsing in clearing the mouth of salt residuals. *Perception,* 1972b, *1,* 209–215.

Orne, M. T. On the social psychology of the psychological experiment: With particular reference to demand characteristics and their implications. *American Psychologist,* 1962, *17,* 776–783.

Panebianco, G. Le variazioni della sensibilita gustativa in gravidanza. *Valsalva,* 1950, *26,* 306–313.

Pangborn, R. M. Influence of hunger on sweetness preferences and taste thresholds. *American Journal of Clinical Nutrition,* 1959, *7,* 280–287.

Pangborn, R. M. In discussion following K. Hoshishima, endocrines and taste. In M. R. Kare and O. Maller (Eds.), *The chemical senses and nutrition.* Baltimore: Johns Hopkins University Press, 1967, p. 149.

Pangborn, R. M., and Simone, M. Body size and sweetness preference. *Journal of the American Dietetic Association,* 1958, *34,* 924–928.

Pfaffman, C., Bartoshuk, L. M., and McBurney, D. H. Taste psychophysics. In L. M. Beidler (Ed.), *Handbook of sensory physiology. Vol. IV, Chemical senses 2: Taste.* Berlin: Springer-Verlag, 1971, pp. 75–101.

Pietras, R. J., and Moulton, D. G. Hormonal influences on odor detection in rats: Changes associated with the estrous cycle, pseudopregnancy, ovariectomy, and administration of testosterone propionate. *Physiology and Behavior,* 1974, *12,* 475–491.

Posner, L. B., McCottrey, C. M., and Posner, A. C. Pregnancy craving and pica. *Obstetrics and Gynecology,* 1957, *9,* 270–272.

Price, R. H. Signal detection methods in personality and perception. *Psychological Bulletin,* 1966, *66,* 55–62.

Pursell, E. D., Sanders, R. E., and Haude, R. H. Sensitivity to sucrose in smokers and non-smokers: A comparison of TSD and percent correct measures. *Perception and Psychophysics,* 1973, *14,* 34–36.

Richter, C. P., and MacLean, A. Salt taste threshold of humans. *American Journal of Physiology,* 1939, *126,* 1–6.

Simone, M., Leonard, S., Hinreiner, E., and Valdés, R. M. Consumer studies on sweetness of canned cling peaches. *Food Technology,* 1956, *10,* 279–282.

Smith, W. I., Powell, E. K., and Ross, S. Food aversions: Some additional personality correlates. *Journal of Consulting Psychology,* 1955a, *19,* 145–149.

Smith, W., Powell, E. K., and Ross, S. Manifest anxiety and food aversions. *Journal of Abnormal and Social Psychology,* 1955b, *50,* 101–104.

Southam, A. L., and Gonzaga, F. P. Systemic changes during the menstrual cycle. *American Journal of Obstetrics and Gynecology,* 1965, *91,* 142–165.

Stevens, S. S. *Psychophysics: Introduction to its perceptual, neural and social prospects.* New York: Wiley, 1975.

Stevenson, H. W., and Allen, S. Adult performance as a function of sex of experimenter and sex of subject. *Journal of Abnormal and Social Psychology,* 1964, *68,* 214–216.

Taggart, N. Food habits in pregnancy. *Proceedings of the Nutritional Society,* 1961, *20,* 35–40.

Thompson, D. A., Moskowitz, H. R., and Campbell, R. G. Effects of body weight and food intake on pleasantness ratings for a sweet stimulus. *Journal of Applied Physiology,* 1976, *41,* 77–83.

Torgerson, W. S. *Theory and methods of scaling.* New York: Wiley, 1958.

Trethowan, W. H., and Dickens, G. Appetite disorders in pregnancy. In N. Morris (Ed.), *Psychosomatic medicine in obstetrics and gynaecology, 3rd international congress, London, 1971.* Basel: Karger, 1972, 126–129.

Valenstein, E. S., Cox, V. C., and Kakolewski, J. W. Further studies of sex differences in taste preferences with sweet solutions. *Psychological Reports,* 1967, *20,* 1231–1234.

Valenstein, E. S., Kakolewski, J. W., and Cox, V. C. Sex differences in taste preferences for glucose and saccharin solutions. *Science,* 1967, *156,* 942–943.

Vaschide, N. Mesure de la sensibilité gustative chez l'homme et chez la femme. *Comptes Rendus Hebdomadaires des Séances de l'Academie des Sciences,* 1904, *139,* 898–900.

Wade, G. N. Sex hormones, regulatory behaviors, and body weight. In J. S. Rosenblatt, R. A. Hinde, E. Shaw, and C. Beer (Eds.), *Advances in the study of behavior,* Vol. 6. New York: Academic Press, 1976, pp. 201–279.

Wade, G. N., and Zucker, I. Hormonal and developmental influences on rat saccharin preferences. *Journal of Comparative and Physiological Psychology,* 1969a, *69,* 291–300.

Wade, G. N., and Zucker, I. Taste preferences of female rats: Modification by neonatal hormones, food deprivation, and prior experience. *Physiology and Behavior,* 1969b, *4,* 935–943.

Wade, G. N., and Zucker, I. Hormonal modulation of responsiveness to an aversive taste stimulus in rats. *Physiology and Behavior,* 1970, *5,* 269–273.

Wallen, R. Sex differences in food aversions. *Journal of Applied Psychology,* 1943, *27,* 288–298.

Wallen, R. Food aversions of normal and neurotic males. *Journal of Abnormal and Social Psychology,* 1945, *40,* 77–81.

Woodworth, R. S., and Schlosberg, H. *Experimental psychology.* New York: Holt, 1960.

Wright, P., and Crow, R. A. Menstrual cycle: Effect on sweetness preferences in women. *Hormones and Behavior,* 1973, *4,* 387–391.

Zucker, I. Hormonal determinants of sex differences in saccharin preference, food intake and body weight. *Physiology and Behavior,* 1969, *4,* 595–602.

15

The Context and Consequences of Contemporary Sex Research: A Feminist Perspective

Leonore Tiefer

1. Introduction

Practicing scientists rarely take the opportunity to reflect on the philosophies—scientific, political, and personal—which guide our work. Moreover, sophistication in such analysis is neither part of our training nor is it an expected or rewarded part of our professional competence.

Nevertheless, it is a valuable exercise, from time to time, to attempt to recognize the philosophical context that guides and defines day-to-day work in science. This conference and celebration, with its emphasis on assessing the contemporary state of our knowledge about sexual behavior and projecting current research trends into the future, provide a signal opportunity for such philosophical reflection.

Trained as a scientist, I do not pretend expertise in the forms of analysis I shall attempt, although the insider's point of view provides illumination not available to other analysts. I see this as a first attempt to step back from practical professional activities to survey some of the contextual features that define sex research.

I would like to consider some of the current aspects and future directions of sex research from the perspectives offered by contemporary femi-

Leonore Tiefer • Department of Psychology, Colorado State University, Fort Collins, Colorado. Present address: Department of Psychiatry, Downstate Medical Center, State University of New York, Brooklyn, New York.

nism. In recent years, feminist analyses have illuminated our understanding of a variety of social institutions and processes. I propose to focus on some of our research on animal and human sexuality, and on biological theories about behavioral differences between the sexes.

The feminist perspective allows exploration of the ways in which ideologies of a variety of types serve to perpetuate a patriarchal vision of social institutions, sex roles, and human nature. I believe that one can use this point of view profitably to examine both the underlying assumptions about scientific research and particular examples of research and theory.

I hope to develop two distinct lines of argument. First, with regard to research on animal, and more particularly, on human sexuality, I will suggest that there is a certain narrowness plaguing our research. Our considerations of female sexual behavior, male reproductive physiology, the subjective aspects of sexuality, and the nonerotic components of sexuality have been impoverished, it seems to me, by both excessive reliance on one rigid scientific model, and on the masculine experience as the primary point of reference. Furthermore, I will suggest that to some degree our excessive reliance on one definition of legitimate scientific research stems from an unfortunate confounding of the norms of science with the masculine behavioral role in Western society.

Second, feminists note with concern a growing imbalance in contemporary psychology toward biological explanations of mental phenomena and suggest that, whereas such a research focus has considerable value and promise, its political consequences require careful analysis and cautious extrapolation of results beyond the particular experimental conditions utilized. Failure to observe such caveats seems to me to disregard the two-way connections between social ideology and scientific research.

Finally, taking into account the merits of these various criticisms as well as the strengths of contemporary lines of research, I hope to educe some positive suggestions for future research and theory.

2. *Feminist Perspectives on the Crisis in Contemporary Science*

Making use of its specialized methodologies, science made enormous advances in understanding natural phenomena. Those very methodologies, however, have come under intense criticism in the last few decades as being inappropriate for the comprehension of complex interactions and as being unable to fulfill the promise of a value-free orientation. Among the allegations frequently made are that science forces the investigator to fractionate whole systems into isolated parts (Ovenden, 1975); to conduct experiments in artificially controlled settings that cannot duplicate natural interactions (Cronbach, 1975); and to enforce an illusion of objective, impersonal observation (Bruner, 1970).

Many critics point out, in particular, that the hard-nosed approaches exemplified in the physical sciences can be especially inappropriate for psychology (Maslow, 1966; Farson, 1965; Sanford, 1965; Koch, 1965). For example, Farson (1965, p. 3) warns, "To force the data of human experience into a mold copied from the exact sciences seems to me to be inappropriate, even crippling."

One of the key norms of science has traditionally been its separation from social and political pressures, its reliance on an objective, consensually determined set of methods and values. If there were one word to sum up the unique contribution of science as a way of obtaining knowledge about the world (as opposed to the methods of art or religion, say), that word would be "objectivity." Jesse Bernard (1950, p. 273), for example, concludes a broad discussion of the relationship between the conduct of science and the cultural context in which it is conducted, "Scientific techniques, both physical and social, then, contain their own correctives for class, cultural, or any other kind of bias. Rich man, poor man, beggerman, thief; doctor, lawyer, merchant, chief; Zuñi, Dobuan, Eskimo—all, using scientific techniques, get the same results if they ask the same questions. . . . Science is objective, beyond culture."

But sociologists of science have long been skeptical of this claim, pointing out the interdependence of science and social ideology in the generation of research questions, the methods used, and especially in the choice of alternative explanations for one's findings (Merton, 1963; Barber, 1952; Blume, 1974; Schumpeter, 1949). Gunnar Myrdal (1970, p. 157) comments, "The ordinary social scientist believes that his analysis and inferences are simply founded upon his observation of facts. . . . Now we all know that when we look back at any earlier era, we regularly find that not only popular discussion but also the work of the social scientists were biased in the sense that their approach was influenced by dominating national or group interests in the society they were part of. However, like contemporary social scientists today, they were firmly convinced that their analysis and inferences were founded simply upon their observations of facts."

The current attacks on science emphasize that knowing about the social structure of science—who does it, how do they learn, where do they do it, who pays for it, what are the explicit and implicit demand characteristics of the business, what are the interrelations between the scientific "establishment" and other institutions in society, what are the codes of conduct for scientists, what are the rewards and how are they gotten—predicts inevitable dependence of scientific behavior on the values of the dominant ideology and reinforcements of the dominant sectors of society.

These various sorts of criticism have not gone unanswered by the scientific community. Too often, however, the scientific response has focused only on the element of irrationalism in the critiques, and the responses have merely restated faith in the ultimate self-correcting properties of science (Frankel, 1973; Hebb, 1974; Pirages, 1975). Falsely claiming that "the objectors to modern psychology do not want an objective science, but a sort of self-contemplation" (Hebb, 1974, p. 73), the defenders of the faith have failed to isolate the constructive elements of the criticism.

Many feminist scholars concur with the criticisms of science. Their critique, like that from other ideological perspectives, focuses on two major points: the ultimate inadequacy of the scientific method as currently understood and developed to answer the most interesting questions of nature, and the unacknowledged relationship between scientific behavior and the political and ideological surround.

Feminist scholars note the similarity between the questioned values of science and some characteristics of the Western model of masculinity. The glorification, for example, of efforts to be independent, analytic, unemotional, and in control in both arenas seems a coincidence not to be disregarded (Carlson, 1972; Nash, 1975; Feldman, 1975). Observations about the similarity of scientific norms and masculine, heterosexual norms have been made before (e.g., Snow, 1956; Maslow, 1966). They do not necessarily say that science evolved its peculiar emphasis because the people doing it were men. Rather, this focus points out that science, like other institutions in society, takes its values from the prevailing patriarchal ethos (Kantor, 1976; Lorber, 1976; Bakan, 1966).

Acceptance of the feminist opinion does not deny the usefulness of research as it is now conducted. Rather, it suggests that *excessive* reliance on traditional methods precludes comprehensive understanding. As Maslow (1966, p. 5) expressed it, "I believe mechanistic science . . . to be not incorrect but rather too narrow and limited to serve as a *general* or comprehensive philosophy."

Second, feminists are acutely aware of the influence of powerful institutions on maintaining the social status quo in society and see the scientific establishment and scientific research as forces that too often in the past have been used to oppress rather than promote the interests of women. These issues will receive greater discussion in a later section of this paper, but suffice it to say here that, whenever the "nature" of people is discussed, especially in the context of a discussion of sex differences, feminists are keenly concerned about the potential for abuse of scientific data. In the nineteenth century, during the birth of both the scientific study of sex differences and the first wave of the feminist movement, this potential for abuse was made manifest by the activities of Francis Galton. "Galton's legacy to the psychological study of sex differences is that he established the field within the broader context of differential psychology, and gave ample evidence that one's attitudes, values, and ideologies may infiltrate one's interpretation of data" (Buss, 1976, p. 285).

Excessive confidence in the traditional methods of science to encompass the complexities of nature and humanity, inattention to the cultural values determining research content and methodological approach, and insensitivity to the role of science in providing material for social policy decisions—these are the elements of the feminist critique of science. Let us see how aspects of our contemporary science of sex research look from this critical vantage point.

3. *Masculine Biases in the Study of Sexual Behavior*

The current theories and data in the area of mammalian sexual behavior, especially human sexual behavior, suffer from a biased and inadequate vision, a state of affairs I tentatively attribute to the narrow definition of sexuality plaguing the research area.

Most sex research in psychology has utilized animal subjects, tested under controlled laboratory conditions, in brief heterosexual pair tests. Usually the animals will be placed together in a relatively small test enclosure. Sometimes one will be tethered, or enclosed in the goalbox of a T maze, or given access to a partner only after performance of some operant response. The traditional measures utilized in the pair test include the timing and frequency of mounts, intromissions and ejaculations by the male, and of lordoses by the female. Occasionally, some measure of "proceptive" behavior by the female (approach, solicitation, darting, or some species-appropriate behavior) will be tabulated. Based on research, theories about the physiological control of sexual behavior, environmental conditions supporting or inhibiting sexual behavior, and the influence of early life experience on adult sexual behavior are drawn.

The "meaning" of sexual activity in such animal studies is limited to genitally oriented copulation. The only acts considered sexual are genital contacts. Necessarily, but sadly for our understanding of sexuality, no relevance is made of the partners' past histories with one another, enjoyment of the various moves, subjective involvement in the processes, reminiscences of previous erotic experiences, fantasies, etc. Likewise, the most prominent studies of human sexuality (Kinsey, Pomeroy, and Martin, 1948; Kinsey, Pomeroy, Martin, and Gebhard, 1953; Masters and Johnson, 1966) also adopt a genital definition for sexual expression in their observations, tabulations, and discussions of the vicissitudes of human sexual processes.

For example, examine the conclusions, widely cited, that males and females are differentially sexually responsive to psychological experiences and stimuli in the arousal of sexual desire and the achievement of sexual satisfaction. In the 1953 volume, Kinsey and his coauthors declare:

> The sexual responses and behavior of the average male are, on the whole, more often determined by the male's previous experience, by his association with objects that were connected with his previous sexual experience, by his vicarious sharing of another individual's sexual experience and by his sympathetic reactions to the sexual responses of other individuals. The average female is less often affected by such psychological factors. (p. 687)

What are the data which substantiate such conclusions and how were they collected? Kinsey's conclusions were based on the differential response of sampled males and females in our society to questions administered in a personal interview such as, "How often are you aroused by commercial movies, burlesque shows, seeing others engaged in sex acts, your own fantasies about the opposite sex, writings or drawings about sexual activity, etc.?"

Data so collected, showing differential responses by the two sexes, seem to me to warrant a conclusion like, "More males than females report being aroused by observing in real life or in symbolic depiction certain sexual acts involving either other humans or animals." To claim, however, that these data warrant a conclusion like "the sexual responses of males are more often determined by their previous experience" seems totally *unwarranted*. Perhaps Kinsey's careful choice of words, "The average female is less often affected by *such* psychological factors" should be taken to mean that he felt women were affected by *other* psychological factors. Unfortunately, Kinsey's failure to specify any other psychological factors resulted in his original qualification being overlooked.

One might suggest assessing, for example, the nature of the romantic involvement and commitment of the partner, the presence of overt reassurance and reinforcement by the partner, the woman's own body image, the woman's feelings of self-worth both within and independent of the particular relationship, the woman's previous experience with forced sexual activity, the woman's attitudes toward sex in exchange for advantages, etc. Studying the involvement of such factors as these requires both the development of new techniques of study and a broader definition of what is a sexual stimulus and a sexual situation than we see in Kinsey's questions of Masters and Johnson's observations.

In fact, recent studies directly measuring sexual arousal through some sort of vaginal plethysmograph or on-the-spot self-report suggest no differences in rate and frequency of female arousal in response to pictures and stories depicting sexual activities as compared with similar measures made in men (Schmidt and Sigusch, 1973). But such research, too, focuses on genital reaction as *the* measure of sexual involvement and offers less illumination than would research on what men and women themselves consider sexual stimuli, and their responses (overt and covert) to such stimuli.

We do have intimations that nongenital elements in the definition of sexuality may be appropriate in studying women's experience. Some women contend that sex without orgasm can be perfectly satisfying, that extravaginally induced orgasm may be more intense than intravaginally induced orgasm although the latter may in some ways be more pleasurable, and that looking at the beloved or anticipating a reunion or recalling an emotionally intense interaction can be erotically, though not genitally, ecstatic.

We could simply not define such phenomena as "sexual," and relegate them to the scientifically inaccessible world where reside such other elusive phenomena as "love." But that would be wrong, because what is sexual ought to be for the scientist what a person calls sexual, and people will continue to call nongenital experiences sexual except when they are filling out questionnaires or answering questions in situations in which the demand characteristics are such that the definitions are made by others.

Rossi (1973) points out how the "phallic fallacy" of male sex researchers demeans nongenital sexuality. Oftentimes, women are labeled "immature" because for them the greatest pleasure in sexual encounters comes from what is labeled "foreplay," rather than what is labeled "sexual intercourse." The genital intercourse bias of sex research has precluded study of the psychological and physiological roles and functions of such "foreplay" acts. Perhaps such behaviors even affect phenomena like luteal hormone (LH) release or sperm transport, effects overlooked by otherwise zealous researchers!

A recent survey of knowledge concerning the behavioral coordination of male and female animals during sexual encounters attempted to identify the interactional elements in the situation (Larsson, 1973). The three sections in the review were titled: (1) behavioral coordination between the male and the female during coitus; (2) reproductive function of the female as determined by exteroceptive stimulation from the male; and (3) the behavior of the male as determined by the physiological condition of the female. The survey, extensive as it was, had little discussion of behavioral variables in the female and their effect on physiological functions in the male.

Although Larsson concludes, "The behavior exhibited by one partner thus generates physiological changes in the other, and an organized pattern of behavioral and physiological responses emerges" (Larsson, 1973, pp. 48–49), the concern of sex researchers with male behavior and female physiology results in less being known about the interactional elements than if the assumptions governing earlier research had been more carefully scrutinized. Not only does this bias weaken our theories, it has practical consequences. For example, it contributes to the imbalance in applied research on contraceptive methods toward those affecting females. I do not deprecate animal research on sexuality, although I believe that the evidence is overwhelming that sociocultural forces not relevant for animals definitively determine human expectations, definitions, and patterns of behavior and sexual experience (Petras, 1973). This inevitably limits the generalizations derived from animal sexuality research.

To increase their value, animal sudies of the future must attempt a greater balance than before, with emphasis being placed on a careful analysis of female patterns of behavior. Beach's (1976) recent attention to the "proceptive" components of female sexual behavioral patterns is an important step in this direction. He asserts:

> The female's tendency to display appetitive responses finds little opportunity for expression in laboratory experiments which focus exclusively on her receptive behavior, or upon the male's execution of his coital pattern. The resulting concept of essentially passive females receiving sexually aggressive males seriously misrepresents the normal mating sequence and encourages a biased concept of feminine sexuality. (p. 115)

We need to remedy the imbalances of the past with studies of the female experience of the Coolidge effect, criteria for individual preferences in choice of sexual partners, the forms of situational inhibition and facilitation to which female mammals are responsive, the aspects of sexual activity that are reinforcing for female mammals, the existence of anything remotely resembling the human orgasm in either male or female nonhuman mammals, etc.

In the last section of this chapter I will suggest some new approaches to the study of human sexuality, but suffice it to say here, as McCormack (1976) has done in a study of political behavior, that sex differences may not be differences in degree, they may well be differences in kind. So long as sex as men experience it is the only kind of sex deemed worthy of study, so long will we deprive ourselves of true understanding.

4. Biology and Research on Sex Differences in Humans

4.1. Biological Theories about the Origins of Group Differences in Humans

Because of the tremendous influence of science on social attitudes and social policy, feminists are particularly concerned with research alleging biological bases for human sex differences. Bardwick (1976) correctly claims that "Radicals seem generally opposed to the idea of physiological determinism and [are] angered at the idea that physiological differences between the sexes might be significant in accounting for psychological and social differences between women and men" (p. 95), but she fails to credit the realistic sources of the misgivings.

In controversial areas, areas in which social policy decisions will have important effects on individual lives and opportunities, the impact of scientific data is an ingredient in policy making not to be underestimated. History suggests that our society has two responses available to the demonstration of differences in capacity between social groups. The first is to exclude from certain social and political opportunities those considered unfit to bear the responsibilities of full citizenship. This choice is adopted when the differences are seen to result from the inexorable consequences of biological givens, to be "natural," and, therefore, predominantly immutable. This is the reason given why 10-year olds may not now vote, and it used to be the reason given why women were not permitted to vote.

The second response is to provide a variety of compensatory, affirmative experiences for those demonstrating lesser abilities to render them more competitive with the higher-achieving group. This choice is adopted when the original differences are seen to result from the inequities of opportunity or encouragement.

There is an interesting relationship, historically, between those advocating one or the other of these positions, a relationship that will take us a bit further into the connection between science and social phenomena. As Eisenberg (1972, p. 124) expresses it, "Some readers may object to 'politicizing' what should be a 'scientific' discussion. My contention is that it is necessary to make overt what is latent in treatises on the 'innate' nature of man."

Pastore (1949) has shown, in a most interesting and undercited study of 24 prominent scientists who published in the field of nature–nurture differences, the almost perfect correlation between the advocacy of liberal or radical philosophies and an environmentalist outlook on group differences on the one hand, and political conservatism and support for a hereditarian explanation for group differences on the other. "This inner relationship suggests that it would be as reasonable to classify the nature–nurture controversy as sociological in nature as to classify it as scientific in nature." (Pastore, 1949, p. 177).

Pastore observes that the period providing his data, the latter part of the nineteenth century and the first half of the twentieth, was marked by widespread social and political changes. He notes, "In this atmosphere of impending social change, the position of the hereditarian would be to favor the *status quo* since he would contend that the essential incorrigibility of man's inherent nature was at the basis of social evils." (1949, p. 2) Pastore then concludes, "It is probable . . . that in most cases the scientists were not aware of the specific impact of their political loyalties upon their scientific thinking" (p. 182) but that the connection cannot be fortuitous. He attempts to analyze whether the scientific statements preceded the political advocacy of a particular position or vice versa, but whether such a chicken-and-egg question can ever be answered, the striking correlation for 22 of his 24 scientists is noteworthy.

The political aspect of biological theories of the causation of human differences has been made stunningly manifest in the recent controversy over the book by Edward O. Wilson (1975) titled *Sociobiology* (Wade, 1976). The social policies that follow from an acceptance of Wilson's position are apocalyptically laid out by a study group formed of Wilson's colleagues at Harvard (Sociobiology Study Group, 1975). They allege that the "biological determinism" advocated by Wilson is a position where "Aggression, competition, extreme division of labor, the nuclear family, the domination of women by men, the defense of national territory, individualism, are over and over again stated to be the result of 'human nature' [and that] present social arrangements are either unchangeable or if altered will demand continued conscious social control because the changed conditions will be 'unnatural'" (p. 1).

Feminists, striving to "free the human personality from the restricting prison of sex-role stereotyping and to develop a conception of mental health that is free from culturally imposed definitions of masculinity and femininity" (Bem, 1975) understandably regard, therefore, scientific trends supporting theories about the biological bases of sex differences and sex roles with alarm, and with skepticism.

4.2. The Biologizing of Contemporary Psychology

For years I have watched a developing trend in contemporary psychology. As Carlson (1972) puts it, "The biologizing of general psychology is well underway" (p. 27). In every field it seems that physiological models and research strategies are held to offer the best route to understanding the traditional subjects of psychological inquiry.

Differences in cognitive processing strategies are attributed to differential functioning of the right and left cerebral hemispheres (e.g., Hellige, 1975). Differences in dietary habits are thought to result from variations in ventromedial hypothalamic functioning (Schachter, 1973). Efforts to alter patterns of emotional responsiveness emphasize biofeedback techniques, asserting that the autonomic foundations of emotion make direct biological intervention the most efficacious route to altered mental experiences (Brown, 1974). The presence of an extra Y chromosome in some men is purported by some to have a direct influence on later development of criminal behavior (Owen, 1972). The most exciting developments concerning the etiology and treatment of psychopathology focus on biochemical theories and research (e.g., Friedhoff, 1976). Other psychological phenomena, too, from attention to intelligence, from childhood behavioral irregularities to nonverbal communication of emotion are alleged to be most usefully understood from the perspective of underlying biological contributions.

Let us look, then, at some research in the area of human sex differences and explore the implications of the current "biologizing."

4.3. Prenatal Contributions to Human Psychological Sex Differences

The commanding paradigm in the study of animal sex differences has been, for the last 20 years or so, the relationship of prenatal hormone influences to later sexual dimorphism. Originating from the powerful studies of embryonic sexual differentiation reported by Jost in the 1950s, hundreds of experiments have been conducted on a multitude of species to elaborate the extent to which prenatal androgenization contributes to the development of behavioral sex differences in copulation, aggressiveness, activity patterns, feeding behavior, learning capacity, emotionality, and, recently, a variety of human behaviors as measured by observation or self report.

Thousands of words have been devoted to creating finer and finer distinctions between organizational and activational effects of hormones, to ascertaining precisely the neural mechanisms affected by critical period androgens, to defining the temporal parameters of the androgenic effect, to identifying which androgens are involved in various species, to revealing the contributions of postnatal hormonal interactions with the earlier critical androgens, to clarifying whether the primary effect is on those tissues that will mediate later hormonal effects or will directly process stimulus input and direct particular behavioral output in response, and so forth. It is a classic example of research begetting more research and more hypotheses than ever could have been imagined.

One recent statement of the presumed effects of prenatal hormones on human psychological development derives from the work of Anke Ehrhardt and her colleagues (Ehrhardt and Baker, 1974a,b). Ehrhardt compares the behaviors of girls prenatally androgenized to some degree by the excessive output of their abnormal adrenal glands with the behaviors of their presumably unaffected female siblings. She also looks at boys suffering from the adrenogenital syndrome and compares them with their presumably unaffected female siblings. She also looks at boys suffering from the andrenogenital syndrome and compares them with their presumably unaffected brothers. She concludes that statistical differences are observed between unaffected and affected girls in energy expenditure, sex of preferred playmates, fight initiation, daydreams about wedding and marriage, outlook toward career possibilities, outlook toward pregnancy and motherhood, interest in dolls and infants, and interest in functional versus dressy feminine clothing, jewelry, and cosmetics. She interprets the "boyish" behavior and indentification of the adrenogenital girls as being a result of the prenatal androgenic stimulation, and unlikely to be due to parental attitudes or other mediating psychological factors.

Although Ehrhardt states her conclusions carefully and takes great pains to qualify their implications, nevertheless her interpretation of her own work requires that she believe, "If prenatal exposure to androgen modifies behavior in genetic females and in the described clinical conditions, one may assume that similar hormonal factors contribute to the development of temperamental differences between males and females in general" (Ehrhardt and Baker, 1974a, p. 49). Ehrhardt and Baker go on to emphasize, "If prenatal hormone levels contribute to sex differences in behavior, the effects in human beings are subtle and can in no way be taken as a basis for prescribing sex roles" (1974a, p. 49), a disclaimer which suggests that they, like contemporary feminists, perceive the connection between biological explanations and social policy prescriptions. But, because of an overemphasis on the alleged direct behavioral consequences of the biological treatments underlying the research, they are led to conclude that the observed group differences in behavior are due to the biological differences in treatment.

By contrast, I believe that there are alternative explanations for their results. I begin by assuming that biological influences can cause behavioral consequences by circuitous routes, the endpoints of which relate only indirectly to the original biological interventions. The first factor to consider is the potential continuing ambivalence by the parents as to the results of the adrenogenital excesses. Bing and Rudikoff (1970), for example, focus on two sets of parents of adrenogenital girls and show how their behavior toward their daughters is profoundly altered by the medical problem, albeit in very different ways.

The second factor to consider is the likely possibility that prenatal androgen does have some direct effect on factors contributing to increased activity level and energy expenditure, but not on any behaviors more complex than these. Were this true, the adrenogenital girls would likely ally themselves with other children with the same high energy levels and identify with their (culturally imposed boyish) norms of fantasies about adult careers, "boys'" toys, and clothing preferences. The effects of peer pressure and identification, with no countermanding by the parents (and it is important to note here that "there was usually very little parental pressure toward more femininity" (Ehrhardt and Baker, 1974a, p. 49) would be sufficient to account for the differences in behavior, fantasy and expectations for adult roles.

To allege that hormones have a direct effect on desire for functional clothes, intention to have a career, choice of playmates, engaging in fights, etc., makes little sense when one looks at the behavior of boys in cultures where male role behavior is very different than in our own. Are we to think that the androgens found in these other cultures are of different types, or that their effects are somehow thwarted or that the effects of socialization are somehow greater in those other cultures?

The only significant difference between boys with the adrenogenital syndrome and their unaffected brothers is a greater expression of energy expenditure (Ehrhardt and Baker, 1974a). The lack of any differences in play, fantasy, or adult aspirations suggests that their behavior is adequately predictable from a knowledge of the socialization rules, and thus that the only behavioral influence of the prenatal androgen (either in normal or excess amounts) is on energy expenditure and socially mediated consequences of external genital morphology.

In a recent review of the literature on prenatal hormone effects on human behavior, Reinisch (1976, p. 82) observers, "IQ is raised only in children whose genitalia are also affected by hormones and not in children who show no genital masculinization." Although the enhanced IQ levels in prenatally androgenized subjects is not thought to be a reliable finding (Ehrhardt and Baker, 1974b), Reinisch's observation is consistent with my hypotheses, unless of course she is suggesting that people think with their genitals! That is, those individuals whose response to the prenatal hormone exposure includes genital masculinization are somehow treated differently and therefore behave differently from those not so affected.

Although my major concern in this section is to point out the failure to consider nonbiological explanations for human sex differences in clinical research where biological variables are manipulated, it is not amiss to mention that some aspects of my alternative explanation may be appropriate even for our understanding of the effects of prenatal androgen on behavior development in the rhesus monkey. Although some studies detail ways in which the mother's behavior cannot completely account for sex differences in monkey play behavior (e.g., Rosenblum, 1974), nevertheless differences in behavior toward male and female offspring on the part of the parent can go a long way toward explaining the sex differences in outcome (Jensen, Bobbitt, and Gordon, 1967; Mitchell, 1968).

A careful analysis of data recently presented with regard to the necessary dosage and duration of prenatal androgen treatment given rhesus monkeys in order to masculinize their behavior reveals that no androgenic treatment over and above that necessary for masculinization of the genitalia is required for "masculinization" of play behavior (Phoenix, 1974). Conceivably, the mothers' differential treatment of offspring, dependent on observations of the external genitalia, is responsible in large part for the sex differences in their subsequent behavior.

In the published excerpts of the discussion following Phoenix's and

Ehrhardt's papers (Friedman, Richart, and Vande Wiele, 1974), Ehrhardt asked Phoenix whether he thought maternal behavior might have been influenced by the presence of male genitalia in the pseudohermaphroditic offspring. He agreed that it might well have been, but the dicussion then turned in a different direction, and it is unclear to what extent Phoenix would have been willing to attribute the changed offspring behavior to changed maternal behavior (Friedman *et al.*, 1974, p. 79). I raise this alternative interpretation of the animal data mostly to show that researchers who manipulate biological variables seem at times blind to nonbiological interpretations of their own results, and that whereas such "linear" thinking might have no particular harm in animal experimentation, it clearly is to be avoided in theorizing about human beings.

4.4. Summary of Biologizing Comments

Postulating and investigating biological contributions to the genesis and expression of human behavior is obviously an important scientific activity, and I am advocating neither its cessation nor its curtailment. I am, however, advocating extreme caution in drawing conclusions about the impact of physiological factors at our present state of inadequate knowledge, because of the politically conservative uses to which such conclusions may be put.

Roberts (1974) suggests that the rejection of a true hypothesis regarding genetic contributions to race differences is socially a far less severe error than erroneously accepting a false hypothesis regarding such contributions, and he therefore advocates using the nonscientific values to set the statistical parameters for such research. I doubt if this suggestion will catch on, but I present this as one way in which one might come to terms with this social–scientific interface.

Our thinking about the causes of human behavior profoundly affects the ways in which we attempt to assign roles to people and the ways in which we attempt to alter people's behavior. Adams (1964) documents the tremendous changes that occurred in the treatment of mental patients in the nineteenth century as professionals adopted a medical model of interpersonal disturbance. "What we choose to believe about the nature of man has social consequences. Those consequences should be weighed in assessing the belief we choose to hold, even provisionally." (Eisenberg, 1972, p. 124).

It seems that many currently recognize the uses to which biologically based information can be put. I cited Ehrhardt to that effect earlier (Ehrhardt and Baker, 1974a,b). Gastonguay (1975) attempts to ride both sides of the same fence in his proposal for a sociobiology unit in high school science. Erring in overstating what is known, drawing far too broad conclusions, he suggests, "It appears that most higher societies depend on some form of hierarchy. This is a natural outcome of species variation; individuals possess differing strength, leadership ability and so on" (1975, p. 484). But his article concludes with this caveat: "However, as we attempt to quantify human behavior in this way, we must use extreme caution, for there are always those who would use such knowledge in order to promote a social stratification of opportunity" (p. 486). He seems to have waived his own opportunity to display such caution.

Alone, such caveats are insufficient when not supplemented by environmental-based hypotheses about the development of the behavior in question. If the only information supplied, and the only interpretations offered focus on physiological determinants, then the warning to exercise caution will go unheeded. Mary Brown Parlee (1973) shows, for example, how assumptions about the physiological etiology of the premenstrual syndrome in women dictated many early attempts to "cure" the problem, caveats about alternative possibilities to the contrary notwithstanding. As long as science holds the unquestioned respect of people in control of policymaking, scientists must bend over backward not to provide premature or erroneous prescriptions, especially in controversial areas such as that I have been discussing.

5. *Suggestions for Future Sex Research*

Even a partial acceptance of the foregoing comments indicates the need for an expansion of the methods we employ in the study of human sexuality, in terms of choosing questions for research, designing our studies, and drawing sensible and sensitive conclusions. Following the structure of earlier sections, let me first discuss the area of sexuality, and then consider research on sex differences.

5.1. *Toward a Nonsexist Understanding of Human Sexuality*

The need to encompass the whole of human sexuality in our theories requires development of methods for understanding the subjective side of sexuality, both for the affective and cognitive domains. Gagnon (1975, p. 140), for example, has recently predicted that future sex research will have to deal with "the calculus of pleasures" as well as more traditional observations and tabulations.

The thought of returning to reliance on introspection in our studies strikes dismay into psychologists aware of the failures of early Structuralist attempts, but several authors have recently pointed out that we can avoid the early errors without abandoning the method altogether. Radford (1974) suggests that introspection as a scientific method has its problems, but that they are not insurmountable, and that information garnered through this method can be of inestimable value in complementing other data.

Levine (1974) notes that students of perceptual processes have managed to incorporate qualitative self report into a viable scientific methodology by the use of multiple cross checks and overlapping intersecting indicators of particular pieces of behavior and states of awareness. The scientific necessity for control does not only mean the use of control groups as part of an experimental design. He encourages a compilation and classification of the various types of data-collection methods and the inferences one may legitimately draw from each. Instead, then, of always using the same style research for each interesting question, one would be able to choose the best research technique for each particular question.

Carlson (1971, p. 208) has proposed such a classification for personality research, enumerating the strengths of "experimental methods of laboratory psychology, the correlational methods of differential psychology, and the clinical methods stemming from the tradition of French psychiatry and Viennese psychoanalysis." She claims that previous discussions of the various psychological methods have seen them as competing, whereas "the three approaches presented here are conceived of as complementary. . . . Tentative knowledge gained from any one approach must ultimately be weighed by alternative methods" (p. 209).

It is important to incorporate our own experiences as sexual human beings into the hypotheses we generate, recognizing of course the limitations introduced by such bias. Nevertheless, each of us would probably attest to the fact that our experiences of sexual arousal and pleasure have undergone numerous changes throughout our lives, yet our current literature seems oblivious to this common experience. Studying the changes in one's perception of sexuality and oneself as a sexual being from childhood, through adolescence, and the phases of adulthood will open vast regions of insight heretofore neglected.

There have been some studies attempting to incorporate subjective emphases. Hessellund (1976) included quotations about sexual fantasies along with quantitative data in his report about masturbation habits of married persons. Vance and Wagner (1976) had college students write their subjective descriptions of orgasm and then asked raters to try to separate those written by males from those written by females. Kanter, Jaffe, and Weisberg (1975) include numerous quotes of personal feelings and recollections in their study of intimate relationships in communal households. Other recent quasianthropological studies of homosexuals, pederasts, and transsexuals have attempted a balance of objective and self-report data.

Others have advocated this sort of research without exploring precise methods. Hochschild (1976, p. 284), for example, suggests that we "should attend to the actor's own definition of his or her feelings in order to find out how emotion vocabularies are used, what inner experiences they refer to, and what social situations or rules call them forth or squash them out." Bell (1975) recommends that researchers involve themselves more in the lives of their homsexual research subjects in order to more fully appreciate the meanings of sexuality in their subjects' experience.

Some specific methods described by Carlson (1971) include short-term longitudinal studies, soliciting anonymous brief introspective accounts of selected constructs during data-collecting situations organized entirely for other purposes, and the reporting of much phenomenological data now incidentally collected but rarely appearing in the literature.

In addition to the introduction of subjective emphases in sex research, we need to begin to see sexuality as encompassing behaviors and attitudes that are profoundly influenced by classes of social rules which are themselves nonerotic. For example, Blumstein and Schwartz (1974) have taken great care to explore the impact of group identification on both self-labeling and sexual activity participation in bisexual women. Moyle (1975) shows how the functions of sexuality in a society (Samoa, in his example) are illuminated by a study of sexual themes in art, poetry, games, and stories. Similarly, studies in our own culture of graffiti, song lyrics, motion picture themes, styles of social dancing, fashion, cosmetic and jewelry adornments, etc., will illuminate our understanding of sexuality.

How we define "sexuality" is one of the critical elements in the decisions about what and how we research. "To label nude scenes of coitus as sexual but scenes that do not expose the genitals as romantic may be to reveal a phallic fallacy to which male researchers are particularly prone" (Rossi, 1973, p. 164). A genitally oriented definition is, according to some writers (e.g., Francoeur and Francoeur, 1974), consistent with and reflective of a particular "hot" sex (in the McCluhan sense of "hot") framework, which may well be inappropriate to describe the experiences and values of future generations.

The challenge of sexuality research in the future is exploring the multiple meanings for sexual expression, and the ways they reflect individual and societal needs. A feminist perspective will broaden and increase the value of sex research by encouraging attention to a wide spectrum of aspects of sexuality, and by reminding us of our ultimate obligation to contribute to the enrichment of human sexual experience.

5.2. Toward a Nonsexist Study of Sex Differences

The inappropriate extrapolation of scientifically studied sex differences is of great concern to feminists. We are particularly "concern[ed] that any evidence of physiological influences will be taken as grounds for the perpetuation of sex inequalities in family, political, and occupational roles"

(Rossi, 1973, p. 151). For example, Kinsey and his colleagues (1953), having demonstrated to their own satisfaction that the sexuality of males and females is differentially affected by "psychologic stimuli" continue: "Since there are differences in the capacities of females and males to be conditioned by their sexual experience, we might expect similar differences in the capacities of females and males to be conditioned by other, nonsexual types of experience." (p. 712).

I would like to suggest a number of directions for future research on sex differences that ameliorate some of these problems. The first suggestion is to do less research specifically designed to explore differences between the sexes. As Michael Lewis (1975) notes, "It seems to me that research in sex differences . . . is probably motivated by implicit variables having more to do with sociopolitical than scientific issues. Remember that for any important psychological variable there is greater within-group than across-group variance." (p. 333).

Moving beyond the study of sex differences to a true study of individual differences can enable us to discover more about the intrasex variation in abilities and predispositions, and perhaps to begin tracing some biological contributions for intrasex differences. Ehrhardt and Baker (1974a, p. 50) look forward to a time when "it can be documented that prenatal hormone levels are among the factors that account for the wide range of temperamental differences and role aspirations within the female, and possibly also within the male sex." I am not sure why such factors are more likely to contribute to female than to male intrasex differences, but such a research focus would be a useful direction.

Maccoby and Jacklin (1974) point out the reluctance of editors to publish studies showing the absence of sex differences, a reluctance that obviously must recede if our ultimate understanding of the range of sex similarities and differences is not to be severely impaired. Those studies that do focus on sex differences would do well to heed Block's (1976) analysis of Maccoby's and Jacklin's book. Block suggests that differences in the "pattern" of performance and traits are likely to be obtained, and that considering sex (or gender) as a moderating variable may be most fruitful for future research and theory.

As a second contribution, I feel that psychologists must gain a greater awareness of cross-cultural differences in sex-role behaviors before tendering any generalizations regarding the origins of sex differences. It seems absurd to claim that some set of behaviors is importantly influenced by male hormones when that same set of behaviors may represent the epitome of femininity in some other culture. It is no accident that Women's Studies, that branch of scholarship infused with a feminist perspective, is determinedly interdisciplinary. We have seen how the narrowness of vision suffered by one discipline can result in serious oversimplification of complex issues, a situation remedied by a focus on interdisciplinary cross-checking and cross-fertilization. I am not suggesting that the psychological point of view, that perspective which emphasizes intraindividual processes and determinants of behavior and mental life, is inappropriate or socially regressive, only that the individual perspective, uninformed by insights from the other social sciences, is likely to err repeatedly.

Finally, the issue of the social responsibility of the scientist is an important one in the area of sex differences research. The unfortunate use of biological theories of group difference to justify inequitable social treatment stands as a large signpost urging caution in the generation and generalization of biologically based hypotheses. I am not recommending that a board of feminist scholars ought to pass on the ideological legitimacy of every research study proposing to investigate biological variables and sex differences. That would be an undemocratic outrage. Rather, I am recommending that an awareness of the uses to which scientific conclusions are put in contemporary society ought to encourage a redoubling of efforts to produce careful observations, well-designed studies, theories that take into account the nature of the present state of evidence, and the likelihood of practical misuse. Attending to the "sociology of psychological knowledge" (Buss, 1975) will prevent the excesses of the past while encouraging sensitivity to the consequences of scientific research in the present and future.

6. Conclusion

In her 1973 Presidential address to the American Psychological Association, Leona Tyler (1973) discusses many of the reasons why the original hope that society would benefit importantly from psychological research has dimmed during the last decades. She suggests that the choice of research projects has come to depend more on "such things as the availability of a new technique or instrument, the recognition of an unanswered question in a report of previous research, a suggestion from a friend or advisor, or just the fact that a federal program has made grant funds available" (p. 1025) than on commitment to a research issue because of its potential theoretical or humanitarian importance. She urges us to reconsider our research directions, and devote ourselves to making choices that will illuminate our understanding and enrich the quality of human life.

Studies of sexuality and of maleness and femaleness are potentially of great moment both for our understanding of what it is to be human and for enriching our experience of what it is to be human. But before we can provide such research, we have to analyze more carefully the factors underlying our choice of method, of conceptual definitions, of research population, our theoretical assumptions, and the potential uses of our data and conclusions. The feminist perspective provides one angle from which to attempt such analysis.

ACKNOWLEDGMENTS

The author gratefully acknowledges the assistance of Charles Tiefer, a former biochemist now at Harvard Law School, during the writing of this chapter.

References

Adams, H. B. "Mental illness" or interpersonal behavior? *American Psychologist*, 1964, *19*, 191–197.

Bakan, D. *The duality of human existence*. Chicago: Rand McNally, 1966.

Barber, B. *Science and the social order*. New York: Free Press, 1952.

Bardwick, J. M. Psychological correlates of the menstrual cycle and oral contraceptive medication. In E. J. Sachar (Ed.), *Hormones, behavior, and psychopathology*. New York: Raven Press, 1976.

Beach, F. A. Sexual attractivity, proceptivity and receptivity in female mammals. *Hormones and Behavior*, 1976, *7*, 105–138.

Bell, A. P. Research in homosexuality: Back to the drawing board. *Archives of Sexual Behavior*, 1975, *4*, 421–431.

Bem, S. L. Beyond androgyny: Some prescriptions for a liberated sexual identity. Keynote address for APA–NIMH conference on the Research Needs of Women. Madison, Wisconsin, 1975.

Bernard, J. Can science transcend culture? *Scientific Monthly*, 1950, *71*, 268–273.

Bing, E., and Rudikoff, E. Divergent ways of parental coping with hermaphroditic children. *Medical Aspects of Human Sexuality*, 1970, *4*, 73, 77, 80, 83, 88.

Block, J. H. Issues, problems, and pitfalls in assessing sex differences: A critical review of *The Psychology of Sex Differences*. *Merrill-Palmer Quarterly*, 1976, *22*, 283–308.

Blume, S. S. *Toward a political sociology of science*. New York: Free Press, 1974.

Blumstein, P. W., and Schwartz, P. Lesbianism and bisexuality. In E. Goode and R. Troiden (Eds.), *Sexual deviance and sexual deviants*. New York: William Morrow and Co., 1974.

Brown, B. *New body, new mind*. New York: Harper and Row, 1974.

Bruner, J. S. Reason, prejudice, and intuition. In A. Tiselius & S. Nilsson (Eds.), *The place of value in a world of facts*. New York: Wiley, 1970.

Buss, A. R. The emerging field of the sociology of psychological knowledge. *American Psychologist*, 1975, *30*, 988–1002.

Buss, A. R. Galton and sex differences: An historical note. *Journal of the History of the Behavioral Sciences*, 1976, *12*, 283–285.

Carlson, R. Where is the person in personality research? *Psychological Bulletin*, 1971, *75*, 203–219.

Carlson, R. Understanding women: Implications for personality theory and research. *Journal of Social Issues*, 1972, *28*, 17–32.

Cronbach, L. J. Beyond the two disciplines of scientific psychology. *American Psychologist*, 1975, *30*, 116–127.

Ehrhardt, A. A. and Baker, S. W. Fetal androgens, human central nervous system differentiation, and behavior sex differences. In R. C. Friedman, R. M. Richart, and R. L. Vande Wiele (Eds.), *Sex differences in behavior.* NewYork: Wiley, 1974a.

Ehrhardt, A. A., and Baker, S. W. Prenatal androgen, intelligence, and cognitive sex differences. In R. C. Friedman, R. M. Richart, and R. L. Vande Wiele (Eds.), *Sex differences in behavior.* New York: Wiley, 1974b.

Eisenberg, L. The human nature of human nature. *Science,* 1972, *176,* 123–128.

Farson, R. E. Can science solve human dilemmas? In R. E. Farson (Ed.), *Science and human affairs.* Palo Alto, California: Science and Behavior Books, 1965.

Feldman, J. The savant and the midwife. *Impact of Science on Society.* 1975, *25,* 125–136.

Francoeur, A. K., and Francoeur, R. T. Hot and cool sex—closed and open marriage. In R. T. Francoeur and A. K. Francoeur (Eds.), *The future of sexual relations.* Englewood Cliffs, New Jersey: Prentice-Hall, 1974.

Frankel, C. The nature and sources of irrationalism. *Science,* 1973, *180,* 927–931.

Friedhoff, A. J. (Ed.). *Catecholamines and behavior, Vol. 2. Neuropsychopharmacology.* New York: Plenum Press, 1976.

Friedman, R. C., Richart, R. M., and Vande Wiele, R. L. *Sex differences in behavior.* New York: Wiley, 1974.

Gagnon, J. H. Sex research and social change. *Archives of Sexual Behavior,* 1975, *4,* 111–141.

Gastonguay, P. R. A sociobiology of man. *American Biology Teacher,* 1975, *37,* 481–486.

Hebb, D. O. What psychology is about. *American Psychologist,* 1974, *29,* 71–79.

Hellige, J. B. Hemispheric processing differences revealed by differential conditioning and reaction time performance. *Journal of Experimental Psychology: General,* 1975, *104,* 309–326.

Hessellund, H. Masturbation and sexual fantasies in married couples. *Archives of Sexual Behavior,* 1976, *5,* 133–148.

Hochschild, A. R. The sociology of feeling and emotion: Selected possibilities. In M. Millman and R. M. Kanter (Eds.), *Another voice: Feminist perspectives on social life and social science.* New York: Doubleday, 1975.

Jensen, G. D., Bobbitt, R. A., and Gordon, B. N. Sex differences in social interaction between infant monkeys and their mothers. In J. Wortis (Ed.), *Recent advances in biological psychiatry.* New York: Plenum Press, 1967.

Kanter, R. M. Women and the structure of organizations: Explorations in theory and behavior. In M. Millman and R. M. Kanter (Eds.), *Another voice: Feminist perspectives on social life and social science.* New York: Doubleday, 1975.

Kanter, R. M., Jaffe, D., and Weisberg, D. K. Coupling, parenting, and the presence of others: Intimate relationships in communal households. *The Family Coordinator,* 1975, *24,* 433–452.

Kinsey, A. C., Pomeroy, W. B., and Martin, C. E. *Sexual behavior in the human male.* Philadelphia: Saunders, 1948.

Kinsey, A. C., Pomeroy, W. B., Martin, C. E., and Gebhard, P. H. *Sexual behavior in the human female.* Philadelphia: Saunders, 1953.

Koch, S. The allures of ameaning in modern psychology. In R. E. Farson (Ed.), *Science and human affairs.* Palo Alto, California: Science and Behavior Books, 1965.

Larsson, K. Sexual behavior: The result of an interaction. In J. Zubin and J. Money (Eds.), *Contemporary sexual behavior: Critical issues for the 1970s.* Baltimore: Johns Hopkins University Press, 1973.

Levine, M. Scientific method and the adversary model. *American Psychologist,* 1974, *29,* 661–677.

Lewis, M. Early sex differences in the human: Studies of socioemotional development. *Archives of Sexual Behavior,* 1975, *4,* 329–335.

Lorber, J. Women and medical sociology: Invisible professionals and ubiquitous patients. In M. Millman and R. M. Kanter (Eds.), *Another voice: Feminist perspectives on social life and social science.* New York: Doubleday, 1975.

Maccoby, E. E., and Jacklin, C. N. *The psychology of sex differences.* Stanford: Stanford University Press, 1974.

McCormack, T. Toward a nonexist perspective on social and political change. In M. Millman and R. M. Kanter (Eds.), *Another voice: Feminist perspectives on social life and social change.* New York: Doubleday, 1975.

Maslow, A. H. *The psychology of science: A reconnaissance.* New York: Harper and Row, 1966.

Masters, W. H., and Johnson, V. E. *Human sexual response.* New York: Little, Brown and Co., 1966.

Merton, R. K. *The sociology of science.* Chicago: The University of Chicago Press, 1963.

Mitchell, G. D. Attachment differences in male and female infant monkeys. *Child Development,* 1968, *39,* 611–620.

Moyle, R. M. Sexuality in Samoan art forms. *Archives of Sexual Behavior,* 1975, *4,* 227–247.

Myrdal, G. Biases in social research. In A. Tiselius and S. Nilsson (Eds.), *The place of values in a world of facts.* New York: Wiley, 1970.

Nash, J. A critique of social science models of contemporary society: A feminist perspective. *Annals of the New York Academy of Science,* 1975, *260,* 84–100.

Ovenden, M. W. Intimations of unity. In N. H. Stenek (Ed.), *Science and society: Past, present and future.* Ann Arbor: Univerity of Michigan Press, 1975.

Owen, D. The 47, XXY male: A review. *Psychological Bulletin,* 1972, *78,* 209–233.

Parlee, M. B. The premenstrual syndrome. *Psychological Bulletin,* 1973, *80,* 454–465.

Pastore, N. *The nature–nurture controversy.* New York: King's Crown Press, Columbia University, 1949.

Petras, J. *Sexuality in society.* New York: Allyn–Bacon, 1973.

Phoenix, C. H. Prenatal testosterone in the nonhuman primate and its consequences for behavior. In R. C. Friedman, R. M. Richart, and R. L. Vande Wiele (Eds.), *Sex differences in behavior.* New York: Wiley, 1974.

Pirages, D. The unfinished revolution. In N. H. Stenek (Ed.), *Science and society: Past, present, and future.* Ann Arbor: University of Michigan Press, 1975.

Radford, J. Reflections on introspection. *American Psychologist,* 1974, *29,* 245–250.

Reinsich, J. M. Effects of prenatal hormone exposure on physical and psychological development in humans and animals: With a note on the state of the field. In E. J. Sachar (Ed.), *Hormones, behavior, and psychopathology.* New York: Raven Press, 1976.

Roberts, M. On the nature and condition of social science. *Daedalus,* 1974, *103* (3), 47–64.

Rosenblum, L. A. Sex differences, environmental complexity and mother–infant relations. *Archives of Sexual Behavior,* 1974, *3,* 117–128.

Rossi, A. S. Maternalism, sexuality, and the new feminism. In J. Zubin and J. Money (Eds.), *Contemporary sexual behavior: Critical issues for the 1970s.* Baltimore: Johns Hopkins University Press, 1973.

Sanford, N. Will psychology study human problems? *American Psychologist.* 1965, *20,* 192–202.

Schachter, S. Some extraordinary facts about obese humans and rats. *American Psychologist,* 1971, *26,* 129–144.

Schmidt, G., and Sigusch, V. Women's sexual arousal. In J. Zubin and J. Money (Eds.), *Contemporary sexual behavior: Critical issues for the 1970s.* Baltimore: Johns Hopkins University Press, 1973.

Schumpeter, J. A. Science and ideology. *American Economic Review,* 1949, *39,* 345–359.

Snow, C. P. The two cultures. *New Statesman,* 1956, *52,* 413–414.

Sociobiology Study Group. *Sociobilogy—A new biological determinism.* Cambridge, Boston: SESPA/Science for the People, 1975.

Tyler, L. Design for a hopeful psychology. *American Psychologist,* 1973, *28,* 1021–1029.

Vance, E. B., and Wagner, N. N. Written descriptions of orgasm: A study of sex differences. *Archives of Sexual Behavior,* 1976, *5,* 87–98.

Wade, N. Sociobiology: Troubled birth for new discipline. *Science,* 1976, *191,* 1151–1155.

Wilson, E. O. *Sociobiology: The new synthesis.* Cambridge, Mass: Harvard University Press, 1975.

16

Inside the Human Sex Circus: Prospects for an Ethology of Human Sexuality

Gordon Bermant

> The relationship between people's self-understanding in everyday life and the theorist's description or explanation of behavior brings us up against a riddle every bit as vexing as the puzzles that spring directly from the rivalry of the rationalist and the historicist method. If we disregard the meanings an act has for its author and for the other members of the society to which he belongs, we run the risk of losing sight of what is peculiarly social in the conduct we are trying to understand. If, however, we insist on sticking close to the reflective understanding of the agent or his fellows, we are deprived of a standard by which to distinguish insight from illusion or to rise above the self-images of different ages and societies, through comparison. Thus, subjective and objective meaning must somehow both be taken into account.
>
> Roberto M. Unger (1976)

The purpose of this paper is to explore briefly the significance of the ethological frame of reference for human sexual behavior. The intent is exploratory and the conclusions are tentative. The method is to expand the frame of reference at certain points in the discussion, in search of a context rich enough to capture difficult intuitions but lean enough to prevent psychological flatulence.

Gordon Bermant • The Federal Judicial Center, Dolley Madison House, 1520 H Street, N.W., Washington, D.C.

1. The Narrowest Ethological Frame of Reference

The first task is to be specific about what is implied by the phrase "ethological frame of reference." If interpreted too broadly, the phrase becomes synonomous with "biosocially oriented approaches to the study of behavior" and loses its historical significance. If interpreted too narrowly, the phrase might come to mean, for example, "the views expressed by Tinbergen or Lorenz, or contained in the text of Eibl-Eibesfeldt (1975)," and lose its heuristic richness. Ethology is now in a period of diversification; hence attempts to capture its essence are likely to fail. The following characterization of ethology, which has struck some perceptive critics as unduly "classical" (a gentle way to say "old-fashioned," perhaps), is therefore offered only as an initial heuristic device, certainly not as a definition of ethology.

A minimally adequate account of the ethological frame of reference would include at least the following distinctive conceptual and methodological features:

1. Commitment to analysis at the species level. While avoiding errors of typological thinking described by Mayr (1963) and Hirsch (1967), ethology has been concerned primarily with behavioral invariances at the level of the species.

2. Commitment to analysis of the relations between the causes and the evolutionary significance of behavior.

3. Commitment to unobtrusive observation in natural settings accompanied, when appropriate, by experiments designed to insure the ecological validity of their conclusions. Ethology combines the methods of natural history and laboratory science to provide a balanced analysis of behavior.

To the extent that these features specify the ethological frame of reference, the prospects for an ethology of human sexual behavior are dim. With regard to the first ethological commitment, we already know a good deal that ethology could tell us about human sexual behavior described at the level of the species. To a first approximation the ethnography of sexual behavior has been completed. The outcomes of anthropological field studies have been well analyzed (Ford and Beach, 1951), suggesting that an ethological frame of reference would not provide dramatic new insights.

With regard to the second ethological commitment, there appears to be little in human sexual behavior *per se* that invites speculation or research concerning an interaction between its immediate causation and its evolutionary significance. This is *not* to say that there are no interesting questions about the relation between the control of mating opportunities in human populations and the future of the species, for there obviously are such questions. It is rather to say that most of these questions are not ethological in an interesting sense. Exceptions might be questions about the origins and maintenance of aversions to marriages across color lines, whenever they exist.

With regard to the third ethological commitment, unobtrusive observations, positive and normative considerations limit the usefulness of traditional ethology. Even if it were possible to observe the sexual behavior of large numbers of people unobtrusively, such observations would be immoral.* The observations that have been done, for example by Masters and Johnson (1966), while meeting the ethical criterion of informed consent of the observed, do not meet the ethological criterion of unobtrusiveness.

These rather negative conclusions arise from a strict delimitation of the concept of sexual behavior and from a strict definition of the sort of work that counts as ethology. But there are at least two examples in the recent literature that take a broader perspective and provide interesting ethological hypotheses about the origins of sex-related behaviors. I will discuss each of these briefly.

2. *The Work of Wickler and Beach*

Based on observations of several vertebrate species, Wolfgang Wickler proposed that any behaviors used in pair-bond maintenance initially functioned as brood-tending or reproductive behaviors. We might expect to find elements in our own pair-bond behavior which, though ritualized (Blest, 1961), are identifiable as elements of child-tending or infantile behavior. For example, Wickler shows that in several unrelated preliterate societies and in the rural areas of some modern societies, there is direct mouth-to-mouth passage of premasticated food from mother or grandmother to infant. These cases suggest the functional precursor of the mouth-to-mouth kiss. As an intermediate stage he cites the case of ritual feeding among the Ituri Pygmies of Central Africa. A man who has killed a large animal distributes some of the meat by slicing and placing it in his mouth and transferring it directly to the mouth of the man next to him. Still further removed from the infant-feeding function is the passage of chewing tobacco between the mouths of heterosexual friends in certain Central European valleys. Ac-

* One particularly astute critic asked why I made this point about morality. The answer is that morality is one determinant of the limits of inquiry in all science, behavioral and social science in particular. There are certain observations that will not be made because they ought not to be made without the free and informed consent of the observed. Yet we have two good reasons to believe that obtaining consent will markedly decrease the generality of the resulting data. First, persons who readily give consent are likely to be special members of the population in ways relevant to the behavior being observed. And second, persons who give consent only after reflection may behave atypically when they know they are being observed.

ceptance by the woman of tobacco from the man was a sign of her romantic interest in him. Finally, there is the kiss during which partners put their tongues in each other's mouths. In summary, Wickler concludes that mouth kissing "is a ritualized brood-tending action, and that at least in some of its forms it goes back to the mouth-to-mouth feeding of small children by their mother" (Wickler, 1972, p. 245).

Wickler makes a related point in discussing the social significance of the female breast. He argues that the fondling and sucking of the breast during adult sexual behavior is a ritualized form of infantile suckling which produces in women very pleasant sensations in both circumstances.

The validity of these alleged relationships depends, finally, on historical information that is unobtainable. Wickler bolsters their plausibility by reference to a collection of similar behavioral circumstances in other species. He does not claim that the behavior observed in humans is homologous with, or produced by the same physiological substrate as, the behavior observed in other species. His claim is for analogous relations based on the need in all of the species to establish and maintain pair bonding and complex social structure. This emphasis on argument by analogy is also found in the Nobel Prize acceptance speech of Lorenz, in which he stressed the strength and utility of analogical arguments in biology (Lorenz, 1974).

The second example of ethological applications to human sexuality was published in 1974 by Frank Beach entitled "Human Sexuality and Evolution." In this essay Beach uses the term "sexuality" to refer to characteristics of masculinity and femininity, as distinct from the physiological characteristics of maleness and femaleness. Sexuality has two major components, which are gender roles and gender identity. Gender roles are the culturally prescribed appropriate behaviors for boys and men, girls and women. As Beach points out, two separable sets of gender roles are prescribed by every human society. Gender identity, on the other hand, is the gender ascribed by the self, e.g., "I am a woman" or "I am a man." The primary hypothesis with which Beach is working is that sexuality is a product of the evolution of our species. But he also emphasizes the gap between human and nonhuman sexuality when he says:

> It is my present feeling that human sexuality is about as closely related to the mating behaviors of other species as human language is related to animal communication, a relationship that is distant indeed. (Beach, 1974, p. 334.)

Beach asserts that "without *sexuality* society in its human form could neither survive in the future nor have evolved in the past" (p. 349). Both the qualitative sex differences involved with primary reproductive function and the quantitative differences associated with male androgenization were exploited by our ancestors as they moved into the hunting and gathering phase of our species' existence. The key is the emergence of the nuclear family: Why did it emerge? Beach suggests that the major causes involved economic interdependence and sharing between the hunting men and the gathering women. He deemphasizes the role of guaranteed continuous male sexual satisfaction as a cause of the emergence and maintenance of the nuclear family. Others have argued that the high rate of copulation observed in our species forms the foundation for the emergence and maintenance of close monogamous pair bonds. In other words, men stick around because of the availability of sex. But Beach argues that the high rate of copulation is a reproductively necessary consequence of the loose relation between ovulation and receptivity. Because receptivity is not indicative of reproductive readiness, copulation must be frequent enough to insure an adequately high probability of fertile matings. This need in turn was served by the emergence of female orgasm, unique in our species. Only in *Homo sapiens* is orgasmic satisfaction available to both participants of a heterosexual pair. It was not intense sexual gratification that kept pairs together, but rather economic interdependence.

Although I have only skimmed the issues that Beach raises and have not elaborated on the evidence he provides, I have shown that he presents ethological arguments expanded by reference to physical and cultural anthropology. An earlier section of his paper is devoted to the ontogeny of human sex and sexuality. Here again, Beach emphasizes the emergence of biologically unique features while describing the interrelatedness of biophysical and social psychological factors in the determination of human conduct. This is the proper goal, for the capacity and need to be influenced by culture is our unique species-specific heritage.

3. *Umwelt, Symbols, Frames, and Childhood*

I would like now to expand the ethological frame of reference again. Two additional characteristics of traditional ethological work are the description of the appropriate *Umwelt* for a species and the analysis of the effects of infantile experience on adult behavior.

The concept of *Umwelt*, or perceptual world of a species, was elaborated in the 1920s by von Uexküll (1921). This idea has not been effectively extrapolated to our own species by biologists. Efforts in this direction have tended to be conjectures about, for example, the existence of human olfactory pheromones; available data are scanty. On the other hand, scholars informed by phenomenology have been alive to the significance of human symbolic interactions for the creation of "reality," that is, the human *Umwelt*. Kenneth Burke put the matter dramatically:

> And however important to us is the tiny sliver of reality each of us has experienced firsthand, the whole overall "picture" is but a construct of our symbolic systems. To meditate on this fact until one sees its full implications is much like peering over the edge of things into an ulti-

> mate abyss. And doubtless that's one reason why, though man is typically the symbol using animal, he clings to a kind of naive verbal realism that refuses to realize the full extent of the role played by symbolicity in his notions of reality. (Burke, 1966, p. 5.)

Erving Goffman bases his analysis of everyday experience on a distinction equivalent to that made between *Umwelt* and environment:

> My aim is to try to isolate some of the basic frameworks available in our society for making sense out of events and to analyze the special vulnerabilities to which these frames of reference are subject . . . I am not addressing the structure of social life [social environment] but the structure of experience individuals have at any moment of their social lives [Umwelt]. (Goffman, 1974, pps. 10, 13).

The importance of symbolicity and subjective frames of reference is no less in our sexual interactions than in the rest of our social lives. Hence, I believe, an account of the realities of human sexual behavior must analyze the symbolic or representational features of our sexual encounters. But this does not place human sexual conduct beyond the reach of ethological analysis, for much of the ethology of nonhuman species has traditionally dealt with the symbolic or communicative functions of behavior. In fact, it has been the elucidation of the affective significance of species-specific gestures, postures, and vocalizations that has been one of ethology's major contributions to biology. The complex forms and meanings of our symbols make the human ethologist's job enormously more difficult. But the work still belongs within the framework of a properly conceived human biology.

Ethology has long been interested in the effects of early experience on adult behavior. In his classical paper "Der Kumpan in der Umwelt des Vogels" (1935), Lorenz provided a theoretical account of the relation between the imprinting process and adult sexual behavior. Lorenz was already sensitive to the possible analogy between imprinting in nidifugous birds and the development of human sexuality. He says:

> Cases are known from human psychopathology in which irreversible fixation of the object of specific instinctive behavior patterns has been observed and which, on purely symptomatic grounds, exhibit a picture entirely similar to that found in birds in which object-imprinting has not taken place in the species-typical manner.

4. *Ethology and Psychoanalysis*

This brief discussion of the *Umwelt* and imprinting sets the stage for a basic point: If an ethology of human sexuality meets the traditional commitments of ethological analysis and is in addition concerned with the human *Umwelt* of symbols and the effects of early experience on adult behavior, then that ethology will be compatible with or similar to much of the program Freud envisioned and pioneered in psychoanalysis. The most promising basis for an ethological theory of human sexuality is psychoanalytic theory; with all its inconsistencies and epistemological embarrassments, psychoanalysis nevertheless provides the richest matrix within which to work. The work of Freud and other major analysts would have to be rediscovered if human sexual ethology were to be built, independently, from the ground up.

The psychoanalytic program, as theory, meets the ethological commitments as already described. First, it is clear that Freud *attempted* to provide generalizations of species-wide significance. Of course, his work has been criticized for ignoring the impact of cultural differences on social–sexual development. But when that criticism is constructively applied, as for example by Malinowski (1927) in arguing for a generalization of the Oedipus complex into the nuclear family complex, then the result is an enlargement of our understanding produced by a constructive cumulation of information and theory.

Second, while psychoanalysis does not emphasize the relation between the stimuli and evolutionary significance of consummatory responses, there is not much in human sexual behavior that appears to demand just this analysis.

Third, while psychoanalytic methods have been more geared to therapy than to theory, nevertheless the psychoanalytic technique is a relatively unobtrusive method for generating large amounts of information about the cognitive and affective structures of members of our species as they are revealed in our species-specific symbol-generating behavior. Freud conceived of the analyst, in his unique position behind the patient, as representing "a neutral sample of all humanity in the patient's emotional life" (Kubie, 1950, p. 64). The "analytic incognito" practiced by the analyst is the human-to-human equivalent of the ornithologist hiding in a duck blind.*

Fourth, it is clear that much of psychoanalytic theory is geared to elucidating the human symbolic *Umwelt*, particularly in regard to the affective significance of symbols. While one may argue about any particular interpretation, the fact that interpretations are required seems indubitable and an important contribution of psychoanalytic theory and technique.

* None of my readers liked this phrase. I left it in because I still believe it. Expertly guided free association is still the most effective technique for discovering how individuals have structured the past symbolically to deal with its influence on their present lives. Competent psychoanalysts are more effective than anyone else in using this technique. Hence, for all its obtrusiveness, and for all the other problems involved in understanding the analyst–analysand relationship, the psychoanalytic method remains the best way known to probe the details of the individual symbolic *Umwelt*.

Fifth, it is equally obvious that psychoanalysis pioneered the investigation of early childhood experience on adult sexual and other emotional behaviors. The first analytic success, the resolution of conversion hysteria through the bringing to consciousness of repressed childhood sexual fantasies, represents a good example of the psychosomatic impact of human symbolic activity. Moreover, since the time of Freud, the idea of infantile and prepubescent sexuality has become widely recognized, to the extent that one may overlook the resistance that met its introduction.

These views contrast with the admonition that psychiatrists should look to ethology for a greater understanding of their own field (e.g., White, 1974). This is a two-way street; those who would practice human ethology should become aware of what has preceded them, psychoanalysis in particular. The assumptions, goals, and insights of analysis, if properly comprehended and utilized by modern ethologists of human sexual behavior, will save time and effort and will speed the development of knowledge and understanding.

5. *Psychology, Morality, and the Idea of Perversion*

The claim being made here is only as strong as one's ability to deliver on it in specific terms. This section of the chapter provides an example from the recent psychoanalytic literature that, I believe, has the heuristic value, the capacity to provide initial hypotheses, that is required for the more empirical or systematic work to which we have been committed. Robert Stoller's arguments in his book *Perversion* (1975) are illuminating, provocative, sometimes facile and uncritical on physiological matters, but surely indicative of the degree of theoretical richness required to account for the reality of individual sexual conduct.*

The first step is to set the background against which the main argument develops. Psychoanalytic accounts of sexual development and behavior emphasize erotically based internal conflicts and their resolution. By internal is meant both cognitive and affective mental representation. Very strong pleasant and unpleasant feelings early in life, and their interpretation by the child within its limited cognitive capacities, constitute the puzzles and partial or total solutions of early sexual development. Adult sexual behavior represents the individual's resolution of the erotic puzzles and threats of infancy and childhood. For Freud and his followers, the successful resolution of erotic sensations generated within a child by maternal contact is the single most important event in the ontogeny of normative

* For another view of Stoller's book, see Money (1976).

sexuality and morality. From the outset, sexual and moral development are closely linked, and this linkage has important consequences in adult life.

It is important to emphasize the relationship between sexuality and morality. Arguments have been made that we need to free ourselves of "moralistic" blinders in order to understand the realities of human sexual behavior. It is often asserted that all forms of sexual gratification are normal variants with different relative frequencies in the human population. So far as it goes, the argument is sound enough. But it really only suggests that we should not blame people for how they achieve sexual gratification, with the usual proviso that the partners in the act, if any, have consented. Perhaps we should not. But neither should we confuse the psychology of sexual morality, particularly its developmental psychology, with the ethical theory of blameworthiness. This is, unfortunately, an easy mistake to make, and it always carries the danger of further confusion between descriptive and normative categories. The danger is profound when dealing with emotionally loaded words like "perversion." Care is required to prevent category errors from blurring the focus of clear argument.

A distinction needs to be maintained between an *ethical theory* of sexual conduct and a *psychological theory* of sexual conduct. The ethical theory includes a description and justification of appropriate sexual practices and partners. It is more-or-less explicit and detailed, formal or informal. It may be part of a larger scheme, for example religious belief, or independent and specific. In any case, it is what the social milieu provides to an individual by way of sanctions, proscriptions, and rationales in sexual matters.

Analysis of the ethical theory is in the form of ethical questions: How is the theory grounded in a more general view of human rights and responsibilities? What calculus is applied in the solution of particular problems? Is the theory consistent? Is it fair? Does application of the theory lead to human happiness? If not, are values other than happiness maintained in those instances? And so on. The theory is normative, and its analysis proceeds on a normative plane. Moral discourse is only partially indebted to or dependent on psychological facts or truths. For example, when analyzing the moral status of some sexual act (e.g., the insertor role in intercourse with a member of a nonhuman species), the incidence or frequency of the act in the population is a datum of no necessary moral force. However, moral theories can be faulted if they demand people to perform acts, including restraints, that they cannot perform by virtue of their human nature.

The psychological theory of sexual conduct, on the other hand, provides descriptions of sexual development and behavior and relates individual variations in these to correlated variations in various biological and social determinants. Analysis of the psychological theory follows the form of scientific analysis generally: Does the theory account for the observed variation? Is it predictive? Does it have clinical significance? Is it more parsimonious, elegant, and useful than its competitors? This is scientific discourse, and hence is only partially dependent on moral considerations. For example, in accounting for the incidence and frequency of some act (e.g., the insertor role in intercourse with members of nonhuman species), the status of that act as "immoral" within the community of interest is to be treated as a datum like any other in the scientific analysis.

The concept of "perversion" can be significant in both ethical and psychological theories of sexuality. In an ethical theory a perversion will be that which has, literally, become misdirected or gone wrong. In the everyday ethical theories that inform the sexual conduct of most Americans, for example, the traditional paradigm of sexual behavior is a married couple having face-to-face intercourse with accompanying desire for conception. Of course, we have markedly broadened the range of acceptable behavior in recent years, particularly in regard to contraceptive behavior, so only a very small proportion of the population would still maintain that contracepted sex is "perverse." Nevertheless, as the behavior in question moves, along any dimension, from the paradigm case, the likelihood that *someone* will find it "perverse" increases.

Professional philosophers have recently attempted to analyze the concept of perversion (Nagel, 1969; Slote, 1975; Solomon, 1975). Their goal is to provide a coherent account of what, if anything, we ought to label as sexually perverse. Not surprisingly, they do not agree. For Nagel (1969), perversion begins with deviation from an ideal form of sexual attraction in which sexual desire is enhanced in each member of a pair by awareness of the desire of the other. A necessary condition for perversion is the independence of one's desire or behavior from a need for awareness of the reciprocal desire of the object of sexual gratification. In Solomon's view, sexual perversion follows from the significance of sexual behavior as a medium of communication:

> [Perversion] does not involve special deviation in sexual activity, peculiar parts of the body, special techniques, or personality quirks. As a language, sex has at least one possible perversion: the nonverbal equivalent of lying, or *insincerity*. And, as an art, sex has a possible perversion in *vulgarity* (Solomon, 1975, p. 286).

And finally, Slote (1975) believes a moral concept of perversion to be inapplicable to human sexual behavior because of the impossible precondition of determining what "natural" sexual behavior is.

The differences among these three views merit serious consideration by all students of human sexuality, whether scientists or philosophers. For present purposes, they exemplify the normative analysis of human sexuality, wherein questions of sexual worth, value, and morality are appropriately considered.

A psychological theory of sexual behavior, on the other hand, is not the place to consider questions of sexual naturalness or perversion, *except insofar as these matters are influential in determining the sexual behavior of the subjects of psychological investigation*. The psychological theory must account for these notions as determinants of conduct and how they structure the sexual *Umwelten* of individuals. Because sexuality and morality develop, as it were, hand-in-hand during childhood and adolescence and even into adulthood, developing ethical theories are bound to be determinative of sexual attitudes and behavior, and vice versa. The general psychological task of description and analysis can proceed without making moral pronouncements about what perversion is; but it can go nowhere at all without saying what "perversion" is, and how it is coped with, within the developing ethical theories of the subjects of psychological investigation. It is to this end that psychoanalytic techniques can be most useful, for as mentioned already they are uniquely suited to working through the kind of memorial material that reflects, however imperfectly, the organization of moral sentiments within an individual. It is in this latter sense of the concept of perversion that I continue the discussion of Stoller's work.

5.1. Stoller's Argument

Stoller describes the moral aspect of sexual conduct in terms of the individual's acceptance of responsibility for his or her own sexuality. He is critical of current trends in sex research that ignore the possibility that people do in fact hold themselves accountable for their forms of sexual gratification:

> Erotic deviance is as specifically human as are murder, humor, fantasy, competitive sports, art, or cooking. This observation is so grossly manifest that one wonders why it has no force in the speculations of modern sex researchers. *Almost every notable study on human sexual behavior since Freud has tried to prove that a person does not create his own deviance* but rather it is thrust on him. . . . The new research . . . seems aimed, at tearing down the conflict theory; no other aspect of Freud's system has created such resistance, perhaps because Freud believed perversion is motivated, that is, a person somehow, in his depths, feels in part responsible for his perversion. (p. 36, italics in original.)

One need not sympathize entirely with the polemics of Stoller's criticism in order to agree with him that sexual behavior is *meaningful* to people, and that explanations of the behavior ought to account for its meaning to those people. The concept of perverse sexual behavior has to do with what the behavior means to the person who is doing it.

In important ways, therefore, categories of sexual meaning will cross-classify physical descriptions of sexual behavior. Analysis of human sexual behavior needs to include both the psychic and social contexts in which the behavior occurs.

This cross-classification takes interesting forms in Stoller's treatment of perversion. For him, perversion is

> a fantasy, usually acted out but occasionally restricted to a daydream (either self-produced or packaged by others, that is, pornography). It is a habitual, preferred aberration necessary for one's full satisfaction, primarily motivated by hostility. By "hostility" I mean a state in which one wishes to harm an object. . . . The hostility in perversion takes form in a fantasy of revenge hidden in the actions that make up the perversion and serves to convert childhood trauma to adult triumph. To create the greatest excitement, perversion must also portray itself as an act of risk-taking. (p. 4)

Stoller analyzes various forms of sexual behavior in the light of this definition of perversion: heterosexual intercourse, masturbation with and without recourse to pornography, exhibitionism, voyeurism, homosexual behavior, fetishism, transvestism, transexualism, both sides of the prostitution transaction, bestiality, frankly sadistic and masochistic sexual behavior, and rape. Stoller's analyses produce unintuitive results that remain provocative yet plausible. For example, he argues that masturbation, which among sexual dysfunction therapists and some feminists has achieved normative, almost hallowed, status, is sometimes perverse, in particular when it is not performed *faute de mieux* and depends upon pornographic material for sustaining a powerful setting for orgasm. Of course, the sexual dysfunction therapists, feminists, and Stoller may all be correct in their separate views of masturbation. Presumably the orgasmic context Stoller describes as perverse is not what masturbation's proponents have in mind.

Through an analysis of transvestite pronography, Stoller demonstrates the relativity of objective correlates of prurience: material that will lead the transvestite to masturbation may bore or puzzle the general public. But a careful consideration of the material reveals the themes definitive of perversion: hostility, risk taking, and childhood trauma repetitively reenacted as adult triumph. Stoller sees in each pornographic genre "the communicated sexual fantasy of a dynamically related group of people" (p. 115). While the hostile content may sometimes be obvious, particularly in pictorial sado-masochistic pornography, it is not always so. The key to the most general conception of pornography lies in the relationship between its manifest content and the dynamic brought to it by its user. To my knowledge, only psychoanalytic theory is rich enough to provide substantive interpretations of this sort. Other approaches to pornography, for example the social script model of sexual conduct suggested by Gagnon and Simon (1973), seem to me inadequate to deal with the differences in various pornographic formats except by *post hoc* reasoning.

Time does not permit a full exposition of Stoller's analyses of the various forms of sexual gratification. But I do want to mention a theme that recurs in his book and is of great interest to us here, namely, the extent of difference between the sexes in the expression of various forms of sexual behavior including perversion. Ford and Beach (1951) noted that there were pronounced sex differences in the incidence and frequency of masturbation, homosexual behavior, and other noncopulatory forms of sexual gratification. Across virtually all the cultures for which information was available, men engaged in a wider variety of forms of sexual gratification, more often, than women did. Ford and Beach also noted that the same sex difference in reproductively sterile sexual behaviors exists in nonhuman species. They reviewed several plausible interpretations of this class-wide invariance but remained cautious about general conclusions. Less cautious than they, twenty years later I elaborated on one of their ideas in an argument against a well-known psychoanalytic theory of homosexual development (Bermant, 1972). I argued then that the greater accessibility of the penis than the clitoris to accidental prepubertal stimulation was the invariant mammalian feature that might account for greater diversity of sexual stimuli for males in adulthood. There are some rather obvious tests of this idea in both anthropological and comparative animal settings, but they are as yet unpursued.

I was therefore particularly interested in Stoller's account of the psychodynamics of sex differences in normative and perverse sexual development. I will only mention here that castration fears, penis envy, and masculine protests, i.e., the analytic constructs whose plausibility is very unintuitive, figure in the theory. I find Stoller's account strained, but now, unlike before, I believe it is on the right track. We very much need to know more about what sexuality means to children and how those meanings are differentially acquired. We are going to have to acquire this information at a time of rapid social change, and we are going to have to interpret all our conclusions in light of the politics of sex and sex-role differences.

5.2. *Psychoanalysis and Symbolic Interactions*

The notion of *sexual script* (Gagnon and Simon, 1973) is a useful addition to our thinking about sexual behavior. Sexual scripts are the more-or-less routinized or normative procedures which guide individuals through various forms of sexual encounter within a society. There are two dimensions to these procedures, the interpersonal and the intrapersonal. The interpersonal dimension is the "organization of mutually shared conventions that allows two or more actors to participate in a complex act involving mutual dependence" (1973, p. 20), while the intrapersonal dimension contains the "motivational elements that produce arousal or at least a commitment to the activity (p. 20). The emphasis in this approach is on the external script and its power to guide conduct.

Gagnon and Simon offer a social environmentalist approach as an alternative to the psychoanalytic position. They criticize the nativism in the analytic account for overemphasizing the importance of biological sexual drive. They also minimize the causal relevance of childhood sexual experiences for adult sexual practice. Aware of the significance of sexual activity for the expression of nonsexual meanings, they emphasize symbolic activity in contrast to the reverse symbolization so basic to Freudian theory. In these and other ways, they present their position as an explicit foil to the psychoanalytic perspective. However, the two positions are so different that they are often talking about different things. When they do converge, it is possible for us to make some preliminary estimates of their relative explanatory and heuristic values. As mentioned earlier regarding pornography, the psychoanalytic theory of Stoller seems the fuller and more useful. I believe the same is true for serious psychosexual conditions like transsexualism.* On the other hand, certain relatively minor aspects of sexual conduct, for example the control of visual access to our bodies by postures and forms of clothing, seem more explainable through the notion of social scripts (Levin, 1975). Thus, it is difficult to account in psychoanalytic terms, but not in terms of social scripts, for the tendency of women to hide their underwear from the sight of the investigating gynecologist (Henslin and Biggs, 1971).

The proper approach to human sexuality includes a synthesis of these two approaches rather than a forced choice between them. Our task is to understand human sexuality from the inside out as well as from the outside in. We are all inside the human sex circus, some in the center ring, some in the side show, some pretending to be only in the audience, but all inside nevertheless. Every sexual act, like every other act, is a *personal statement*, and our theory of sexual acts should help us clarify the kinds of personal statements we are making. Each of us brings unique complexity to sexual expression; something like psychoanalytic theory and method are required to parse the personal statements each of us makes. But we take our sexual instruction from the social environment, and something like symbolic interactionist theory is required to parse the lessons the environment provides. I will close this chapter with a very tentative exploration of a current sexual issue that exemplifies the interactions of psychodynamics with larger social forces.

* Stoller argues that some cases of transsexualism arise without any history of sexual trauma and are not, therefore, neurotic or perverse.

6. Conclusion: The Public Expression of Sexual Fantasies

Sexual descriptions in literature, explicit sexual films, and similar erotic artifacts are concrete realizations of fantasy. They can serve a number of functions, but the most obvious is educative: they show to other people what the possibilities for fantasy are. To the extent that they become popular, they also make a claim for legitimacy, if only because we tend to legitimate the familiar. Different segments within our society will be affected by different forms of expression. The sources for our sexual scripts differ in this regard as in others. We are dependent to varying degrees on the information supplied by particularly influential sources of social instruction in the mass media. There is no reason *a priori* to believe that the sexual scripts developed and legitimated through all these channels will be the same. Yet I believe there is a theme that can now be seen as emerging at several levels of concretized sexual fantasy, from National Book Award novels to the most obvious pornography.

I offer the following thesis: It is possible to trace an evolution in popular consciousness during the past several years that has partially legitimated progressively greater degrees of perverse sexual expression, e.g., sexual expression with covert or overt hostile content.

If one looks at the sexual themes prevalent in novels coming from our most highly regarded authors, the evolution is demonstrable. During the 1960s the avant-garde issues were adultery or promiscuity and the artistic legitimation of masturbation and oral–genital sex. Roth's *Portnoy's Complaint* (1969) and Updike's *Couples* (1968) gained their sexual punch through humorous or graceful accounts of these activities. In part their success as socializers is measurable by the critical as well as popular esteem the novels achieved.

If Updike and Roth represent the high road to sexual socialization, then commercial pictorial pornography represents the low road, and here there is little doubt that oral–genital sex was the first variant to be stressed. And the middle road, I believe, is to be found in the views of sex manual authors. Here again, telling people oral sex and masturbation are really "OK things to do" was the first major task of the new wave of sex behavior specialists that swelled in the 1960s.

While I am not maintaining that oral sex is perverse, I am claiming that its depiction in pictorial pornography tends to emphasize the power of men over women and in that sense has a definite hostile element. The perversity is relatively covert but present nevertheless.

If oral–genital sex has achieved an increasing measure of social acceptability as a topic of public fantasy, and therefore as an element in common sexual scripts, anal sex and concern with anal function generally are not so prevalent. Yet one can find in novels of the National Book Award level an increased concern with these issues. In *Portnoy's Complaint*, there is a scene clearly intended to display "kinkiness," in which a man lies under a glass table while a woman defecates on it and a second woman watches. And in Mailer's *American Dream* (1964) oral–anal sex is an emotionally explosive issue. There is also exposition of anal sex in commercial pornography, but the frequency is substantially lower than for oral sex. And again, among serious nonfictional commentators, there is a tendency to view the anus and its product as the next taboo to be conquered. Arno Karlen, in his *Sexuality and Homosexuality* (1971) suggests facetiously that the only thing left to do on stage, given that nudity and sex have become commonplace, is for the players to turn to the audience and defecate. Also in this genre is the banal treatment of scatological sexual behavior appearing in the mass-circulation magazine *Hustler*.

One does not have to be a card-carrying Freudian to agree that the emotional ambience of scatological sexuality is sufficiently great as to warrant concern for its possible hostile content. I am not saying that all forms of anal sexual gratification are perverse, but I do believe that anally oriented sexual behavior can easily carry very hostile messages.

The next stage in this evolution has been, I believe, the increased portrayal of frank sadistic and masochistic sexuality. Not surprisingly, some of this has been tied to anal sexuality, as in the example from *Portnoy's Complaint* already described and, much more explicitly, in an extraordinary scene in Pynchon's *Gravity's Rainbow* (1973). Karlen (1971) has noted that sado–masochistic sex was a not uncommon theme in the New York underground cinema of the 1960s, but now, of course, it figures as the theme of nationally distributed pornographic films. I believe the newspaper advertising of these films can be shown to reflect increasingly hostile content. Obviously, by this point no symbolic interpretations are required: the hostility is patent.

Patent, but not complete. There is another stage left, and we can see some examples of it. While I can think of no fictional representation equivalent to those mentioned above, I can refer to a well-known book we might all wish were fiction, namely Bugliosi's *Helter Skelter* (1974), the account of the Manson Family and its murders. There is reference there to the filming of sexual scenes culminated by murder and dismemberment. There is some reason to believe that these bizarre events were not isolated to the Manson craziness. A report in the *Crime Control Digest* for October 6, 1976, cites both New York City Police and FBI sources in regard to the increased circulation of what are being called "Snuff" or "Slasher" films in which a woman is killed by stabbing after participating in various sexual activities. The authenticity of the murders has not been verified, and there are informal reports that the slayings were simulated. But the intent to appeal to violent sexual fantasy is the same. These films are in either case devoid of artistic merit and distributed for the commercial possibilities inherent in a taste for sexual violence.

This is a grim account, and it does not please me to close my paper on the sour note it strikes. I do not present this material to alarm or shock. There is enough sensitivity to these issues in the general community that I feel no need to take up the cudgel against all forms of pornography because of the turn to hostility and violence. It would be an obvious mistake to read the direction of causality from violence in sexual pornography to increased violence in the society in general. Indeed it may be that the pornographers need to move toward hostility and violence in order to match the competition in the so-called nonpornographic marketplaces of R-rated movies and some television programs. My speculations on this issue would be no better informed than yours.

My intent has been to exemplify an approach to human sexuality in full social context that finds its sources of instruction in both psychodynamic and sociological frames of reference. To reduce the scope of theoretical inquiry will be to do injustice to the goal of ethology, which is to understand the behavior of organisms in their actual environments. We live inside ourselves as well as in society, and I believe our ethological theories must reflect these joint realities.

ACKNOWLEDGMENTS

A number of colleagues offered comments to help me revise this manuscript after I presented a version in Berkeley. Rebecca Field was particularly generous with her time, ethological knowledge, and editorial skill. I also appreciate the comments of Frank Beach, Julian Davidson, Don Dewsbury, Steve Glickman, Tom McGill, Ben Sachs, and Robert Stoller.

References

Beach, F. A. Human sexuality and evolution. In W. Montagna (Ed.), *Reproductive behavior.* New York: Plenum Press, 1974. pp. 333–365.

Bermant, G. Behavior therapy approaches to modification of sexual preferences: Biological perspective and critique. In J. Bardwick, (Ed.), *Readings in the psychology of women.* New York: Harper and Row, 1972.

Blest, A. D. The concept of ritualization. In W. H. Thorpe and O. L. Zangwill (Eds.), *Current problems in animal behavior.* Cambridge, England: Cambridge University Press, 1951.

Bugliosi, V., and Gentry, C. *Helter skelter: The true story of the Manson murders.* New York: Norton, 1974.

Burke, K. *Language as symbolic action.* Berkeley: University of California Press, 1966.

Eibl-Eibesfeldt, I. *Ethology,* 2nd. ed. New York: Holt, Rinehart, and Winston, 1975.

Ford, C., and Beach, F. A. *Patterns of sexual behavior*. New York: Harper, 1951.
Gagnon, J. H., and Simon, W. *Sexual conduct*. Chicago: Aldine, 1973.
Goffman, E. *Frame analysis*. Cambridge, Massachusetts: Harvard University Press, 1974.
Henslin, J. M., and Biggs, M. A. Dramaturgical desexualization: The sociology of the vaginal examination. In J. M. Henslin, (Ed.), *Studies in the sociology of sex*. New York: Appleton-Century-Crofts, 1971.
Hirsch, J. *Behavior-genetic analysis*. New York: McGraw-Hill, 1967.
Karlen, A. *Sexuality and homosexuality*. New York: Norton, 1971.
Kubie, L. *Practical and theoretical aspects of psychoanalysis*. New York: International Universities Press, 1950.
Levin, R. J. Facets of female behavior supporting the social script model of human sexuality. *Journal of Sex Research*, 1975, *11*, 348–352.
Lorenz, K. Der Kumpan in der Umwelt des Vogels. [Translated as "Companions as factors in the bird's environment"], in K. Lorenz, *Studies in animal and human behavior*. Cambridge, Massachusetts: Harvard University Press, 1970.
Lorenz, K. Analogy as a source of knowledge. *Science*, 1974, *185*, 229–234.
Mailer, N. *American dream*. New York: Dial, 1964.
Mailinowski, B. *Sex and repression in savage society*. Cleveland: World Publishing Co., 1927.
Masters, W. H., and Johnson, V. E. *Human sexual response*. Boston: Little, Brown, and Co., 1966.
Mayr, E. *Animal species and evolution*. Cambridge, Massachusetts: Harvard University Press, 1963.
Money, J. Perversion or paraphilia. A review of R. Stoller, *Perversion. Contemporary Psychology*, 1976, *21*, 528–529.
Nagel, T. Sexual perversion. *The Journal of Philosophy*, 1969, *66*, 5–17.
Pynchon, T. *Gravity's rainbow*. New York: Viking, 1973.
Slote, M. Inapplicable concepts and sexual perversion. In R. Baker and F. Elliston (Eds.), *Philosophy and sex*. Buffalo, New York: Prometheus Books, 1975.
Solomon, R. Sex and perversion. In R. Baker and F. Elliston (Eds.), *Philosophy and sex*. Buffalo, New York: Prometheus Books, 1975.
Stoller, R. *Perversion*. New York: Pantheon, 1975.
von Uexküll, J. *Umwelt und Innenwelt der Tiere*. Berlin: Springer, 1921.
Unger, R. M. *Law in modern society*. New York: Free Press, 1976.
Updike, J. *Couples*. New York: Knopf, 1968.
White, N. *Ethology and psychiatry*. Toronto: University of Toronto Press, 1974.
Wickler, W. *The sexual code*. New York: Doubleday, 1972.

Appendix I

Bibliography of Frank A. Beach

1. 1937 The neural basis of innate behavior. I. Effects of cortical lesions upon the maternal behavior pattern in the rat. *Journal of Comparative Psychology, 24* (3): 393–436.
2. 1938 A new method for marking small laboratory animals. *Science, 87:* 420.
3. 1938 The neural basis of innate behavior. II. Relative effects of partial decortication in adulthood and infancy upon the maternal behavior of the primiparous rat. *Journal of Genetic Psychology, 53:* 109–147.
4. 1938 Techniques useful in studying the sex behavior of the rat. *Journal of Comparative Psychology, 26* (2): 353–359.
5. 1938 Sex reversals in the mating pattern of the rat. *Journal of Genetic Psychology, 53:* 329–334.
6. 1939 Maternal behavior of the pouchless marsupial *Marmosa cinera. Journal of Mammalogy, 20* (3): 315–322.
7. 1939 The neural basis of innate behavior. III. Comparison of learning ability and instinctive behavior in the rat. *Journal of Comparative Psychology, 28* (2): 225–262.
8. 1940 Effects of cortical lesions upon the copulatory behavior of male rats. *Journal of Comparative Psychology, 29* (2): 193–244.
9. 1941 Effects of brain lesions upon running activity in the male rat. *Journal of Comparative Psychology, 31* (1): 145–178.
10. 1941 Effects of lesions to corpus striatus upon spontaneous activity in the male rat. *Journal of Neurophysiology, 4:* 191–195.
11. 1941 Apparatus for inflicting subcortical lesions in the rat brain. *Science, 93:* 383–384.
12. 1941 Copulatory behavior of male rats raised in isolation and subjected to partial decortication prior to the acquisition of sexual experience. *Journal of Comparative Psychology, 31* (3): 457–470.
13. 1941 Female mating behavior shown by male rats after administration of testosterone propionate. *Endocrinology, 29* (3): 409–412.
14. 1941 Instinct and intelligence. *Transactions of the New York Academy of Science, Series II, 4* (1): 32–36.
15. 1942 Analysis of the stimuli adequate to elicit mating behavior in the sexually inexperienced male rat. *Journal of Comparative Psychology, 33* (2): 163–207.

16. 1942 Comparison of copulatory behavior of male rats raised in isolation, cohabitation, and segregation. *Journal of Genetic Psychology, 60:* 121–136.
17. 1942 Effects of testosterone propionate upon the copulatory behavior of sexually inexperienced male rats. *Journal of Comparative Psychology, 33* (2): 227–247.
18. 1942 Execution of the complete masculine copulatory pattern by sexually receptive female rats. *Journal of Genetic Psychology, 60:* 137–142.
19. 1942 Coauthor: Priscilla Rasquin. Masculine copulatory behavior in intact and castrated female rats. *Endocrinology, 31* (4): 393–409.
20. 1942 Sexual behavior of prepuberal male and female rats treated with gonadal hormones. *Journal of Comparative Psychology, 34* (3): 285–292.
21. 1942 Copulatory behavior in prepuberally castrated male rats and its modification by estrogen administration. *Endocrinology, 31* (6): 679–683.
22. 1942 Central nervous mechanisms involved in the reproductive behavior of vertebrates. *Psychological Bulletin 39,* (4): 200–226.
23. 1942 Analysis of factors involved in the arousal, maintenance and manifestation of sexual excitement in male animals. *Psychosomatic Medicine, 4* (2): 173–198.
24. 1942 Male and female mating behavior in prepuberally castrated female rats treated with androgens. *Endocrinology, 31* (6): 673–678.
25. 1942 Importance of progesterone to induction of sexual receptivity in spayed female rats. *Proceedings of the Society for Experimental Biology and Medicine, 51:* 369–371.
26. 1943 Coauthor: Thelma Weaver. Noise-induced seizures in the rat and their modification by cerebral injury. *Journal of Comparative Neurology, 79* (3): 379–392.
27. 1943 Effects of injury to the cerebral cortex upon the display of masculine and feminine mating behavior by female rats. *Journal of Comparative Psychology, 36* (3): 169–198.
28. 1943 Interindividual behavior among animals. *Transactions of the New York Academy of Science, Series II, 6* (1): 14–18.
29. 1944 Effects of injury to the cerebral cortex upon sexually-receptive behavior in the female rat. *Psychosomatic Medicine, 6* (1): 40–55.
30. 1944 Relative effects of androgen upon the mating behavior of male rats subjected to forebrain injury or castration. *Journal of Experimental Zoology, 97* (3): 249–295.
31. 1944 Experimental studies of sexual behavior in male mammals. *Journal of Clinical Endocrinology, 4* (3): 126–134.
32. 1944 Responses of captive alligators to auditory stimulation. *American Naturalist, 78:* 481–505.
33. 1945 Angry mosquitos. *Science, 101:* 610.
34. 1945 Coauthor: A. Zitrin. Induction of mating activity in male cats. *Annals of the New York Academy of Science, 46:* 42–44.
35. 1945 Bisexual mating behavior in the male rat: Effects of castration and hormone administration. *Physiological Zoology, 18* (4): 390–402.
36. 1945 Hormonal induction of mating responses in a rat with congenital absence of gonadal tissue. *Anatomical Record, 92* (3): 289–292.
37. 1945 Play in animals. *Encyclopedia Brittanica.*
38. 1945 Current concepts of play in animals. *American Naturalist, 79:* 523–541.
39. 1946 Coauthor: A. Marie Holz. Mating behavior in male rats castrated at various ages and injected with androgen. *Journal of Experimental Zoology, 101* (1): 91–142.
40. 1947 A review of physiological and psychological studies of sexual behavior in mammals. *Physiological Review, 27* (2): 240–305.
41. 1947 Evolutionary changes in the physiological control of mating behavior in mammals. *Psychological Review, 54* (6): 279–315.

42. 1947 Hormones and mating behavior in vertebrates. Recent Progress in Hormone Research. *Proceedings of the Lauren. Conference, 1:* 27–63.
43. 1949 Coauthor: Roslyn S. Pauker. Effects of castration and subsequent androgen administration upon mating behavior in the male hamster (*Cricetus auratus*). *Endocrinology, 45* (3): 211–221.
44. 1949 Coauthor: A. Marie Holz-Tucker. Effects of different concentrations of androgen upon sexual behavior in castrated male rats. *Journal of Comparative Physiology and Psychology, 42* (6): 433–453.
45. 1949 Coauthor: R. W. Gilmore. Response of male dogs to urine from females in heat. *Journal of Mammalogy, 30* (4): 391–392.
46. 1949 A cross-species survey of mammalian sexual behavior. *Psychosexual Development in Health and Disease.* New York: Grune & Stratton. 52–78.
47. 1949 Psychology of the dog. *Dog Encyclopedia.* (Davis, Ed.)
48. 1949 Coauthor: G. Levinson. Diurnal variations in the mating behavior of male rats. *Proceedings of the Society for Experimental Biology and Medicine, 72:* 78–80.
49. 1950 Coauthor: G. Levinson. Effects of androgen on the glans penis and mating behavior of castrated male rats. *Journal of Experimental Zoology, 114* (1): 159–171.
50. 1950 Commentary on *Corydon* by André Gide. New York: Farrar, Straus and Co.
51. 1950 The Snark was a Boojum. *American Psychologist, 5* (4): 115–124.
52. 1950 Sexual behavior in animals and men. The Harvey Lectures, Ser. XLIII: 254–279.
53. 1951 Effects of forebrain injury upon mating behavior in male pigeons. *Behaviour, IV* (1): 36–59.
54. 1951 Instinctive behavior: Reproductive activities. In *Handbook of Experimental Psychology.* (S. S. Stevens, Ed.), New York: John Wiley.
55. 1951 Body chemistry and perception. In *Perception: An Approach to Personality.* (R. R. Blake and G. V. Ramsey, Eds.), New York: Ronald Press.
56. 1952 "Psychosomatic" phenomena in animals. *Psychosomatic Medicine, 14* (4): 261–270.
57. 1952 Sex and species differences in the behavioral effects of gonadal hormones. Ciba Foundation Colloquia on Endocrinology. *III.* London: J. & A. Churchill.
58. 1952 Mechanisms of hormonal action upon behavior. Ciba Foundation Colloquia on Endocrinology. *III.* London: J. & A. Churchill.
59. 1952 Coauthor: J. Kagan. Effects of early experience on mating behavior in male rats. *Journal of Comparative Physiology and Psychology, 46* (3): 204–208.
60. 1953 Animal research and psychiatric theory. *Psychosomatic Medicine, 15* (5): 374–389.
61. 1953 Coauthors: William F. Heindenreich III. and Charles F. Alexander, Jr. Survival of mating behavior in male rats after thyroid-parathyroidectomy. *Endocrinology, 52* (6): 719.
62. 1953 Animals in psychological research. I. *Bulletin of the National Society for Medical Research, 7* (5): 2–6.
63. 1953 Animals in psychological research. II. *Bulletin of the National Society for Medical Research, 7* (6): 2–6.
64. 1954 Coauthor: J. Jaynes. Effects of early experience upon the behavior of animals. *Psychological Bulletin, 51* (3): 239–263.
65. 1954 The individual from conception to conceptualization. In *Psychology and the Behavioral Sciences.* Pittsburgh: University of Pittsburgh Press.
66. 1954 Ontogeny and living systems. Group Processes. Transaction of the First Conference, Josiah Macy Foundation.
67. 1955 Coauthors: A. C. Goldstein and O. Jacoby. Effects of electroconvulsive shock on

sexual behavior in male rats. *Journal of Comparative Physiology and Psychology, 48:* 173–179.

68. 1955 Coauthor: J. W. Kislak. Inhibition of aggressiveness by ovarian hormones. *Endocrinology, 56* (6): 684–693.
69. 1955 Coauthors: A. Zitrin and J. Jaynes. Neural mediation of mating in male cats. II. Contribution of the frontal cortex. *Journal of Experimental Zoology, 130* (3): 381–402.
70. 1955 The de-scent of instinct. *Psychological Review, 62* (6): 401–410.
71. 1956 Coauthor: Lisbeth Jordan. Effects of sexual reinforcement upon the performance of male rats in a straight runway. *Journal of Comparative Physiology and Psychology, 49* (2): 105–110.
72. 1956 Coauthors: Myron W. Conovitz, Frederick Steinberg, and Allan Goldstein. Experimental inhibition and restoration of mating behavior in male rats. *Journal of Genetic Psychology, 89:* 165–181.
73. 1956 Coauthors: A. Zitrin and J. Jaynes. Neural mediation of mating in male cats. I. Effects of unilateral and bilateral removal of the neocortex. *Journal of Comparative Physiology and Psychology, 49* (4): 321–327.
74. 1956 Coauthors: A. Zitrin and J. Jaynes. Neural mediation of mating in male cats. III. Contributions of occipital parietal, and temporal cortex. *Journal of Comparative Neurology, 105* (1): 111–125.
75. 1956 Coauthor: J. Jaynes. Studies of maternal retrieving in rats. I. Recognition of young. *Journal of Mammalogy, 37* (2): 177–180.
76. 1956 Coauthor: J. Jaynes. Studies of maternal retrieving in rats. II. Effects of practice and previous parturitions. *American Naturalist, 40* (851): 103–109.
77. 1956 Coauthor: J. Jaynes. Studies of maternal retrieving in rats. III. Sensory cues involved in the lactating females' responses to her young. *Behaviour, 10* (1–2): 104–125.
78. 1956 Coauthor: Lisbeth Jordan. Sexual exhaustion and recovery in the male rat. *Quarterly Journal of Experimental Psychology, 8* (3): 121–133.
79. 1956 Characteristics of masculine "sex drive." In *Neb. Symp. on Motivation:* 1956 (M. R. Jones, Ed.). Lincoln: U. Nebr. Press, 1–32.
80. 1958 Neural and chemical regulations of behavior. In *Biological and Biochemical Bases of Behavior.* (Harry F. Harlow and Clinton N. Woolsey, Eds.). Wisconsin: University of Wisconsin Press, 1958, 263–286.
81. 1958 Evolutionary aspects of psychoendocrinology. In *Behavior and Evolution.* (Anne Roe and George Gaylord Simpson, Eds.). New Haven: Yale University Press, 81–102.
82. 1958 Normal sexual behavior in male rats isolated at fourteen days of age. *Journal of Comparative Physiology and Psychology, 51:* 37–38.
83. 1959 Coauthor: H. Fowler. Effects of "situational anxiety" on sexual behavior in male rats. *Journal of Comparative Physiology and Psychology, 52* (2): 245–247.
84. 1959 Coauthor: H. Fowler. Individual differences in the response of male rats to androgen. *Journal of Comparative Physiology and Psychology, 52* (1): 50–52.
85. 1959 Coauthor: R. Rabedeau. Sexual exhaustion and recovery in the male hamster. *Journal of Comparative Physiology and Psychology, 52* (1): 56–61.
86. 1959 Coauthor: R. Whalen. Effects of ejaculation on sexual behavior in the male rat. *Journal of Comparative Physiology and Psychology, 52* (4): 249–254.
87. 1959 Coauthor: R. Whalen. Effects of intromission without ejaculation upon sexual behavior in male rats. *Journal of Comparative Physiology and Psychology, 52* (4): 476–481.

88. 1960 Experimental investigations of species-specific behavior. *American Psychologist, 15* (1): 1–19.
89. 1961 Coauthors: R. Whalen and R. Kuehn. Effects of exogenous androgen on sexually responsive and unresponsive male rats. *Endocrinology, 69* (2): 373–380.
90. 1961 Sex differences in the physiological bases of mating behavior in mammals. In *Physiology of Emotions* (Alexander Simon, Ed.). Illinois: Charles C Thomas, 151–162.
91. 1962 Coauthors: H. Fowler, S. Hicks, & Constance D'Amato. Effects of fetal irradiation on behavior in the albino rat. *Journal of Comparative Physiology and Psychology, 55* (3): 309–314.
92. 1963 Coauthor: R. Kuehn. Quantitative measurements of sexual receptivity in female rats. *Behaviour, 21* (3–4): 282–299.
93. 1963 Reproductive anatomy and mating behavior of an intersex rat. *Journal of Mammalogy, 44* (1): 57–61.
94. 1963 Coauthors: J. Wilson and R. Kuehn. Modification in the sexual behavior of male rats produced by changing the stimulus female. *Journal of Comparative Physiology and Psychology, 56* (3): 636–644.
95. 1963 Coauthor: J. Wilson. Mating behavior in male rats after removal of the seminal vesicles. *Proceedings of the National Academy of Science, 49* (5): 624–626.
96. 1963 Coauthor: J. Wilson. Effects of prolactin, progesterone and estrogen on reactions of nonpregnant rats to foster young. *Psychological Reports, 13*, 231–239.
97. 1964 Biological bases for reproductive behavior. In *Social Behavior and Organization among the Vertebrates* (Wm. Etkin, Ed.). Chicago: University of Chicago Press, 117–142.
98. 1965 Coauthor: N. G. Inman. Effects of castration and androgen replacement on mating in male quail. *Proceedings of the National Academy of Science, 54* (5): 1426–1432.
99. 1965 Experimental studies of mating behavior in animals. In *Sex Research: New Developments* (J. Money, Ed.) New York: Holt, Rinehart, 113–134.
100. 1965 Retrospect and prospect. In *Sex and Behavior* (F. A. Beach, Ed.). New York: Wiley, 535–569.
101. 1966 Coauthors: W. H. Westbrook & L. G. Clemens. Comparisons of the ejaculatory response in men and animals. *Psychosomatic Medicine, 28* (5): 749–763.
102. 1966 Ontogeny of "coitus-related" reflexes in the female guinea pig. *Proceedings of the National Academy of Science, 56* (2): 526–532.
103. 1966 Perpetuation and evolution of biological science. *American Psychologist, 21* (10): 943–949.
104. 1966 Sexual behavior in male rat. *Science, 153* (3737): 769–770.
105. 1967 Cerebral and hormonal control of reflexive mechanisms involved in copulatory behavior. *Physiological Review, 47* (2): 289–316.
106. 1967 Coauthor: B. J. LeBoeuf. Coital behavior in dogs. I. Preferential mating in the bitch. *Animal Behavior, 15* (4): 546–558.
107. 1967 Coauthor: T. W. Ransom. Effects of environmental variations on ejaculatory frequency in male rats. *Journal of Comparative Physiology and Psychology, 64* (3): 384–387.
108. 1967 Maternal behavior in males of various species. *Science, 157* (3796): 1591.
109. 1968 Coauthor: G. Bignami. Mating behavior in the chinchilla. *Animal Behavior, 16* (1): 45–53.
110. 1968 Coauthors: C. M. Rogers & B. J. LeBoeuf. Coital behavior in dogs. II. Effects of estrogen on mounting by females. *Journal of Comparative Physiology and Psychology, 66* (2): 296–307.

111. 1968 Coital behavior in dogs. III. Effects of early isolation on mating in males. *Behaviour, 30* (2-3): 218–238.
112. 1968 Coauthor: A. Merari. Coital behavior in dogs. IV. Effects of progesterone in the bitch. *Proceedings of the National Academy of Science, 61* (2): 442–446.
113. 1968 Coauthor: W. H. Westbrook. Dissociation of androgenic effects on sexual morphology and behavior in male rats. *Endocrinology, 83* (2): 395–398.
114. 1968 Coauthor: W. H. Westbrook. Morphological and behavioral effects of an "antiandrogen" in male rats. *Journal of Endocrinology, 42:* 379–382.
115. 1968 Factors involved in the control of mounting behavior by female mammals. In *Perspectives in Reproduction and Sexual Behavior: A Memorial to William C. Young* (M. Diamond, Ed.) Bloomington: Indiana University Press, 83–131.
116. 1969 Coauthor: G. Eaton. Androgenic control of spontaneous seminal emission in hamsters. *Physiology and Behavior, 4* (2): 155–156.
117. 1969 Coauthors: R. G. Noble & R. K. Orndoff. Effects of perinatal androgen treatment on responses of male rats to gonadal hormones in adulthood. *Journal of Comparative Physiology and Psychology, 68* (4): 490–497.
118. 1969 Locks and beagles. *American Psychologist, 24:* 971–989.
119. 1969 Coital behavior in dogs. VII. Effects of sympathectomy in males. *Brain Research, 15:* 243–245.
120. 1970 Coauthor: A. Merari. Coital behavior in dogs. V. Effects of estrogen and progesterone on mating and other forms of social behavior in the bitch. *Journal of Comparative Physiology and Psychology, 70,* 1–22.
121. 1970 Coital behavior in dogs. VI. Long-term effects of castration upon mating in the male. *Journal of Comparative Physiology and Psychology, 70*, 1–32.
122. 1970 Coital behavior in dogs. VIII. Social affinity, dominance and sexual preference in the bitch. *Behaviour, 36,* 131–148.
123. 1970 Coital behavior in dogs. IX. Sequelae to "coitus interruptus" in males and females. *Physiology and Behavior, 5:* 263–268.
124. 1970 Coauthor: L. Nucci. Long-term effects of testosterone phenylacetate on sexual morphology and behavior in castrated male rats. *Hormones and Behavior, 1,* 223–234.
125. 1970 Coauthor: R. E. Kuehn. Coital behavior in dogs. X. Effects of androgenic stimulation during development on feminine mating responses in females and males. *Hormones and Behavior, 1,* 347–367.
126. 1970 Coauthors: A. Zitrin, J. D. Barchas, & W. C. Dement. Sexual behavior of male cats after administration of parachlorophenylalanine. *Science, 170* (3960): 868–870.
127. 1970 Some effects of gonadal hormones on sexual behavior. In *The Hypothalamus* (L. Martini, M. Motta, and F. Fraschini, Eds.). New York: Academic Press, 617–640.
128. 1970 Hormonal effects of socio-sexual behavior in dogs. In *Mammalian Reproduction* (M. Gibian & E. J. Plotz, Eds.). Berlin: Springer-Verlag, 437–466.
129. 1971 Coauthor: R. H. Sprague. Effects of different forms of testosterone on mating in castrated rats. *Hormones and Behavior, 2,* 71–72.
130. 1971 Coauthor: L. Nucci. Effects of prenatal androgen treatment on mating behavior in female hamsters. *Endocrinology, 88,* 1514–1515.
131. 1971 Hormonal factors controlling the differentiation, development and display of copulatory behavior in the ramstergig and related species. In *Biopsychology of Development* (L. Aronson & E. Tobach, Eds.). New York: Academic Press, 249–296.
132. 1972 Coauthors: R. E. Kuehn, R. H. Sprague, & J. J. Anisko. Coital behavior in dogs.

XI. Effects of androgenic stimulation during development on masculine mating responses in females. *Hormones and Behavior, 3,* 143–168.

133. 1974 Behavioral Endocrinology and the Study of Reproduction. *Biology of Reproduction, 10,* 2–18.
134. 1974 Effects of gonadal hormones on urinary behavior in dogs. *Physiology and Behavior, 12,* 1005–1013.
135. 1974 Coauthor: R. K. Orndoff. Variation in the responsiveness of female rats to ovarian hormones as a function of preceding hormonal deprivation. *Hormones and Behavior, 5,* 201–205.
136. 1974 Human Sexuality and Evolution. In *Reproductive Behavior* (W. Montagna and W. A. Sadler, eds.) Plenum Press, pp. 333–365.
137. 1974 F. A. Beach. Autobiography. In *A History of Psychology in Autobiography,* Vol. VI. (G. Lindzey, ed.) Prentice-Hall, Inc., pp. 31–58.
138. 1975 Behavioral Endocrinology: An Emerging Discipline. *American Scientist, 63:* 178–187.
139. 1975 Hormonal Modification of Sexually Dimorphic Behavior. *Psychoneuroendocrinology, 1,* 3–23.
140. 1975 Variables Affecting "Spontaneous" Seminal Emission in Rats. *Physiology and Behavior, 15,* 91–95.
141. 1976 Sexual attractivity, proceptivity, and receptivity in female mammals. *Hormones and Behavior, 7,* 105–138.
142. 1976 Prolonged hormone deprivation and pretest cage adaptation as factors affecting the display of lordosis by female rats. *Physiology and Behavior, 16,* 807–808.
143. 1976 Cross-species comparisons and the human heritage. *Archives of Sexual Behavior, 5:* 469–485.
144. 1976 Coauthors: Bonnie Stern, Marie Carmichael, and Edoma Ranson. Comparisons of sexual receptivity and proceptivity in female hamsters. *Behavioral Biology, 18,* 473–487.
145. 1977 Coauthor: Michael G. Buehler. Male rats with inherited insensitivity to androgen show reduced sexual behavior. *Endocrinology, 100,* 197–200.
146. 1977 Confessions of an Imposter. *Pioneers in Neuroendocrinology,* Vol. II. (In press)
147. 1977 Coauthors. J. Anisko, A. Johnson and I. Dunbar. Hormonal control of sexual attraction in pseudohermaphroditic female dogs. *Journal of Comparative Physiology and Psychology* (In press).

Book and Film Reviews

1. 1943 Bird psychology. A review of *Bird Display: An introduction to bird psychology.* E. A. Armstrong. In *Ecology, 24* (4): 503–505.
2. 1951 *Bees: Their Vision, Chemical Senses and Language.* K. von Frisch. In *Psychological Bulletin, 48* (4): 365–366.
3. 1952 Symposia of the Society of Experimental Biology. IV. Physiological Mechanisms of Animal Behavior. In *Psychological Bulletin, 49:* 358.
4. 1952 *Sexual Behavior in Penguins.* L. E. Richdale, 1951, University of Kansas Press. In *Quarterly Review of Biology*
5. 1953 *Sexual Behavior in Western Arnhem Land.* E. M. Berndt and C. H. Berndt. 1951 Viking Publication in Anthropology #16. *American Anthropology, 55,* 601–602.
6. 1953 *Postural Development of Infant Chimpanzees. A Comparative and Normative*

Study Based on the Gesell Behavior Examination. Austin H. Riesen and Elaine F. Kinder, Yale University Press, 1952.

7. 1954 *The Herring Gull's World.* N. Tinbergen. Frederick A. Praeger, 1953. In *Psychological Bulletin,* 1952.
8. 1957 *Learning and Instinct in Animals.* W. H. Thorpe. In *Nature,* 1957.
9. 1957 *Studies of the Psychology and Behavior of Animals in Zoos.* H. Hediger. In *Contemporary Psychology.*
10. 1958 *Instinctive Behavior: The Development of a Modern Concept.* Claire H. Schiller, Ed., & Trans. New York: International Universities Press, 1957. In *Contemporary Psychology, 3* (7): 177–179.
11. 1960 *The Sage of Sex. A Life of Havelock Ellis.* Arthur Calder-Marshall. New York: G. P. Putnam. In *American Journal of Psychology, 74* (1): 151–154.
12. 1960 Nature and Development of Affection. H. F. Harlow and R. Zimmerman, Penn. State university. (Film). In *Contemporary Psychology, 6* (1): 27.
13. 1962 *The Encyclopedia of Sexual Behavior* (2 vols.). A. Ellis and A. Abarbanel (Eds.): New York: Hawthorne Books, 1961. In *Contemporary Psychology 7* (1): 16–17.
14. 1962 *Sex and Internal Secretions* (3rd ed.) W. C. Young (Ed.) Baltimore: Williams & Wilkins Co., 1961. In *Contemporary Psychology* 70–71.
15. 1964 *Impotence and Frigidity.* Donald W. Hastings, M.D. Boston: Little, Brown Co., 1963. In *Contemporary Psychology, 9,* (10), 396.
16. 1965 *Comparative Psychology: Research in Animal Behavior.* Stanley C. Ratner and M. Ray Denny. Homewood, Ill.: Dorsey Press, 1964. In *Contemporary Psychology, 10* (8), 345–346.
17. 1965 *Maternal Behavior in Mammals.* Harriet L. Rheingold, New York: Wiley, 1963. In *Contemporary Psychology, 10* (2), 64–65.
18. 1965 *A Model of the Brain.* J. Z. Young. Oxford: Clarendon Press. In *Scientific American,* April, 147–150.
19. 1966 *Human Sexual Response.* William H. Masters and Virginia E. Johnson. Boston: Little, Brown. In *Scientific American,* August.
20. 1970 *Endocrinology and Human Behavior.* Richard P. Michael, Ed. New York: Oxford University Press. In *Journal of Nervous and Mental Disease, 150* (2).
21. 1972 *Dr. Kinsey and the Institute for Sex Research.* Wardell B. Pomeroy. New York: Harper and Row. *Science, 177,* 416–418.
22. 1973 *Man and Woman: Boy and Girl.* John Money and Anke A. Ehrhardt. Johns Hopkins University Press, Baltimore and London. 1972.

1a. 1947 Physiological and psychological factors in sex behavior. S. B. Wortis et al. In *Annals New York Academy of Sciences,* 1947, Vol. XLVII, Art. 5, pp. 603–664.

Biographical Memoirs

1. 1959 *A Biographical Memoir of Clark Leonard Hull, 1884–1952.* National Academy of Sciences Biographical Memoirs, *XXXIII:* 125–141.
2. 1961 *A Biographical Memoir of Karl S. Lashley, 1890–1958.* National Academy of Sciences Biographical Memoirs, *XXXV:* 164–204.

Books

1. 1948 *Hormones and Behavior.* Paul B. Hoeber, Inc. New York.
2. 1952 Coauthor: C. S. Ford. *Patterns of Sexual Behavior.* Harper Bros., and Paul B. Hoeber, Inc., New York.

3. 1965 (Editor) *Sex and Behavior.* Wiley, New York. Reprint Edition. 1974, Robert E. Krieger Publ. Co.
4. 1977 (Editor) *Human Sexuality in Four Perspectives.* Johns Hopkins University Press, Maryland.

Nontechnical Publications

1. 1935 First steps in establishing the school newspaper. Quill & Scroll, *10:* 10–16.
2. 1936 Are you printing what your subscribers want? *Scholastic Editor, 15:* 4–21.
3. 1942 Why don't we behave like apes? *Science Digest,* October, 1942.
4. 1944 The Saga of Oscar, the musical 'gator. *Natural History, 53* (1): 456–459.
5. 1947 Of course animals can think. *Natural Hisotry, 56* (3): 116–118, and in *Science Digest,* February, 1948.
6. 1947 Open the door, Richard! *Natural History, 56* (7): 326–332.
7. 1947 Payday for primates. *Natural History, 56* (10): 448–451, and in *Naturen, 10,* 1948, and in *Science Digest,* May, 1948.
8. 1947 Do they follow the leader? *Natural History, 56* (8): 356–359.
9. 1947 Singing lessons for birds. *Fauna, 9* (4): 105–108, and in *Science Digest,* April, 1948.
10. 1947 Brains and the beast. *Natural History, 56* (6): 272–275.
11. 1948 Can animals reason? *Natural History, 57* (3): 112–116.
12. 1948 Dogs at the automat. *Fauna, 10* (8): 85–87.
13. 1949 Psychology of the dog. *The Dog Encyclopedia.*
14. 1950 Beasts before the bar. *Natural History, 59* (8): 356–359.

Appendix 2

Predoctoral and Postdoctoral Students of Frank A. Beach

Norman T. Adler, predoctoral 1963–67
Joseph Anisko, predoctoral 1969–74
Gordon Bermant, postdoctoral 1962–64
Giorgio Bignami, postdoctoral 1963–64
Lynwood G. Clemens, predoctoral 1961–66
Julian Davidson, postdoctoral 1962–63
R. Denniston, postdoctoral 1952–53
Donald A. Dewsbury, postdoctoral 1965–66
Richard Doty, postdoctoral 1971–73
Ian Dunbar, predoctoral 1971–74; postdoctoral 1974–
G. Gray Eaton, predoctoral 1966–69
Joyce Fleming, predoctoral 1967–70
Harry Fowler, predoctoral 1956–57
Stephen Glickman, postdoctoral 1962–64
Allan C. Goldstein, predoctoral 1953–57
Lawrence Harper, predoctoral 1961–66
Carolyn Hoffberg, predoctoral 1955–59
Julian Jaynes, predoctoral 1946–50
Ann Johnson, predoctoral 1969–75
Jerome Kagan, predoctoral 1950–54
Knut Larsson, postdoctoral 1958–59
Fred Leavitt, postdoctoral 1966–67
Burney Le Boeuf, predoctoral 1962–66
Edmund D. Lowney, postdoctoral 1956–57
Thomas E. McGill, postdoctoral 1959–60
Ariel Merari, predoctoral 1965–69
Ralph Noble, predoctoral 1967–70
Robert K. Orndoff, predoctoral 1972–76
Ronald Rabedeau, predoctoral 1956–60
Timothy Ransom, predoctoral 1965–72
Charles M. Rogers, predoctoral 1948–52
Burton S. Rosner, postdoctoral 1950–51

Benjamin Sachs, predoctoral 1962–66
Marvin Schwartz, predoctoral 1952–55
Jeffrey Stern, predoctoral 1965–69
Leonore Tiefer, predoctoral 1966–69
William H. Westbrook, predoctoral 1961–68
Richard E. Whalen, predoctoral 1956–60
James Wilson, predoctoral 1957–66
Dale Wise, predoctoral 1966–69
Arthur Zitrin, predoctoral 1939–41; postdoctoral 1970–71

Author Index

Subject Index

9 780306 310843